T0325731

Wireless Public Safety Networks 1

Series Editor
Pierre-Noël Favennec

Wireless Public Safety Networks 1

Overview and Challenges

Edited by

Daniel Câmara
Navid Nikaein

First published 2015 in Great Britain and the United States by ISTE Press Ltd and Elsevier Ltd

ISTE Press Ltd
27-37 St George's Road
London SW19 4EU
UK

www.iste.co.uk

Elsevier Ltd
The Boulevard, Langford Lane
Kidlington, Oxford, OX5 1GB
UK

www.elsevier.com

For information on all our publications visit our website at http://store.elsevier.com/

British Library Cataloguing-in-Publication Data
A CIP record for this book is available from the British Library
Library of Congress Cataloging in Publication Data
A catalog record for this book is available from the Library of Congress
ISBN 978-1-78548-022-5

Printed and bound in the UK and US

Contents

Preface

This book series is dedicated to the public safety networks (PSNs) field. PSNs are the networks established by the authorities for handling disaster scenarios. PSNs are the kind of networks established by the authorities to either warn the population about an imminent catastrophe or coordinate teams during the crisis and normalization phases. A catastrophe can be defined as an extreme event causing a profound damage or loss as perceived by the afflicted people. PSNs have the fundamental role of providing communication and coordination for emergency operations.

The series is organized as follows; this first book presents an overview of PSNs and communication systems with a particular focus on current and future challenges

---

Preface written by Daniel CÂMARA and Navid NIKAEIN.

and trends. The second book[1] provides a system view, focusing on specific enabling technologies that can be exploited in the context of current and next-generation PSNs. The third book[2] describes applications and services dedicated to PSNs.

Emergency management phases

Disasters can be of different types: natural disasters, such as hurricanes, floods, droughts, earthquakes and epidemics, or man-made disasters, such as industrial and nuclear accidents, maritime accidents and terrorist attacks. In both cases, human lives are in danger and the telecommunication infrastructures may be seriously affected or even no longer operational.

The disaster management involves three main phases, shown in Figure 1:

– *Preparedness*: at this phase all the equipment and people should be ready to enter in action, if needed. It consists of training, equipment maintenance, hazards detection and education;

– *Crisis*: this phase goes from the break-out point (decision to respond), to the immediate disaster aftermath, when lives can still be saved. Crisis is understood as the society's response to an imminent disaster; it is different from the disaster itself;

– *Return to normal situation*: this phase consists of the building and maintenance of temporary communication mechanisms/structures while the regular mechanisms are being repaired or rebuild.

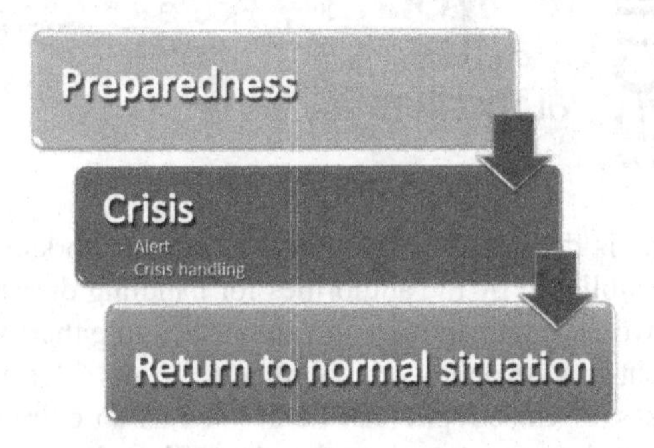

Figure 1. *Disaster management phases*

1 Câmara D., Nikaein N., Wireless Public Safety Networks 2, ISTE Press, London and Elsevier, 2016.

2 Câmara D., Nikaein N., Wireless Public Safety Networks 3, ISTE Press, London and Elsevier, 2016.

Crisis subphases

In a situation of crisis, the involved parties can be classified in the following way, taking into account also the degree of mobility they need:

– local authority(ies); fixed: the group in the administrative hierarchy competent to launch a warning to the population and the intervention teams;

– citizens; either mobile or fixed: non-professional people involved in the crisis;

– intervention teams; mobile: professionals (civil servants or militaries) in charge of rescuing citizens in danger, preventing hazard extension or any time-critical mission just after the break-out of the crisis; in charge of caring for injured people once the crisis is over;

– risk management center; fixed: group of experts and managers in charge of supervising operations. The risk management center works in close cooperation with local authorities;

– health centers; fixed: infrastructure (e.g. hospital) dedicated to caring injured citizen and backing intervention teams as for this aspect of their mission.

Alert phase

It is important to properly manage this critical phase as it is the moment where a quick response is the most efficient in terms of lives and goods saved. This means notifying professionals and people of the incoming hazard.

Warning makes sense if there is a delay between the very break-out of the hazard and the damages it could cause. This leaves time for people to escape and avoid the endangered area. Warning the population is typically the local authorities' responsibility since they are the only ones who can clearly appreciate the danger depending on local circumstances. Deciding that the situation is critical may be taken at governmental, national level. This is the case, for example, for earthquakes in all European countries.

Crisis handling phase

Coordination of intervention teams begins when the crisis breaks out. The local authorities alert them just before the population and then transfer the supervision to the risk management center. Later on, intervention teams still receive instructions from their local authorities, from the risk management center and the health center.

Intervention teams send back information to local authorities, risk management center and health centers about the situation and request for help. They typically use a specific purpose network deployed especially to attend to the needs of that particular event. Usually, the same network is used for receiving instructions and returning feedback.

Reference architecture

Public safety, or emergency networks, have as an objective to provide more information to authorities and to the population, and to help on the organization of the relief efforts in the case of a catastrophe. Figure 2 presents a generic reference architecture where we have the three main PSN communication systems, i.e. situation awareness, rapidly deployable and emergency alert ones:

– *situation awareness* is related to the elements responsible for providing an overall real-time picture of events with an assessment of the consequences on the population and on property. This kind of system helps authorities on the decision-making process;

– *rapidly deployable* systems allow the exchange between elements of the relief effort and also grant a communication channel with the support teams, in control rooms. In this way, the control teams can access field information and coordinate the relief efforts;

– *emergency alert* grants authorities the means to warn the population about possible dangers and the status of a given disaster in an efficient and fast manner, through several media routes simultaneously.

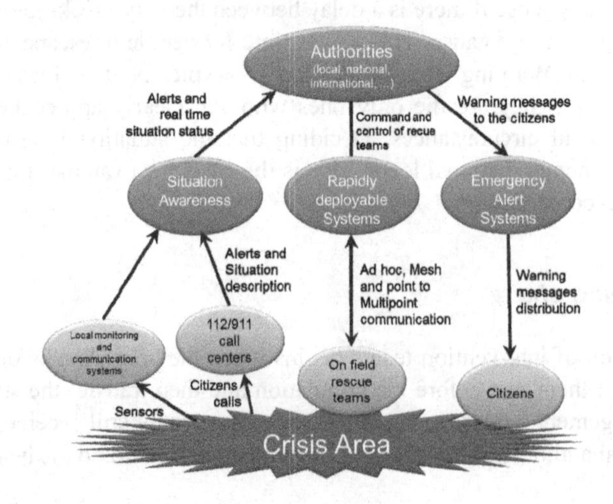

Figure 2. *Different elements of public safety systems*

Technologies

Different technologies can be used for PSNs: some were specifically designed to be used for the authorities in case of emergency, but others are standard ones. Figure 3 presents some of the possible technologies and how they can be organized to provide communication for the teams on the terrain. This series will discuss many of these technologies and how they are used on the field.

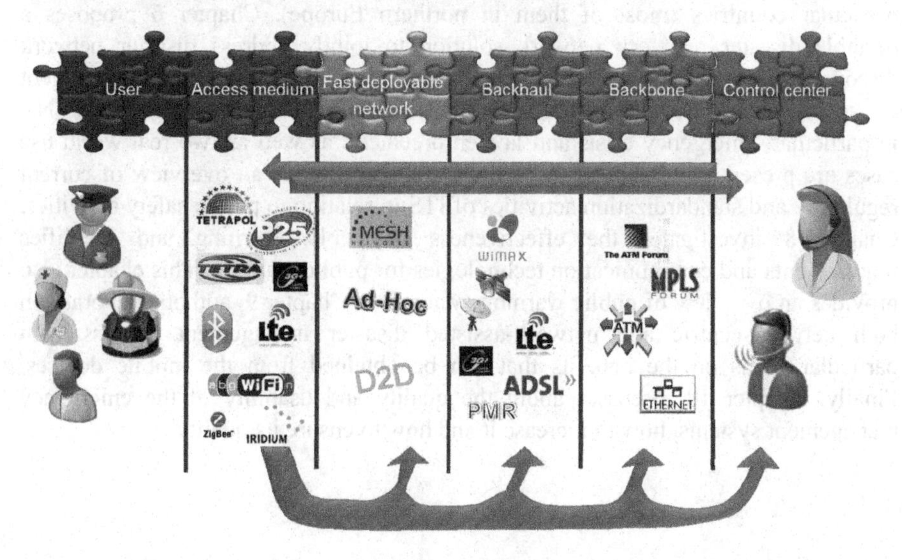

Figure 3. *Possible interconnections between technologies and actors*

Book overview

The aim of this first edited book is to provide a big picture on different elements of PSNs and dig into few challenges and trends in view of current and next-generation PSNs.

The book is organized into 10 chapters. Chapter 1 presents various techniques and technologies that can be used to build a PSN, and the main design choices to be made. The role of communication technologies in PSNs is analyzed in Chapter 2. Furthermore, the challenges to efficiently deploy, operate and interoperate present and future technologies and how these technologies can help public safety agencies to meet their expectations are described. The evolution of public protection and disaster relief (PPDR) systems in terms of new services and capabilities, by leveraging emerging broadband wireless technologies such as long term evolution (LTE) in addition to existing legacy PMR systems as TETRA and TETRAPOL are discussed in Chapter 3. Direct communication technologies, as one of the most important PSN requirement to establish direct communication links between terminals without the presence of any form of infrastructure, and proximity services are presented in Chapter 4.

Chapter 5 analyzes the interoperability issues in PSN, with particular focus on European networks, and points out some agreements and treaties active between particular countries (most of them in northern Europe). Chapter 6 proposes a portable disaster recovery network solution to jointly address disaster network discovery and search and rescue networking problem. The role of intelligent transport system (ITS) and related technologies to augment the capabilities of PSNs, in particular, emergency crisis and law enforcement, as well as two real world use cases are presented in Chapter 7. The authors also provide an overview of current regulatory and standardization activities of ITS in relation to public safety activities. Chapter 8 investigates the effectiveness of public warning and identifies requirements and communication technologies for public warning. This chapter also provides an overview of public warning systems. In Chapter 9, authors elaborate on both network-centric and network-assisted disaster management process with particular focus on the benefits that can be obtained from the mobile devices. Finally, Chapter 10 discusses about the quality and usability of the emergency management systems, how to increase it and how to ensure its quality.

1

Public Safety Network: an Overview

A public safety network (PSN) is a communication network used by emergency services, such as police, fire brigades and medical emergency services, to prevent or respond to incidents that endanger people and/or property. Over the last few years, in order to enhance their efficiency, public safety organizations and first responders have increasingly turned to use personal devices and networked applications. Consequently, wireless systems, hand-held computers and mobile video cameras are increasingly present in PSNs. These devices are an efficient way to increase services' efficiency, visibility and ability to instantly collaborate with central command, co-workers and other agencies.

The need to access and share the new vital flow of data and images should take into account the complete information chain, i.e. from the information acquisition by

Chapter written by Tullio Joseph TANZI and Jean ISNARD.

sensors, to its use by the relief teams directly in the field. Moreover, PSNs must incorporate the command-chain level to ensure an efficient organization of the relief effort.

This chapter presents various techniques and technologies that can be used to build up PSNs, and the reason driving the different choices. Some examples of existing systems are discussed.

1.1. Introduction

The media remind us daily, that we live in a world where risks are ever present. In addition to natural disasters [EMD 15] (hurricanes, cyclones, typhoons, earthquakes, tsunamis, landslides, floods, forest fires, etc.), there are also risks generated by human activity (armed conflict, industrial accidents, transportation accidents, socio-economic movements, etc.). A relatively new phenomenon is the risk of falsification of information related to new technologies supposed to protect us [PER 07]. A society which desires to survive must identify each type of risk and evaluate the necessary means to prevent and manage a situation.

All modern communication and information technologies concerning location, remote sensing, advanced technology terminals, multimedia, video, advanced techniques of computer sciences and databases, security protocols, signal processing, communication and networking protocols, etc., represent a valuable toolbox to support the modern engineer, especially one that evolves in the field of risk management [TAN 09].

Communication is an indisputable contribution for those who are in charge of risk management. They authorize the constitution of real-time management systems by promoting the exchange and share of information. Safety networks used for the Public Protection and Disaster Relief (PPDR) are a good example. The topic of setting up networks for national and international coverage and broadband will stay with us for still a long time. Indeed it has several aspects: in addition to technical and operational components, inputs of improvement from various organizations working on the subject have to be taken into account.

It has to be remembered that PPDR related radiocommunication matters are an issue of sovereignty of States and that PPDR requirements may vary to a significant extent from one country to another. For example, Europe will consider future harmonization of PPDR only if the action is flexible enough to consider different national circumstances such as the PPDR scenarios, the amount of available spectrum and the type of network which may be a dedicated, a commercial or a hybrid solution.

In addition to communications, spatial analysis techniques represent a powerful contribution to decision-making in a critical situation (e.g. crisis management); one of the aims is to be able to manage and control the risk. The difficulties concern both information collection and distribution to men on the ground [WIL 10].

Encountered constraints are important. They are mainly due to the fact that critical systems are using a multiplicity of sources of heterogeneous data from various geographically dispersed sensors. In addition to the geolocation and mobility tracking of sensors, the huge quantity of data to be processed represents another source of substantial problems. In fact, only an interdisciplinary approach can help to solve all these difficulties [TAN 09].

The use of these data permits to obtain an accurate synthesis of the observed situation. Data fusion of sensors allows us to elaborate meta-indicators describing the evolution of the observed phenomena. Spatial data mining and supervised or non-supervised classification capabilities are also used to extract relevant information for decision-making.

A reliable and secure network, in an environment often devoid of infrastructure and sometimes hostile, has to distribute information and therefore must be designed above this layer of embedded networks (to ensure mobility) and compensate for a possible existing infrastructure failure. This network must be flexible to support all types of data (including video). It has to comply with the needs of quality of service (QoS) and offer a high level of safety and security. It must possess dynamic reconfiguration capabilities taking into account the position of sensors and various equipment.

Finally, it is necessary to set up a logical architecture for transport, processing and display of information.

Regarding information transport: which are the required networks' characteristics and in which way information is displayed (dashboards, synthesis, use of avatars, cartographic support 2D and 3D, etc.)?

The main objective of information processing is to deliver the relevant information to operational actors. In other words the tasks of this type of system are [TAN 98]:

> *To obtain the right information, at the right time, in the right place, at the right person... for the right decision*

This constitutes the specificity of information networks for risk management.

1.2. A multi-level response

Such a system must be able to provide a response at several levels [TAN 98, LAU 98]. Indeed, when we analyze the events provoking a crisis, we are in the presence of successive phases. Figure 1.1 shows the temporal cycle of a catastrophic event. During each of these phases, the events, their consequences, the conditions in which work the operators responsible for the crisis management will vary, and which will request different actions to be performed. Depending on the conditions in which they are, operators will decide on the priority of action to be taken.

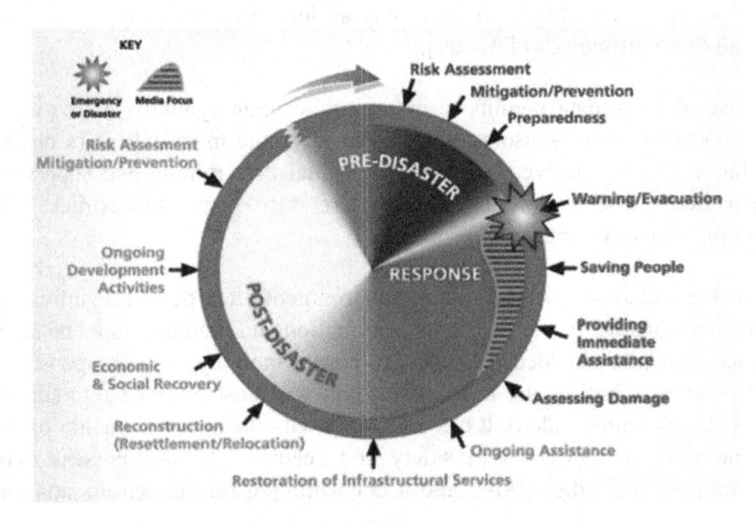

Figure 1.1. *Temporal cycle of a catastrophic event (http: //www.disasterscharter.org)*

The resolution of a crisis means that any solution is implemented in order to avoid and/or minimize negative consequences of the crisis. The search for why and how of the handling of the incident will be analyzed at a later time.

Once a crisis is over and conditions have returned to normal, we have to analyze the event with a view to integrate that gained experience. In this analysis, the time factor is pre-eminent as illustrated in [TAN 98].

1.3. Observation services

Radio-based monitoring services, which are in fact the only services able to operate at any time, are mainly based on instruments, i.e. ground and space radiometers and radars. However, it has to be underlined that data provided by radars are in common use whereas those provided by radiometers (passive sensors) are still confined in the research domain.

1.3.1. *Observations by satellites*

Measurements made by satellite depend on the accessibility of data and periodicity of observation. The problem of accessibility to data has been resolved at the UNISPACE III conference, which was held in July 1999 in Vienna. The European Space Agency (ESA) and the French Space Agency (CNES) initiated the "international charter space and major disasters" which was subsequently signed by most space agencies (see www.disasterscharter.org). This charter, which entered into force in November 2000, allows us to mobilize space resources and related Earth resources (RADARSAT, ERS, ENVISAT, SPOT, IRS, SAC – C satellite NOAA, LANDSAT, ALOS, DMC, etc.) In a crisis, the Parties will make every effort to facilitate access to all data relevant for this area.

At present, a relevant example is a growing demand for very high resolution pictures produced by synthetic aperture radars (SARs) operating in the Earth exploration-satellite service (EESS) (active). The image resolution needed for global environmental monitoring can only be achieved by corresponding transmission bandwidth. Resolution 651 (WRC-12) invites ITU-R to study on options for the extension of the EESS (active) allocation in the band 9,300–9,900 MHz by up to 600 MHz anywhere within the frequency ranges 8,700–9,300 MHz and/or 9,900–10,500 MHz.

Report ITU-R RS.2178 describes in detail the essential role and global importance of radio spectrum use for Earth observations and related applications in general. Moreover, the information on applications provided in this report, concentrates particularly on the description of applications of space-borne SAR[1] requiring high resolution information to less than 0.3 m. This desired resolution can only be achieved if a chirp transmission bandwidth of 1,200 MHz is available. Such a bandwidth requires an extension of the current EESS (active) allocation by

1 Synthetic-aperture radar (SAR) is a form of radar which is used to create images of an object, such as a landscape – these images can be 2D or 3D representations of the object. SAR uses the motion of the SAR antenna over a target region to provide finer spatial resolution than is possible with conventional beam-scanning radars. SAR is typically mounted on a moving platform such as an aircraft or spacecraft, and it originated as an advanced form of side-looking airborne radar (SLAR). The distance the SAR device travels over a target creates a large "synthetic" antenna aperture (the "size" of the antenna). As a rule of thumb one can assume that the larger the aperture is, the higher the image resolution becomes, regardless whether physical aperture or synthetic aperture – this allows SAR to create high resolution images with comparatively small physical antennas.

600 MHz. Such a high resolution will enable unprecedented features for long-term (4D, i.e. 3D space dimensions and one time dimension) global monitoring as well as for environmental monitoring and land-use purposes.

It is to be recognized that the allocation to the EESS around 9,600 MHz combines the advantage of the largest possible bandwidth at the lowest possible frequency regarding propagation conditions. Much lower frequencies cannot provide this large bandwidth, while much higher frequencies increasingly suffer from worsening propagation conditions.

Very high resolution mapping and monitoring is required by some applications that stipulate a substantial socioeconomic benefit, disaster relief and humanitarian aid actions require *ad hoc* access to up-to-date geo-information, including remote areas of the globe. Airborne imaging is very often hampered by remoteness of the area to be observed and by cloudy weather conditions. Today's radar satellites are too limited in resolution to allow adequate infrastructure damage assessment (and consequently a rough estimate of the number of affected people) to assist emergency aid activities. Also, identification of trafficable roads, landing strips or suitable spaces to set-up first aid or refugee camps is limited by the resolution of today's radar sensors.

1.3.2. *Ground-based observations*

Having their own objectives, ground based radars [WES 04] complement the satellite observations. Moreover, they can be installed on sensitive areas. Another technique is the continuous monitoring of GNSS signals (USA GPS, the Russian GLONASS, the European Galileo and the Chinese Beidou). The effect of variations of physical characteristics of the atmosphere (electron density, atmospheric density, temperature, humidity, etc.) on the phase and the amplitude of signals from GNSS satellites gave birth to a new discipline called "GNSS meteorology" [BEV 92, BUS 96]. The radio shadowing technique is used to measure the crossed through atmospheric layers by these signals [JAK 05]. A use of inversion methods allows us to trace the variations of one or more parameters of the environment and to highlight the temporal evolution of weather events such as, for example, the formation of thunderstorms [DOE 08]. A similar approach is used for the detection of seismic origin phenomena such as tsunamis [OCC 08].

In this case, the phenomenon responsible for the disruption of GNSS signals is due to the vertical displacement of the Earth surface, which causes the atmosphere immediately above to follow the same vertical movement as the ground surface. The perturbation is propagating upwards as an atmospheric wave until the ionosphere. At the altitudes of the lower layers of the ionosphere, it produces a change in the electron density (total electronic content or TEC) which, when the initial disturbance is important, leaves a signature on GNSS signals.

1.3.3. *Alert systems, their limits and some current research*

Several types of warning systems have been fielded in recent years or are under development. Needless to say that each second counts to save lives and prevent damages. As a good example of the essential role of data transmission during the initial phase of disaster management, we have chosen the deep-ocean assessment and reporting of tsunamis (DART) system [DIS 13] designed by PMEL for Pacific Marine Environmental Lab. For further information, see http://nctr.pmel.noaa.gov/Dart/dart_ref.html [BER 01, BER 11].

The Pacific Tsunamis warning system, based in Honolulu (Hawaii), managed by the National Oceanic and Atmospheric Administration (NOAA), regroups several national bodies (USA, Federal of Russia, Japan, Chile, etc.). In total, 28 countries are members of the Pacific system. Its objectives are to detect and locate earthquakes in the Pacific to assess their (tsunami) impact and provide information to the countries concerned to warn inhabitants and protect them as much as possible. Countries such as Japan, which is particularly exposed, have taken important measures for risk prevention. In this respect, the ocean bottom seismograph (OBS) system allows detecting earthquakes in the sea using seismographs and instruments that measure the pressure exerted by water. The data is transferred every 20 s by cable to surface stations, then by telephone to the Tsunami Warning Centre of the Japan Meteorological Agency (JMA), located in Tokyo. It allows us to give the alert an hour before the arrival of a tsunami.

To complete this package, buoys are placed strategically on the open sea. They form the DART system. This system was set up in 1995 by the United States, to protect their coasts. It uses the detection of pressure changes in the sea. The system is composed of a buoy and a deep pressure recorder associated with the GOES satellite to transmit information. The buoy has several meteorological sensors – anemometer, barometer and salinity of the water sensor, etc. – and equipment dedicated to the transmission of information to the GOES satellite to then pass the information to the centre of the tsunami alarm located in Hawaii. Figure 1.2 shows the diagram of the DART system.

The key sensor of the DART system is a deep pressure recorder (BPR, for bottom pressure recorder). This BPR is a high precision measuring device that uses a pressure transducer composed of a quartz crystal and a “Bourdon tube”. The changes in frequency of crystal vibration are measured with great precision and then interpreted to determine the size of the tsunami. The DART system is installed around the Pacific Ocean as shown in Figure 1.3.

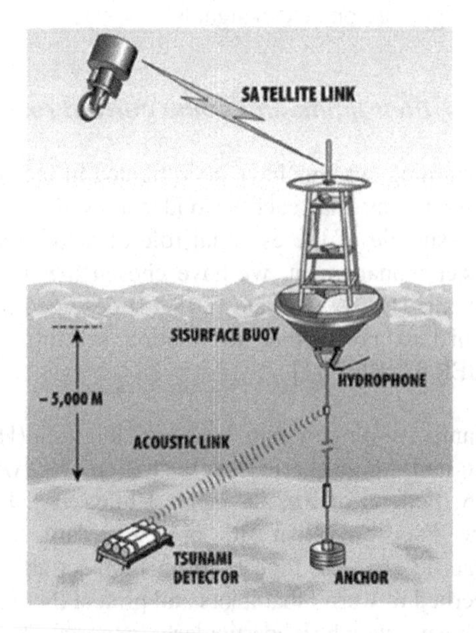

Figure 1.2. *Overview of DART system*

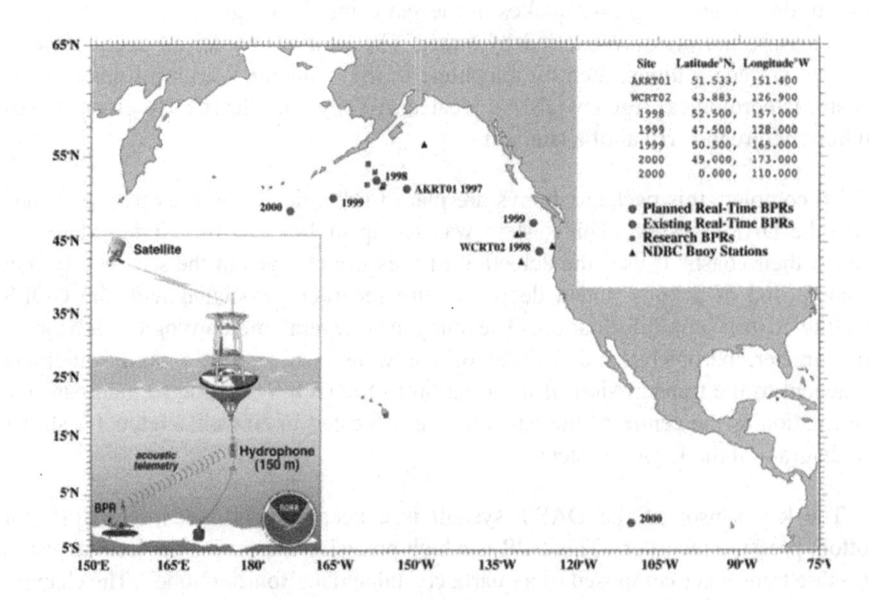

Figure 1.3. *DART stations location*

As shown in Figure 1.4, each DART station includes: (1) a detection system BPR, which is deposited on the seabed to detect pressure changes announcing a tsunami, (2) a surface buoy receiving the information transmitted by the BPR via an acoustic link and transmitting the data to a satellite which relays information to ground stations which are part of the NOAA and PMEL tsunami warning systems.

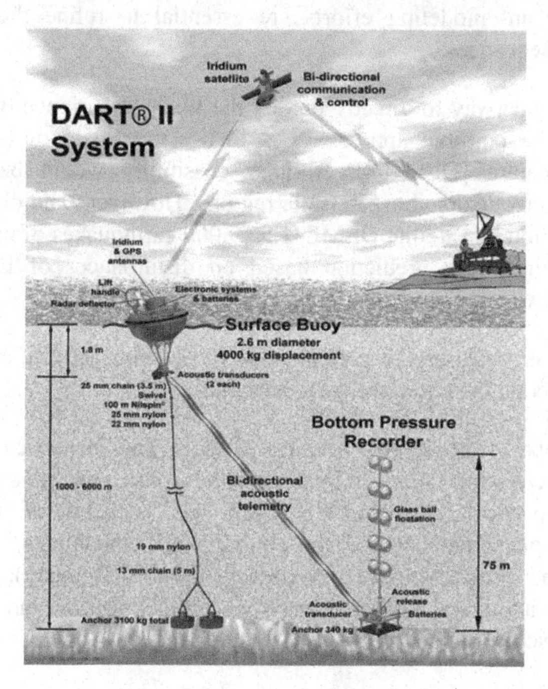

Figure 1.4. *DART II system*

It is evident that risk management crucially needs telecommunication support. “To provide the right information to the right person at the right time, for the right decision” requires the deployment of elaborate solutions necessitating the cooperation of two types of skills:

1) “Trade” skills of those who by their expertise are able to act efficiently and are aware of the risks which crisis situations can provoke.

2) ICT skills necessary to imagine technical solutions in order to assist people in their management tasks.

The operational teams benefit from robust techniques of image processing of information. The latter can be quickly deployed by Mesh networks. The advantage of this type of network is to be very flexible and thus to avoid dark zones.

Before the crisis, the priority is to strengthen all forecast services. Multiple space weather-related services are already in place. They allow issuing alerts: (1) on the risks of "blackout" for HF and VHF communications, (2) on foreseeable disturbances of the GNSS signals distorting the location, (3) on potential degradations of the images produced by SARs and other sensors. In all these domains, significant modeling efforts are essential to refine the forecasts and evaluate the consequences.

Research is underway to directly predict the likely occurrence of a catastrophic event. One of the possible approaches is to use the effect of tsunamis on the ionospheric signature [OCC 08]. Another possibility would be to use some electromagnetic precursors above seismic regions. However, data obtained after 2.5 years of observation by satellite DEMETER (9,000 earthquakes of magnitude above 4.8) does not allow any prediction based on disturbances of EM propagation characteristics of waves above these areas [NEM 08].

Similarly, it is necessary to continuously track solar activity and to take into account its impact on the choice of frequency bands.

During all phases, all remote sensing systems, and in particular the systems based on radiometers and radars on board satellites (which are not sensitive to cloud blankets), are particularly useful for both the selection of information on meteorological phenomena (cyclones, hurricanes, cloudbursts, etc.), and for monitoring of the disaster area (e.g. the recent flood of Bombay). It goes without saying that all additional information from optical sensors and ground-based instruments is also useful (see above GNSS meteorology).

1.4. Communication services

In addition to various operators there are several actors, i.e. government officials, media, etc., entitled to access to some kind of information: it is certainly an additional factor of complexity.

Presently there is no universal communication system accessible at any moment and everywhere in the world. For political (security), economic (cost) or technical (high-latitude areas, isolated areas, etc.) reasons several communication systems are currently used (satellite, HF and VHF communications links, etc.). During the crisis, any redundancy, at least where it exists, can be extremely useful. It allows us to ensure the continuity of communications. But there are still regions in developing countries which have no communication systems, although some initiatives are currently being implemented to install new communication infrastructure [PRA 92, CAR 08, LIE 08] or to deploy *ad hoc* networks during critical periods. Figure 1.5 shows an example of information processing chain.

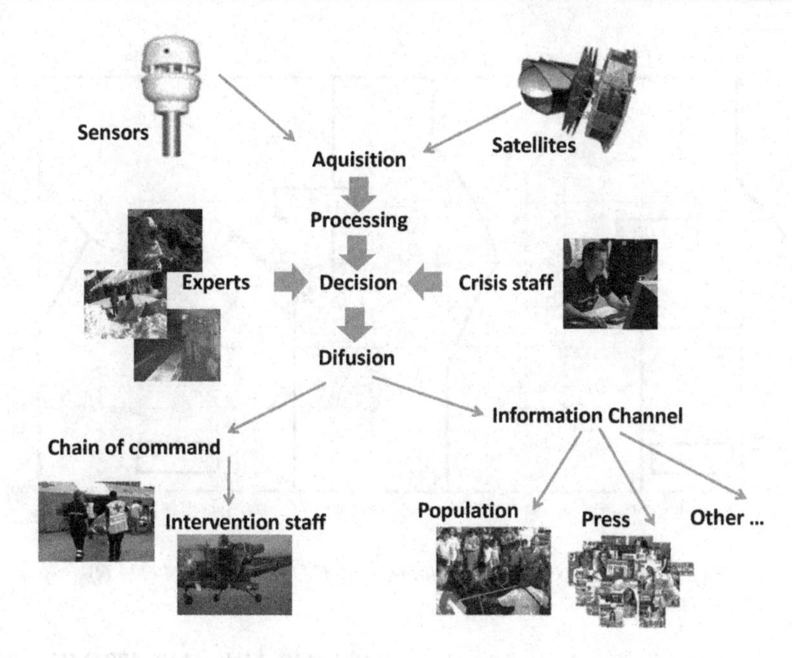

Figure 1.5. *Information processing chain (example) [TAN 09b]*

Several factors are bringing some improvement to the situation. For example, the transition from analog to digital broadcasting is not only contributing to an efficient use of the spectrum but is also creating opportunities for ICT applications and wireless broadcast communication needed in the PPDR framework.

Broadcasters produce a wide range of content and services such as traditional radio and TV, but also time-shifted, on-demand hybrid content and data services. As far as the necessary access to frequency spectrum is concerned the ITU Res. 646 provides information for this purpose as well as the Report ITU-R M.2033 and the ITU-D Handbook on disaster relief.

Work is under way toward regional harmonized frequency bands in the three regions as defined by ITU (see Figure 1.6).

In region 1, the band 380-470 MHz is the band within which the subband 380–385/390–395 MHz, the preferred core harmonized band for permanent public protection activities within certain agreed countries of the region.

In region 2, three bands are under analysis 746–806 MHz, 806–869 MHz and 4,940–4,990 MHz.

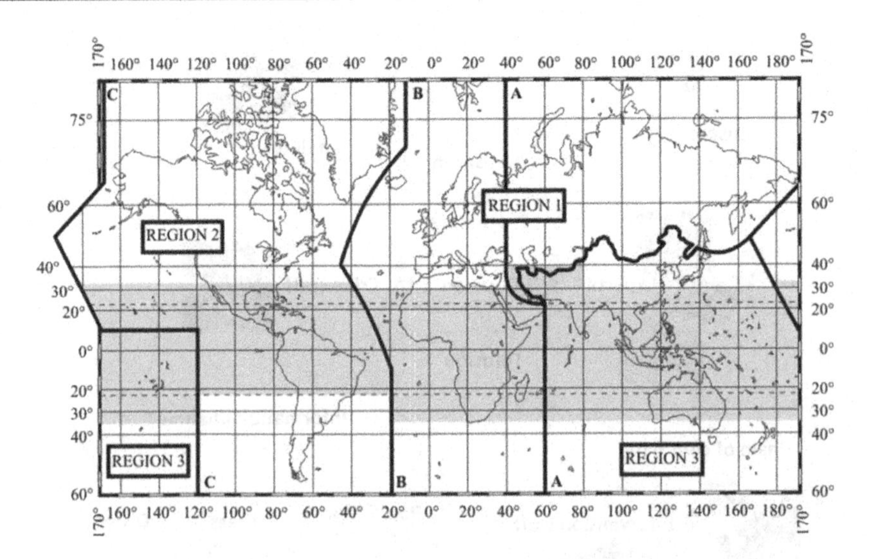

Figure 1.6. *Map of regions defined by ITU (courtesy of ITU-R)*

In region 3, the bands envisaged are 406.1–430 MHz, 440–470 MHz, 806–824/851–869 MHz, 4,940–4,990 MHz and 5,850–5,925 MHz. Some countries have also identified the bands 380–400 MHz and 746–806 MHz for PPDR applications.

New technologies for broadband PPDR applications are being developed in various standards organizations: for example, a joint standardization programme between the European Telecommunications Standards Institute (ETSI) and the Telecommunications Industry Association (TIA), known as Project MESA (mobility for emergency and safety applications) has commenced for PPDR.

IEEE 802.16 specifications (WiMAX – worldwide interoperability for microwave access) are good candidates for broadband land mobile radio (LMR), in particular PPDR application; it is of course a competitor of long-term evolution (LTE). LTE advanced will be a standard in accordance with IMT Advanced 4G defined by ITU. The future 5G (beyond 2020) is expected to be able to provide "ubiquitous" connectivity including naturally audio-visual media services. This will necessitate additional frequency bandwidths probably in high frequency bands.

1.5. Telecommunications during the crisis

During a crisis [TAN 06], the data collected in real-time by satellites, radars and other sensors have to follow chains of processing (signals and images: sorting, comparison, estimation and simulation) to be distributed to various users: to experts for analysis, to authorities responsible for decision-making, to media, finally to population for information.

The implementation of such a system brings to light two aspects of the follow up of risk: the temporal aspect in which it is necessary to find a balance between the time needed to send warning signals of the imminence of the phenomenon and the spatial aspect identifying where the disaster takes place.

For a more efficient risk management, new technologies are presently being developed. For the management of risk before, during and after the potential crisis, new technological needs have arisen: new methodology design of systems based on the telecoms, space and alphanumeric databases, deployment of powerful tools taking into account spatial and temporal aspects and local specificity in terms of risks.

When the risk becomes effective disaster, crisis cells must have simultaneous access to many environmental or regulatory databases. Unfortunately, their structures and software are often different if not incompatible. Figure 1.7 shows an example of multisources architecture.

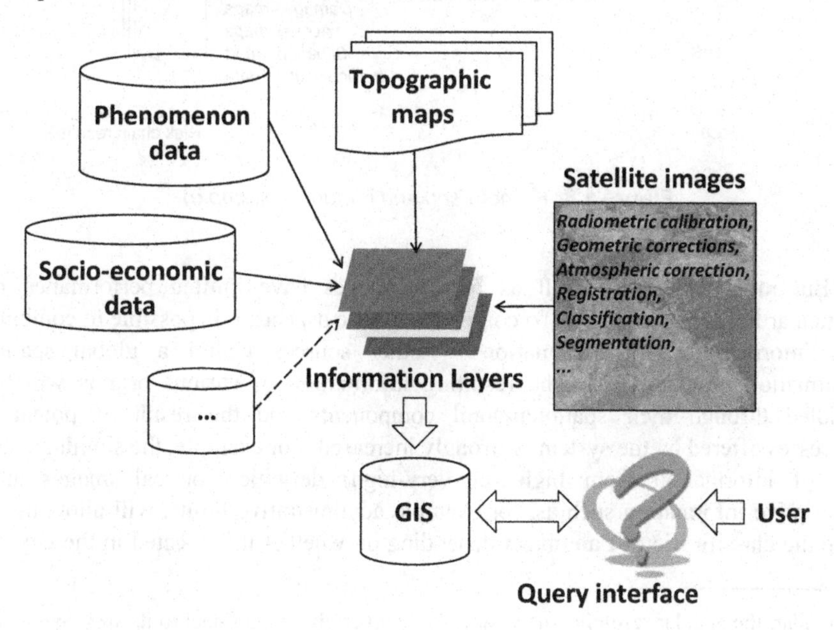

Figure 1.7. *Multisources architecture (example) [TAN 09b]*

The use of remote sensing imagery (see Figure 1.8) to achieve the mapping of the area is of primary importance [ELK 15]. Space images can be acquired shortly after the event and cover a wide area of territory. It can therefore be hoped that shortly after the disaster, a cartography of existing infrastructures based on these images can be obtained. However, they will not achieve total infrastructure mapping. The optical images may be unusable in the presence of an excessive overcast. They must be supplemented by radio images such as those obtained by the Synthetic Aperture Radar[2] (SAR) that are not affected by cloud cover[3].

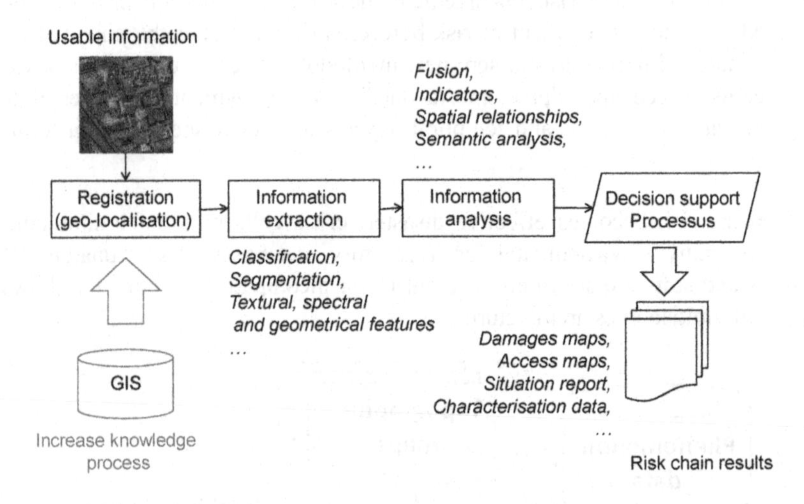

Figure 1.8. *Remote sensing imagery (example)*

But radio images as well as optical images have limited performance in particular limited resolutions. To compensate these aspects, it is possible to combine this information with information of other sources within a global spatial information system (GIS). The resulting information of various origins will be handled through their spatiotemporal components and the result of potential processes offered by the system is strongly increased. For example, the simultaneous use of information from high (or very-high) definition optical images and descriptive information such as, for example administrative limits, will allow us to ease the classification of an object depending on whether it is located in the city or

2 In radar, the angular resolution of an antenna is inversely proportional to its size. Synthetic Aperture Radar (SAR) technique operates the relocation of the antenna to form an antenna "synthesis" of more important dimension, and therefore an angular resolution higher than the same antenna, motionless. The large antenna is reconstructed by signal processing.

3 The use of radar waves makes the observation possible whatever is the weather or even if is the night.

in the forest[4]. Cross checking with textual data will facilitate the definition of the semantics of objects from the optical image depending on the context.

The results produced by the remote sensing process are then transmitted to the rescue teams. The information is used at several levels: strategic, tactical and operational (see Figure 1.9). Once more, an efficient communication network is required to enable the dissemination of information.

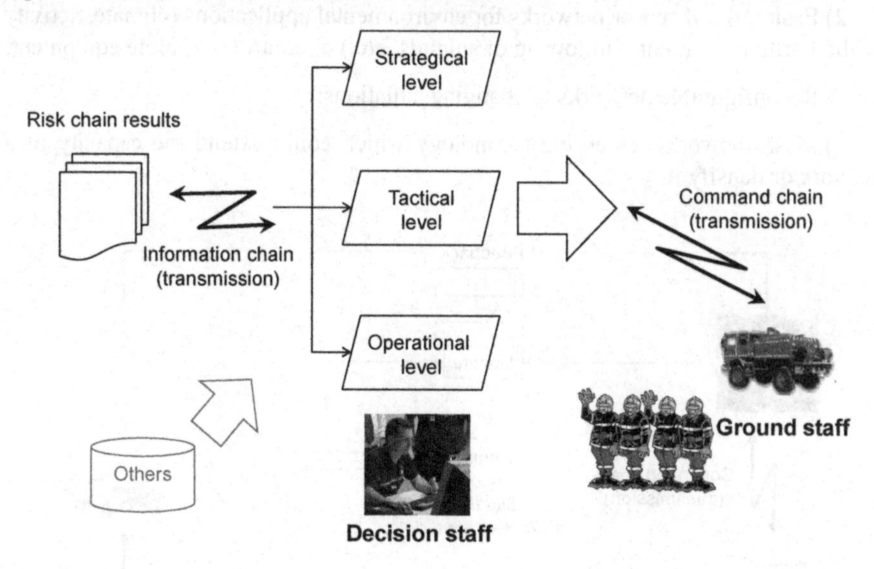

Figure 1.9. *Risk results used in intervention level (example)*

The validity of the acquired and produced information depends on the evolution of the situation in the area affected by the disaster. In any case, it will become rapidly obsolete. It is therefore necessary to set up a communication channel to feedback changes detected by the intervention teams and integrate them in the ensuing updates (see Figure 1.10). It goes without saying that a reliable and efficient communication network is indispensable.

1.6. Ad hoc networks

If no communication equipment has survived the disaster, the deployment of *ad hoc* systems is necessary, all components of the latter communicate in a direct way

4 The coupling with data produced by radar or infrared sensors will allow better classification of information using the characteristics of the observed objects.

by using their own interfaces [CAV 06]. The frequent advantage of the *ad hoc* network is to be self-sufficient and fast to deploy.

The use of *ad hoc* technology offers several advantages such as:

1) Fast installation of emergency services: search and rescue of people after earthquakes, fires, floods, etc;

2) Easier use of sensor networks for environmental applications (climate, activity of the Earth, movement's follow up of animals, etc.) or control of remote equipment;

3) Reconfigurable networks to changing situations;

4) Mesh networks: emerging technology which could extend the capacity of a network or densify it.

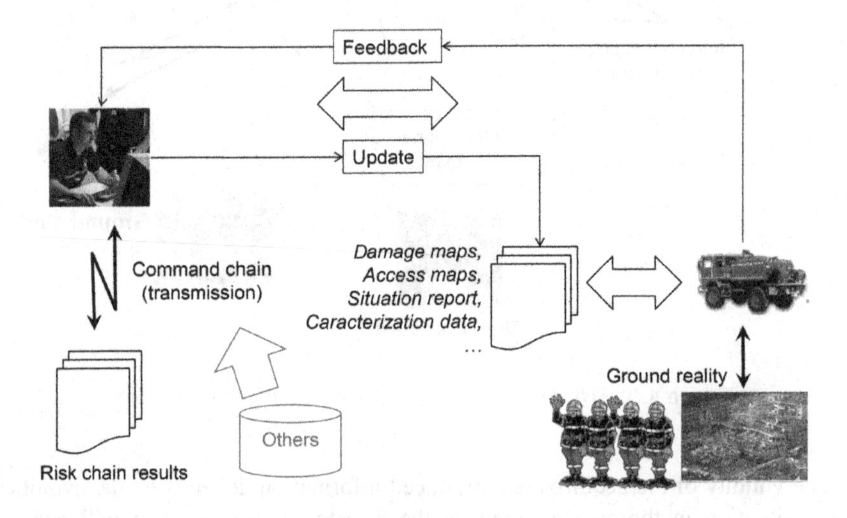

Figure 1.10. *Ground update level (example)*

The information network dedicated to authorities and to media has to be added to take into account their specific aspects. A point which does not always appear in the deployment of *ad hoc* networks during catastrophic events is the allocation of frequencies. A network implemented urgently, in a region where all frequency bands are already allocated has to be able to operate without major interference.

The need to search for free frequency bands [GHO 06] is probably one of the major reasons for attempting to dynamically apply the concepts of reconfigurable radio and cognitive radio [TAN 10].

1.7. Conclusion

The techniques of processing (sorting out and comparison) and communication of information have largely benefited from the development of information science to result in what is today called information and communication technologies (ICTs) [BRO 91, GRA 94]. A largely accepted definition of ICTs can be found in [BAK 85, CAR 90] and [GEN 86]:

> *ICTs are a set of non-human resources dedicated to keeping in stock, processing and manipulating information. They also include the principle of structural organization of these resources as well as their rules of application.*

This definition underlines the two main components: information and communication.

The field of applications of ICTs has been significantly enlarged: natural or anthropic disasters, public health, security, etc. As a result, the context has become very complex: destruction of general infrastructures, in particular of communication networks, etc.

This shows that we have to do with a very sensitive problematical question requiring an immediate and appropriate reply. In this framework, it goes without saying that communication networks play a major role:

1) Before the crisis with regard to the setting-up of a system of anticipation, prevision and alert taking into account the local physical constrains and the intercommunication with all actors concerned.

2) During the crisis to manage rapidly and efficiently all necessary interventions and co-ordinate the various actors.

3) After the crisis to draw conclusions from lessons learned and consequently enhance future ways of intervention.

The main new techniques include the use of embedded data, geolocation, fusion of various sources of information, measuring networks of alert systems (meteorological radars, pluviometry devices, toxicity, etc.). Data integrity and their updating are thus of primary importance.

In fact, to increase the effectiveness of PSF networks in general and PPDR in particular, the following actions are required:

1) Pay attention and take into account the evolution of communication networks.

2) Learn from disaster management experience.

3) Encourage scientific research on indicators of imminent natural disasters (the disaster management resulting from anthropic activities should be approached in a different manner).

The contributions which follow illustrate the broad scope of the matter.

1.8. Bibliography

[BAK 85] BAKOPOULOS Y., Toward a more precise concept of information technology, Working Paper MIT/CISR, vol. 126, p. 32, 1985.

[BER 01] BERNARD E.N., GONZÁLEZ F.I., MEINIG C. *et al.*, "Early detection and real-time reporting of deep-ocean tsunamis", *Proceedings of the International Tsunami Symposium 2001 (ITS 2001)*, NTHMP Review Session, Seattle, WA, pp. 97–108, 7–10 August 2001.

[BER 11] BERNARD E., MEINIG C., "History and future of deep-ocean tsunami measurements", *Proceedings of Oceans'11 MTS/IEEE*, Kona, IEEE, Piscataway, NJ, no. 6106894, p. 7, 19–22 September 2011.

[BEV 92] BEVIS M., BUSINGER S., HERRING T.A. *et al.*, "Remote sensing of atmospheric water vapour using the Global Positioning System", *Journal of Geophysical Research*, vol. 97, pp. 15787–15801, 1992.

[BRO 91] BROUSSEAU E., Les contrats dans une économie d'échange et de production: technologies de l'information et de la communication et coordination inter-entreprises, PhD Thesis, Université Paris Nord, 1991.

[BUS 96] BUSINGER S., CHISWELL S.S.R., BEVIS M. *et al.*, "The promise of GPS in atmospheric monitoring", *Bulletin of the American Mathematical Society*, vol. 77, pp. 5–17, 1996.

[CAR 90] CARRE D., "Info-révolution, usages des technologies de l'information", *Autrement Revue*, Paris, p. 348, 1990.

[CAR 08] CARCELLE X., "An introduction to power line communications", *MySQL Conference and Expo*, Santa Clara, CA, p. 2, 14–17 April 2008.

[CAV 06] CAVIN D., SCHIPER A., Towards reliable communication and agreement in mobile ad-hoc networks: algoritms, simulation and testbed, PhD Thesis, Lausanne, EPFL (http://infoscience.epfl.ch/record/89453), 2006.

[DIS 13] Using science for DISASTER RISK REDUCTION, Report of the UNISDR Scientific and Technical Advisory Group, 2013.

[DOE 08] DOERFLINGER E., "Les applications météorologiques du système de positionnement satellitaire GPS", *Navigation*, vol. 56, no. 223, pp. 15–40, 2008.

[ELK 15] EL KHARKI O., "Panorama sur les méthodes de classification des images satellites et techniques d'amélioration de la précision de la classification", *Revue française de photogrammétrie et de télédétection*, no. 210, pp. 23–38, April 2015.

[EMD 15] EM-DAT, The OFDA/CRED International Disaster Database, available at http://www.emdat.be, Université Catholique de Louvain, Brussels, 2015.

[GEN] GENTHON C., L'industrie électronique mondiale: innovation, intervention étatique, IREP, Université de Grenoble, 1986.

[GHO 06] GHOZZI M., DOHLER M., MAX F. *et al.*, "Cognitive radio: methods for the detection of free bands", *Comptes Rendus Physique de l'Académie des Sciences*, vol. 7, pp. 794–804, 2006.

[GRA 94] GRATACAP A., Impact des TIC sur la globalisation des marchés et la mondialisation des activités de la firme industrielle, PhD Thesis, Université Paris I, 1994.

[JAK 05] JAKOWSKI N., "Radio occultation techniques for probing the ionosphere", *Radio Science Bulletin*, vol. 314, pp. 4–15, 2005.

[LAU 98] LAURINI R. "La télégéomatique, problématique et perspectives", *Revue Internationale de Géomatique*, France, vol. 8, nos. 1–2, pp. 27–44, 25-27 November 1998.

[LIE 08] LIENNARD M., OLICAS CARRION M., DEGARDIN V. *et al.*, "Modeling and analysis of in-vehicle power line communication channels", *IEEE Transaction on Vehicular Technology*, vol. 57, no. 2, pp. 670–679, 2008.

[NEM 08] NEMEC F., SANTOLIK O., PARROT M. *et al.*, "Spacecraft observations of electromagnetic perturbations connected with seismic activity", *Geophysical Research Letters*, vol. 35, no. 5, p. 5, 2008.

[OCC 08] OCCHIPINTI G.A., KOMJATHY A., LOGONNÉ P., "Tsunami detection by GPS: how ionospheric observations might improve the Global Warning System", *GPS World*, pp. 50–56, February 2008.

[PER 07] PERROT P., CHOLLET G., "Biometrics and forensic sciences: a same quest for identification?", *International Crime Science Conference*, London, p. 7, 2007.

[PRA 92] PRADHAN B.D., "Wireless rural communication systems for developing countries", *Wireless Communications, Conference Proceedings, 1992 IEEE International Conference on Selected Topics*, Vancouver, Canada, pp. 254–256, 25-26 June 1992.

[TAN 98a] TANZI T., SERVIGNE S., "A crisis management information system", *Proceedings of the International Emergency Management and Engineering Society*, TIEMEC, Washington D.C., pp. 211–220, 19–22 May 1998.

[TAN 98b] TANZI T., LAURINI R., SERVIGNE S., "Vers un système spatial temps réel d'aide à la decision", *Revue Internationale de Géomatique, France*, vol. 8, no. 3 pp. 33–46. 1998.

[TAN 06] TANZI T., DELMER F., *Ingénierie du risque*, Hermès Lavoisier, Paris, 2006.

[TAN 09a] TANZI T., PERROT P., *Télécoms pour l'ingénierie du risque*, Hermès Lavoisier, Paris, 2009.

[TAN 09b] TANZI T., LEFEUVRE F., "L'apport des radios sciences à la gestion des catastrophes", *Journées scientifiques 2009 d'URSI-France: Propagation et Télédétection*, Paris, France, vol 1, pp. 401–428, 24–25 March 2009.

[TAN 10] TANZI LEFEUVRE F.,, "Radio sciences and disaster management", *Comptes Rendus Physique de l'Académie des Sciences*, Paris, France, vol. 11, no. 1, pp. 114–124, 2010.

[WES 04] WESTWATER R., CREWELL S., MÄTZLER C., "A review of surface-based microwave and millimetre wave radiometric remote sensing of the troposphere", *Radio Science Bulletin*, vol. 310, pp. 59–80, September 2004.

[WIL 10] WILKINSON P.J., COLE D.G., "The role of radio science in disaster management. URSI", *Radio Science Bulletin*, vol. 358, pp. 45–51, December 2010.

2

The Evolutionary Role of Communication Technologies in Public Safety Networks

Existing networks for public safety communications are mostly based on systems such as terrestrial trunked radio (TETRA), TETRAPOL and Project 25. These systems are mainly designed to support voice services. However, public safety communication networks are challenged to expand their scope way beyond their original functions toward more sophisticated devices and support new services including packet data communications. Therefore, governments, public safety agencies and research communities are continuously working together and are making significant progresses toward improving public safety communications capabilities. Next-generation mobile technologies are the enablers for meeting the

---

Chapter written by Karina Mabell GOMEZ CHAVEZ, Leonardo GORATTI, Tinku RASHEED, Dejene Boru OLJIRA, Riccardo FEDRIZZI and Roberto RIGGIO.

new requirements of the public safety community. Public safety and commercial systems are typically designed and deployed to fulfill different needs and have different requirements, which are directly affecting the quality of service of the communications. The unique and vital nature of public safety affects the technical decisions that are necessary to guarantee connectivity for everyone, anywhere and anytime. For this reason, recent enhancements of the 4G long term evolution (LTE) in the field of public safety communications, such as device-to-device and group communications for mobiles in physical proximity, accrue a great opportunity to bring new services and a high-level of technological innovation to the public safety workers or first responders. We describe and analyze the evolution of public safety communication systems from the technical standpoint providing a complete overview of what is available in the market and a glimpse of future trends. Furthermore, we discuss the challenges to efficiently deploy, operate and interoperate present and future technologies for public safety communication networks and describe how these technologies can help public safety agencies to meet their expectations.

2.1. Introduction

Natural disasters and unexpected events can stress the public land mobile network (PLMN) in terms of 1) network elements that could be damaged, or 2) peaks of traffic injected in the mobile network by the users. For example, extremely devastating events like earthquakes or tsunamis as well as large public gatherings such as the Olympic Games put the mobile network under exceptional workload conditions. Therefore, Public Protection and Disaster Relief (PPDR) organizations, or alternatively public safety agencies, are demanded to intervene to provide different kinds of support and provide rescue services whenever needed. In this context, the current dedicated networks for public safety (such as TETRA) suffer from technological limitations and lag far behind recent advances of electronics and telecommunication systems. For all these reasons, several initiatives have started in recent years in terms of both projects and standardization efforts to cover this gap [APC 15]. If on the one hand this is sufficient to provision voice-based services such as one-to-one or one-to-many communications as well as message dispatching, it is not sufficient to satisfy the demands imposed by data communications.

Recently, 4G LTE and its advanced version LTE-A have been selected as the next generation of cellular technology to be exploited for public safety communications [ABS 15, DOM 13]. Public safety users have stricter requirements than commercial users since several situations might jeopardize the lives of rescuers. Extremely low latency for starting a session (even less than one millisecond), dependability, resilience and availability of services are crucial requirements to be considered by the third generation partnership project (3GPP) for the LTE improvements required by public safety operations. We focus on the particular aspects pertaining to public safety communications; the evolution of the public safety networks in terms of communication technologies and presents an in-depth description of the future trends for public safety. Then, an overview of the 4G LTE technology and its improvements for public safety operations is also presented giving particular attention to the device-to-device (D2D) communication mode. This feature is readily available in traditional public safety networks (referred to as direct mode) but is still lacking in the 4G LTE technology. Finally, the main challenges that could be encountered by public safety agencies in terms of deployment and operation of public safety communications are analyzed focusing on disaster scenarios.

The remainder of this chapter is organized as follows. Section 2.2 surveys the relevant literature related to the topic of communication technologies for public safety networks. Section 2.3 shows an overview of the 4G LTE extensions for public safety communications. Section 2.4 analyzes the public safety communications requirements and main challenges. Conclusions are drawn in section 2.5.

2.2. Communication technology evolution for public safety networks

PPDR organizations increasingly rely on telecommunication networks in order to perform their duties. Even though analog systems are still extensively used by public safety workers [BAL 14], due to the recent technological advances, the public safety radios have been upgraded from analog to digital since 1990s. The driver behind the advances in public safety telecommunication systems is mainly the identified requirements [ITU 03] and the need of the support multimedia applications (real-time access, high-resolution maps, floor plans, on-field live video transmissions to a central unit, remote database access, etc.) [FER 13a]. The specific operational and security requirements of public safety networks motivated the development of different standards dedicated to public safety communications. Standards that have been adopted in the context of public safety are TETRA and its evolution TETRA enhanced data services (TEDS) and TETRAPOL, mainly adopted by European countries, and the Association of Public-Safety Communications Officials-International – Project 25 (APCO-25 or P25) mainly spread in the United States of America (USA). In emergencies or during special events, exceptional stress on communication networks occurs temporarily. It is worth to mentioning that, even though some rapidly deployable base stations are available on the market (e.g. MTS1 Base TETRA Station [MTS 15]), current technologies rely on fixed network infrastructures deployed on the basis of a long term planning. For instance, in Great

Britain, the Airwave network, which consists of over 3,000 base stations, was deployed and it is currently used by police forces, fire and ambulance services [AIR 15]. Even if public safety networks can be deployed according to the principle of redundancy to improve communication resilience [AIR 15], it has been recognized that in case of big disasters fixed telecom infrastructures can be seriously damaged hence affecting the operation of PPDR organizations.

2.2.1. *Review of the currently deployed technologies*

TETRA is a standard of the European Telecommunications Standards Institute (ETSI) [ETS 01] that was firstly published in 1995. Nowadays, TETRA is a mature standard open to multi-vendor technology based on fixed network infrastructures or base stations. It is employed in public safety since it is able to provide different services such as: group call, individual call, integrated voice and messaging, security features such as authentication and encryption, direct mode operation (DMO) and advanced radio features employed by public safety operators such as dynamic grouping, emergency call, etc. The TETRA voice codec is based on conjugate structure-algebraic code excited linear prediction (ACELP) technique providing a quality near to the global system for mobile communications (GSM) system at almost half of the bit-rate. Moreover, TETRA codec is able to provide a better quality than GSM in noisy environments. The frequency bands for TETRA products that are typically found in the market today are as shown in Table 2.1. In order to achieve spectrum efficiency, TETRA air interface makes use of time division multiple access (TDMA) technique and the modulation scheme is π/4-shifted differential quaternary phase shift keying (π/4-DQPSK). TETRA uses 25 kHz spaced radio frequency bearers divided into four time slots each. Thus, TETRA provides four communications channels per 25 kHz of band in order to achieve high spectrum efficiency.

TEDS is the evolution of TETRA to support high speed data (HSD) requirements [ETS 07]. In order to support high-speed data services, TEDS standard supports radio frequency bandwidths of 25, 50, 100 and 150 kHz. In the development process, particular attention was paid to the backward-compatibility with TETRA equipments. For this reason, the air interface in TEDS still supports 25 kHz channels with a π/4-DQPSK modulation. However, adaptive channel modulations are employed exploiting the following techniques:

1) π/4-DQPSK (for common TETRA and TEDS control channel).

2) π/8-D8PSK (for early migration requiring modest increase in speed).

3) 4 QAM (for efficient links at the edge of radio coverage).

4) 16 QAM (for moderate speeds).

5) 64 QAM (for high transfer rates).

TETRAPOL is another standard, substantially different from TETRA, which was originally developed for the French police forces [BAL 14, PAS 15]. An international standardization body did not develop the TETRAPOL standard; however, it is defined in publicly available specifications (PAS) at the disposal of any manufacturer willing to create TETRAPOL equipment. TETRAPOL can provide different services similarly to TETRA, such as: group call, individual call, wireless data, integrated voice and data, security with authentication and encryption, dynamic grouping, emergency call, etc. Even if according to the standard TETRAPOL should operate at frequencies between 70 MHz and 520 MHz, at the present time, its commercially available embodiments only support the bands in the range of 380–450 MHz. Unlike TETRA, TETRAPOL utilizes frequency division multiple access (FDMA) technique to achieve spectrum efficiency and Gaussian minimum shift keying (GMSK) modulation. Each TETRAPOL channel is divided into 12.5 kHz or 10 KHz radio channel; thus, it can achieve up to 2 voice channels within a 25 kHz radio channel. Two versions of the air interface are defined: one for operations below 150 MHz (VHF) and the other for operations above 150 MHz (UHF). The air interface in these two cases differs only in the specification of the interleaving and the differential encoding for voice and data frames. Due to its limited data service capabilities, TETRAPOL can be compared to TETRA release 1 (see Table 2.1).

APCO-25 is the standard developed in the USA for digital wireless communications for the public safety agencies. Similarly, to the other standards for public safety, it is able to provide a number of services like: group call, individual call, wireless data, integrated voice and data, security with encryption, dynamic grouping and emergency call. Additionally, APCO-25 maintains a backward compatibility with the analog systems. Thus, APCO-25 equipment can perform a direct communication with analog radios operating at the same frequency (without the network infrastructure). This is a key advantage of P25 over TETRA, since it allows for incremental deployment. As shown in Table 2.1, commercially available APCO-25 devices operate in the 136–174 MHz (VHF), 403–512 MHz (UHF) and 800 MHz bands. APCO-25 is composed of a phased development [APC 15]. In the first phase, the air interface uses FDMA for channel access and 12.5 kHz channels. In the phase 2, it is forecasted to support both FDMA and TDMA providing one voice channel per 6.25 kHz channel. Phase 2 equipment is not backward compatible with phase 1 due to the TDMA access technique; however, TDMA radios in the second phase should be able to operate in phase 1 mode to support interoperability in the upgrade process. At the present time, vendors are producing phase 1 equipment even if they can be upgraded to phase 2.

	Region	Access	Carrier Bandwidth	Channels per Carrier	Frequencies [MHz]	Maximum Data Rate [Kb/s]
TETRA	Europe	TDMA	25 kHz (25/50/100/150k Hz in TEDS)	4	380-400, 410-430, 800	28.8
TETRAPOL	Europe	FDMA	25 kHz (12.5 or 10 kHz channels)	2	380 – 450 MHz	4.8
APCO-25	USA	FDMA	25 kHz	2	136 – 174 (VHF) 403 – 512 (UHF) 800	9.6

Table 2.1. *Features of traditional public safety communications*

2.2.2. *Future trends in public safety communications*

As previously discussed, public safety communication technologies evolved over the past years in order to meet the strict requirements of public safety operations. Hence, public policies for communication systems adopted by PPDR organizations have evolved over many decades as well. In particular, old policies where based on the assumptions that:

1) local agencies should have maximal flexibility at the expense of standardization;

2) commercial carriers have little role to play;

3) public safety should not share spectrum or infrastructure;

4) narrow band voice applications should dominate.

Therefore, primary public safety communications systems are designed based on those policies and being run by thousands of independent local agencies leads to interoperability failures, inefficient use of spectrum, lower dependability and higher costs. Moreover, public safety requirements have changed significantly over time, and technology has changed as well, so there are many reasons to follow the fundamental shift in both policies and technologies. Policy reforms should include expanding the role of commercial service providers, allowing public safety to share spectrum with other players and expanding capabilities beyond traditional voice communications. Therefore, operational and technological innovation trends for public safety communications expected in the coming years are summarized in the following questions.

Are public safety networks going to adopt the commercial broadband data services and applications?

Traditionally, public safety communications systems primarily provide voice services. However, there are many reasons to bring services that could be useful to first responders, including broadband data transfer, real-time video and geolocation, which would enable dispatchers to track the precise location of first responders during an emergency. These new services and applications are already available in commercial broadband cellular networks. Therefore, the adoption of 4G LTE broadband technology in public safety will be central to the discussion in terms of 1) how it will be funded and deployed and 2) how public safety agencies will adopt packet data services for their day-to-day operations. With the amount of infrastructure sites and backhaul costs required to support LTE, most public safety agencies will look to share their infrastructure/communication systems with consumer/commercial operators. Therefore, new standards and technologies to enable sharing of public safety traffic with that generated by commercial users while guaranteeing channel access to public safety, deserves to be investigated. Additionally, discussions about how broadband technology will be leveraged for public safety are getting a lot of attention, especially considering the following aspects:

1) Will it be public or private networks?

2) Will it be using shared or dedicated spectrum?

3) Will it be a shared network?

How will traditional public safety networks evolve in the future?

In parallel to the ongoing discussion around public safety and 4G LTE technology, PPDR organizations will continue to migrate and deploy new P25 and TETRA land mobile radio (LMR) networks. LMR has been the standard for public safety communications for more than 80 years. LMR systems are inherently easier to set because of the smaller number of sites required to cover an area as compared to broadband data systems such as 4G LTE. The main advantages of keeping an LMR system for public safety communications are 1) LMR systems are private therefore sharing issues are not there, and 2) D2D communications require zero infrastructure in traditional LMR. As a result, LMR technologies remain the best tool for mission critical services like direct mode, group call and push-to-talk (PTT) voice communications.

Will it be possible to share communication resources between public safety agencies?

Public safety agencies will increasingly interoperate to share communications resources. Shared resources could be in the form of regional shared P25 and TETRA radio infrastructures, multi-agency command centers and possibly 4G LTE broadband networks in the future. Typical examples are emergency numbers such as

911 (USA), 999 (UK) and 112 (EU), answering points and multi-agency emergency operation centers. Increasing needs for interoperable communications will continue to push agencies to work together to share infrastructure, data and resources. The advantage of reducing the operational costs of maintaining and operating command centers and control rooms will lead more PPDR organizations to work together to build larger, regional, multi-agency centers. Some of these networks will include public–private partnerships and multi-border communication agreements.

Which levels of communication security will be needed in public safety networks?

Public safety agencies use private communication systems such as LMR networks with end-to-end encryption, including over the air encryption keys management. The new trend of sharing infrastructure/communication systems not only between public safety agencies but also with commercial operators will create concern regarding the security of the communications. Accordingly, new security standards and technologies to enable hardware and software encryption solutions like the assured mobile environment (AME) to provide standard-based security solutions need to be further studied.

What will the first responder's equipment look like in the future?

The evolution of smartphone and tablet devices is also spurring the debate around the first responders' equipment evolution, where the relevant questions are:

1) What should a public safety broadband device look like (form factor) [THA 15, ALC 15]?

2) What should it do, and how should it operate?

3) What are the requirements for cold and hot weather operation?

4) What are the requirements for resilience to snow and rain?

5) What are the requirements for gloved operation?

6) Should public safety devices rely on touch screens for their operations?

7) What are the requirements to keep devices operational in the field?

8) Where and when commercial devices can be used?

In the future, agencies will test commercial devices in public safety use cases, where extreme operational conditions are not required. Moreover, the rapid growth of commercial wireless services has led to mass market production and lowered costs. Thus, equipment used in public safety could be much cheaper than what could nowadays possible, if they resemble commercial equipment.

How will drones, robots, cameras and sensors complement the work of the public safety agencies?

The complexity and capabilities of drones and robots is rapidly evolving especially in terms of functionalities and cost reduction. This trend will allow, for example, the use of drones and robots for 1) replacing human intervention in dangerous situations, and 2) monitoring the disaster areas. Typical applications of drones and robots will be for example live video streaming for airborne surveillance in hazardous situations or over incident scenes for monitoring the level of radiations over a disaster zone. Video surveillance cameras and unattended sensors are also becoming the most important tools to extend the eyes and ears of public safety agencies. Consequently, public safety workers are increasingly being equipped with mobile video cameras to improve their efficiency, visibility and ability to instantly collaborate with the central command, coworkers and other agencies.

How will temporal rapid deployment networks complement the fixed public safety networks?

Current public communication technologies are not designed or not suitable to address rapid and temporal deployments of emergency communication systems immediately after a disaster has occurred. In this context, new public safety technologies that can be rapidly rolled out to provide dependable and resilient network connectivity at high data rates over large geographical areas will deserve more attention in the future.

How will the use of social networks influence the public safety operations?

Developing strategies to quantify, capture and find ways to use public and private data to predict and understand the chaos that public safety agencies will face during or after disaster scenarios have to be investigated in the future. Such new strategies will use and combine real-time data flows, aggregate public/private sector video shooting, access to social media and will correlate/link data to command centers and control rooms to improve incident management. Citizens have increasing expectations for the agencies that serve them since they already constantly use social networks for gathering information about transportations, weather conditions, and news between many others. Hence, the ability to capture and share data for managing disaster situations through social networks will be more important than ever. In this respect, the following points needs to be addressed in the future:

1) Investigate the most effective way to deliver messages using telecommunication infrastructures and their impact on the community.

2) Investigate how social behavior is affected after receiving the alarm messages and what are the consequences.

3) Designing alarm messages for effectively delivering the message and avoid creating panic to the recipients.

4) Investigate how social networks can help crisis management.

Will the use of cloud services for public safety applications be the new trend?

Moving to "the cloud" is one of the biggest trends in information technologies and public safety is increasingly more sensitive to this trend. This is to say that cloud information services could range from email and productivity applications to more specific public safety applications. PPDR organizations will be more comfortable with not having application and database hosted on-site. Adoption of smartphone and tablet devices will play an important role as many apps rely on resources in the cloud. Fundamentally, moving to "the cloud" automatically involve also edge computing which is about pushing processing for remotely isolated applications away from the CORE of the data center to the outer edges of the network. Thus, processing some data locally can significantly reduce the amount of data that must be sent over the network to a central data center and alleviate network traffic bottlenecks, as well as application performance concerns. The concept of "cloud services" will be extended to radio and broadband networks services. Cloud services can relay in new networking trends, such as software defined network (SDN) and network function virtualization (NFV), for allowing the effective virtualization and management of wireless and wired communication network resources. SDN creates network abstractions to enable faster innovations while NFV reduces Capital expenditures (CAPEX), operating expenses (OPEX), space and power consumption. NFV is highly complementary to SDN, but not dependent on it (or vice versa). Even if both NFV and SDN functionalities can be implemented independently each other, although the two concepts and solutions can be combined and potentially provide benefit to public safety communications.

How will public safety networks reduce the costs of maintenance and operations?

Traditionally, public safety agencies purchase their own communication technologies. However, communication technology as-a-service is emerging as a new model to reduce the costs to maintain and operate the communication network. Adoptions of cloud services and broadband radio as-a-service will allow public safety agencies to keep their communication system up to date whereby agreements between public safety agencies and vendors to keep their system constantly upgraded.

Finally, future trends for the public safety communications need to follow a clear roadmap in order to allow a harmonized and successful migration toward new and future standard technologies. Figure 2.1 shows a possible roadmap for public safety communications evolution relying on what has been discussed above.

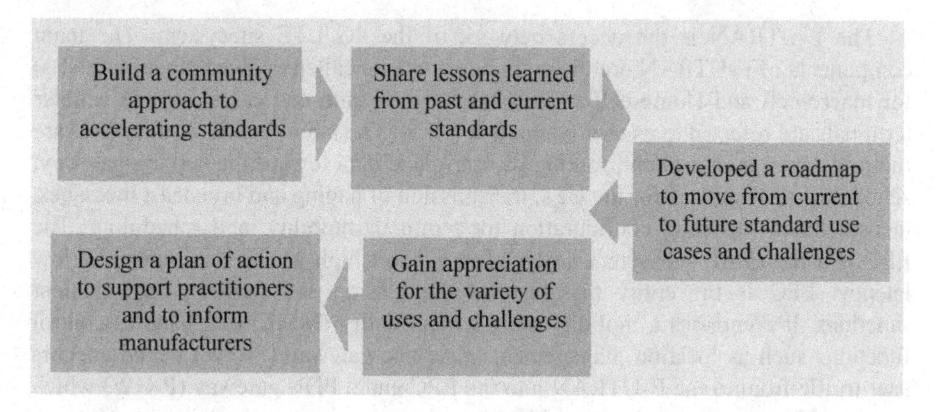

Figure 2.1. *Roadmap for the evolution of public safety communications*

2.3. 4G long term evolution for public safety communications

The adoption of 4G LTE as candidate for public safety communications introduces a unique opportunity to improve the response of PPDR organizations and it will bring high-level applications to first responders. Indeed, 3GPP Rel. 12 incorporates enhanced features to meet the requirements of the public safety community. In fact, establishing common technical standards for commercial and public safety communication networks offers several advantages to both communities. In the following sections, the main changes that 4G LTE will undergo to adapt to public safety uses are discussed.

2.3.1. *LTE standard overview*

4G LTE is the mobile broadband technology standardized by 3GPP to satisfy the increasing demand of mobile broadband services with higher data rates and QoS. The standard adopted orthogonal frequency division multiplexing (OFDM) as the multiple access technology to efficiently support wideband transmissions. Multiple-input multiple-output (MIMO) techniques are used to improve spectral efficiency. The 4G LTE network is characterized by a flexible air interface with low latency and enhanced performance supporting flexible channel bandwidths from 1.4 to 20 MHz in both uplink and downlink directions. Both frequency division duplex (FDD) and time division duplex (TDD) allocation methods are supported to accommodate flexible channel bandwidths in the allocated frequency spectrum. The architecture of the 4G LTE system (and the current release LTE-Advanced) is divided into two main subsystems: 1) the evolved universal mobile telecommunications system (UMTS) terrestrial radio access network (E-UTRAN), and 2) the evolved packet core (EPC).

The E-UTRAN is the access network of the 4G LTE subsystem. The main components of E-UTRAN are the radio base stations called evolved NodeBs (eNBs) for macro-cell and Home-eNBs (HeNBs) for the femto-cell cases and the cellular terminals are referred to as user equipment (UEs). The main functions of an eNB are radio resource management, routing of user plane data toward the serving gateway, scheduling of resources for the UEs, transmission of paging and broadcast messages, measurement, reporting configuration for terminals mobility, and scheduling. The EPC is a flat all-IP subsystem designed to support high packet data rates and low latency. EPC is the entity that handles the session and mobility management functions. It comprises a mobility management entity (MME) that handles control functions such as location management, a service gateway (S-GW) which anchors user traffic from/to the E-UTRAN into the EPC and a PDN gateway (P-GW) which provides IP connectivity to external IP networks.

2.3.2. *LTE-based public safety networks: improvements and challenges*

LTE is the widely deployed global mobile broadband standard featuring greater flexibility to support a wide variety of deployment scenarios and operators' needs. In an effort to harmonize commercial cellular networks and dedicated public safety technologies, as well as employ the inherent features of 4G LTE system to public safety communications, 3GPP has introduced enhancements to the standard LTE. This aims to maintain the current LTE features while incorporating functionalities that are specific to the public safety communications and to maximize the technical commonalities between commercial and public safety sectors and so provision the best cost effective solutions to both communities. The features identified by 3GPP to enrich the inherent facets of the 4G LTE system to support public safety communications are described and analyzed below.

2.3.2.1. *Proximity services (ProSe)*

In 3GPP, *D2D* is defined as the communication between two UEs in physical proximity using 4G LTE air interface to set up a direct link without routing via base station and core network. Proximity-based communication identifies devices in the vicinity and enables optimized communication between them. This essential function needs to be supported in public safety to allow communication between public safety officers even if the network coverage is down or when the device is out of network coverage. For commercial cellular users, it is acceptable to move around to find a location with adequate coverage but this is unacceptable or impossible in mission critical communications since the constraints of the incident scenario. To enable D2D, 3GPP has standardized the concept of proximity service (ProSe) [GPP 12], which is conceived to cater for both mission critical and commercial communication requirements whereby D2D ProSe-enabled services over 4G LTE air interface.

The high-level features of the ProSe network consist of ProSe discovery and ProSe direct communications setup:

1) *The ProSe discovery function* identifies ProSe-enabled devices in proximity either at the level of the radio access network (direct proximity discovery) or at the EPC level [GPP 13]. In direct proximity discovery, the UE would search for nearby UE devices autonomously and require them to participate in the device discovery process to periodically transmit/receive discovery signals. The direct discovery can operate both under network coverage and out of coverage and does not preclude network assistance when available in either a push or pull mechanisms. In a push-based discovery, the UE broadcasts its presence while in pull discovery the UE requests information regarding discoverable proximity devices. In EPC-level proximity discovery, the EPC determines the proximity of UE devices, and a UE device is allowed to start the discovery procedure after it receives its target information from the network. This scheme requires the network to keep track of the candidate D2D UE devices, reducing the discovery burden on the mobile terminals.

2) *The ProSe direct communication setup* enables the establishment of communication paths between two or more ProSe-enabled UEs that are in direct communication range (after the ProSe discovery function phase finished). The ProSe direct communication path could use E-UTRAN or WLAN. For public safety specific usage, two scenarios are considered:

– public safety ProSe-enabled UEs can establish the communication path directly between two or more public safety ProSe-enabled UEs served by the E-UTRAN;

– ProSe direct communication is also facilitated by the use of a ProSe UE-to-network Relay, which acts as a relay between E-UTRAN and UEs not served by E-UTRAN.

Notice that in 3GPP the procedure for direct communication setup is not standardized, however several procedures are proposed in the literature [GOR 13, GOR 14].

2.3.2.2. *Group communication*

Group communication is a service that offers efficient exchange of voice and data between groups of devices in a controlled manner. In order to be able to exploit 4G LTE technology for public safety group communications, such a feature is needed [DOM 13]. To accomplish the effort to enable group communication, the 3GPP study group introduced an enhancement of 4G LTE that enables group communication services called Group Communication System Enabler over 4G LTE (GCSE-LTE) [GPP 14]. In the standard, group communication is represented by the group communication service application server (GCS-AS), which exploits the services offered by the GCSE-LTE. This network entity makes the decision to use either unicast or broadcast mode for sending traffic to the public safety devices (handhelds of the first responders).

2.3.2.3. *Mission critical push-to-talk*

Mission critical PTT voice systems are utilized by public safety allowing users to selectively and sequentially transmit messages to one another, either on a one-to-one or one-to-many basis. A PTT session is a half-duplex communication where one person speaks while the other(s) is (are) only in listening mode with the permission granted on a priority basis. This is a multicast or broadcast service in which a transmitter in a session sends packet data traffic to a dedicated mission critical PTT application server that copies the traffic for all the recipients. Mission critical PTT is under standardization in 3GPP Rel. 13 [GPP 15], which assumes mission critical PTT application server as part of the GCS-AS. GCS is a generic function for voice, video and data whereas mission critical PTT is a voice communication service. The service will include regular group calls, broadcast group calls where no response is expected by initiating user, priority-based group call such as emergency group call that could pre-empt other calls in progress and private one-to-one calls.

To leverage public safety communication based on 4G LTE the enhancements included by 3GPP are under development for full-fledged functionalities. Moreover, new features such as Isolated E-UTRAN operation for public safety are under development to ensure the necessary resilience in case the backhaul connection is cut-off during mission critical operations and enable autonomous security mechanisms. Such new features are items part of the standardization road map toward 3GPP Rel. 13 [GPP 14].

2.3.3. *D2D-LTE main challenges for public safety usages*

Although the foreseen 4G LTE system enhancements for mission critical communication are promising, the standardization faces several challenges which have to be addressed adequately [XIN 14, COM 14, JSA 14]:

1) *UE mobility*: one such a challenge is the mobility of the UEs in D2D communication mode that affects the temporal correlation of shadowing as well as the fast fading, for instance increasing the Doppler spread. Additionally, the mobility of the UEs will determine the lifetime of the D2D communication network.

2) *Propagations aspects*: the antenna height and inter-link correlation are the other design challenges in D2D communication networks. With the antenna height of a UE around 1.5 m, the UE-UE link incurs higher path loss than eNB-UE link of the same size. Small inter-UE distances are expected in D2D communication networks, which can potentially cause much higher correlations in the propagation characteristics resulting in degraded intelligibility of the received signals.

3) *Energy consumption*: other aspects that require sufficient attention are power allocation and control in D2D communication network, which have a direct impact on energy consumption and efficient recourses utilization. For public safety communications, the battery duration of handheld devices is a crucial mission critical requirement especially during dangerous missions.

4) *Spectrum aspects*: spectrum allocation and management is another issue that hinders the evolution of 4G LTE for public safety communications because of political, regulatory and economic influences on the solutions to be adopted [FER 13b]. Due to the dynamics of public safety communications, policies that rely only on exclusive use of dedicated spectrum do not suit well the requirements of PPDR organizations [BUD 07, PEH 07]. A hybrid approach in which spectrum is shared with commercial systems could be followed in order to enable instantaneous and reliable access to sufficient amount of spectrum while simultaneously improving the overall spectrum utilization efficiency [FER 12].

5) *Interference management*: D2D communication network design must carefully manage channels designated exclusively for D2D links, as well as channels jointly used by both UE-to-UE and UE-to-eNB connections. In fact, reservation pair by pair of UEs that participate in different D2D communication networks (UEs physically in proximity) could overlap partially or even completely. In other words, inter-D2D network interference could occur, hence requiring interference management and mitigation techniques. For example, a simple way to circumvent the problem could consist of forming less D2D networks with a higher number of connected UEs in each, which could positively contribute to relieve this problem. The existing literature on D2D interference management addresses channel assignment relying on game theory, D2D admission control; D2D power control and D2D relay selection.

6) *Synchronization issues*: eNB synchronization beacon can be used for D2D network synchronization (when D2D networks are created in coverage of at least one eNB). However, there are two aspects that have to be underlined: i) UEs might be associated to different eNBs that are not synchronized in LTE- FDD, and ii) even when located in the same cell, UEs are located at different distances from the eNB and different timing advance adjustments are required. When D2D communication networks are out-of-coverage of the eNBs, periodic transmission of synchronization signals from UEs may be required. Although the reuse of existing 4G LTE signals like primary/secondary synchronization signals (PSS/SSS) is simple and obvious,

deciding which UE is going to transmit the synchronization signal is definitively more challenging.

7) *Indoor and outdoor localization*: localization algorithms present in the literature are usually based on combination of GPS, mobile, sensors and signal propagation techniques. These algorithms rely on, centralized systems that, combining different techniques, are able to accurately track the temporal localization of the UEs. In the case of D2D networks, where centralized systems are not present, the localization will become a challenge, especially for indoor UEs. GPS, mobile and sensor techniques for indoor D2D networks are not the best option (especially in public safety scenarios) while localization algorithms based on propagation techniques can be used. However, the penetration loss of indoor-to-outdoor and outdoor-to-indoor propagation is usually a function of environment characteristics (building material properties, temperature, rain between others) which make it difficult to be predicted with enough accuracy for provide reliable localization of the UEs.

8) *Security*: in 3GPP, there is a specific group studying security issues for D2D communications design, including routing, security key management and attack identification. The UEs sending and receiving the data must be assured their data is not eavesdropped by malicious devices outside the D2D communication network. Additionally, security issues are one of the most relevant topics to be addressed in public safety communications due to the sensitive nature of data that first responders exchange during mission critical communications.

2.3.3.1. *BS/eNB main challenges for public safety usages*

Standardization updates in protocols and operations are also expected at BS/eNB level. Standardization groups will face several challenges, which are summarized in the following [YUT 12,GOM 14]:

1) *Mobility management*: the multi-technologies standard for public safety communications will challenge the new generation of BS/eNBs. Improving the soft hand-off process between technologies (TETRA, TETRAPOL, 3G, 4G and WiFi), reducing the vehicular penetration loss that affects communication and allowing non-cellular radio's inter-working with LTE for traffic offload will be the focus of advanced mobility management mechanism [YUT 12]. These topics are crucial for supporting public safety communications.

2) *Self configuration, plug and play*: LTE small cells or temporal BS/eNB require a lot of hand crafting, tuning and long setup time for performance. In public safety communication the required setup time of the communication devices are crucial especially in critical missions. Therefore, LTE advanced research areas will focus on introduce of self-configuration, plug and play small or temporal BS/eNB capable of (1) embedding self-configuration and learning mechanisms at different networking levels, and (2) embedding the most fundamental CORE/EPC operations.

The advantages of these approaches are the reduction of the dependency between network entities making the network more robust in case of temporal deployments and backhauling disruption for instance [GPP 14, GOM 14].

3) *Service Level QoS and Service Level Agreement (SLA) Support*: early deployments or cellular networks focused on simple applications of voice and data only. Ensuring the required QoS and SLA demanded by public safety operations will request further differentiation at the flow and packet levels especially for advanced public safety applications. QoS and SLA enforcement techniques for the different types of services need to be introduced at BS/eNB level in order to guarantee high resilience public safety communications.

2.4. Public safety communication requirements and main challenges

PPDR organizations include emergency management agencies, law enforcement agencies, fire departments, rescue squads, emergency medical services (EMS) and many others. All these agencies make use of information and communication technology (ICT) for their day-to-day communications. In the following section, the requirements of public safety communications in terms of deployment, operations and challenges are described.

2.4.1. *Deployment requirements of public safety communication networks*

There are a number of possible ways forward with the goal of developing public safety networks. Allowing first responders to make use of multiple ICT systems will increase the chances that first responders can use more capabilities during their daily operations on the field. The most relevant ICT systems that can be used for public safety operations are summarized below.

1) *Public safety agencies communication infrastructure*: municipal or regional systems based on typical public safety technologies (TETRA, TETRAPOL or P25) play a useful role for public safety operations. Public safety communication networks are used by emergency services organizations, such as police, fire and emergency medical services to prevent or respond to incidents that harm or endanger persons and/or properties. The typical systems used by public safety are TETRA and its evolution TEDS, as well as TETRAPOL in European countries, whilst APCO-25 in the USA. These technologies allow provisioning several necessary services, for example group call, individual call, wireless data, integrated voice and data, security with authentication and encryption, dynamic grouping, emergency call, etc.

2) *WiFi-based municipal systems*: this system can be deployed by municipalities in cities using wireless local area network broadband technology, or just be

strategically placed in hot spot areas. These systems are relatively low-cost, provide high data rates and can serve many needs, including, but not limited to, public safety scenarios. Although this technology is currently not able to support some mission-critical applications over large regions, it can serve well certain uses. These uses include fixed applications, such as transferring data from a remote surveillance camera to a command center and applications where lives do not depend on ubiquitous and instantaneous access such as transferring arrest reports from a police car back to the police station.

3) *Commercial cellular carriers*: a commercial company could provide services to public safety agencies across the nation focusing on services such as broadband connectivity that are not completely available for public safety communications as today. Cellular carriers can compete to offer services to PPDR organizations in a way that the diversity of telecommunication networks can greatly increase service dependability and radio coverage. Cellular carriers also can provide new services that are not offered typically to PPDR organizations. However, it is still necessary to pay attention on whether public safety's service requirements will be met adequately and in perpetuity.

It is important to note that large-scale destruction of communication infrastructures caused by natural disasters or unexpected events might hamper the communication of the public safety agencies over a large area. To fulfill the requirement of rolling out flexible and rapidly deployable resilient communication infrastructures for public safety, several projects are designing and validating innovative holistic system-wide approaches for ensuring dependable communication services based on the following main features: rapid deployment, flexibility, scalability, resilience and provision of interoperable broadband services. Hereinafter some examples are provided to clarify this point.

4) *Ad hoc networks*: *ad hoc* networks are ideally suited for applications where all devices are mobile or they are transported to an emergency scenario when this is needed. Such networks might be set up quickly among portable devices placed in a burning building or among fast-moving police cars (i.e. vehicle-to-vehicle communications). This is also an effective solution where much of the communication is local, for example, to enable public safety devices operating inside an urban subway system to communicate between one another at high data rates. In this case, some forms of D2D communications could be applied in particular to relieve out of network coverage conditions.

5) *Aerial-terrestrial temporal networks*: rapid and temporarily deployable communication networks are an important option to be considered in cases of disaster events. For example, the European FP7 ABSOLUTE project [ABS 15] aims to design a holistic approach that opportunistically combines terrestrial, aerial and satellite communication network capabilities. The proposed architecture, shown in Figure 2.2, relies on the features of 4G LTE technology to efficiently support low

latency and high capacity requirements of future public safety networks. ABSOLUTE architecture is based on the following elements:

Aerial 4G LTE base stations (AeNB): the AeNB subsystem will be deployed by means of tethered Helikites platforms equipped with the 4G LTE payload remote radio head (RRH), batteries and antennas. Optical fiber using CPRI interface connects the RF part (i.e. RRH on the aerial platform) to the base band eNB (BB-eNB) located on the ground. The complete system (RF part and BB-eNB) is capable of acting as an aerial 4G base station and it is connected to a flexible management entity (FME), which is a distributed virtual EPC. The FME is connected to a Ka-band satellite modem in order to provide Internet access for the served devices. The use of AeNBs allows envisioning efficient mobile network planning, more efficient and advanced network management in different mobility patterns, dynamic spectrum access and management as well as the provision of rapidly deployable multi-purpose services.

Portable land mobile unit (PLMU): the PLMU is a standalone and self-sufficient communication node that integrates a WLAN access point, an IP router, a 3G femtocell, a wireless sensor network gateway, a TETRA base station, a Ka-band satellite modem and a 4G eNB. Additionally, the PLMU includes other parts that support its main role as a communication platform such as batteries, power supply and a PC that controls all of the PLMU functionalities.

Multi-Mode User Equipment (MM–UE): the MM–UE purpose is to integrate several technologies in order to provide at any time the necessary communication means for first responders. The MM-UE is able to communicate with AeNB, PLMU and is engaged in D2D communications.

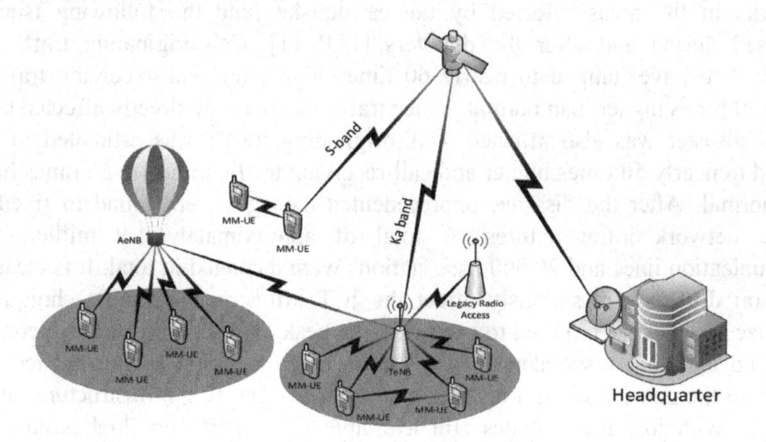

Figure 2.2. *Hybrid aerial-terrestrial-satellite network architecture*

6) *Satellite networks*: satellite systems are another communication option given the possibility to cover vast geographical regions and immunity to earthquakes, hurricanes, and to terrorist attacks. Hence, satellite systems play an important role in sparsely populated areas where terrestrial coverage can be expensive, or in areas where terrestrial systems have been destroyed by a disaster. It must be noticed that they are generally not the first choice in all cases in which terrestrial networks are available. The time it takes for a signal to travel to/from a satellite is inherently problematic in many applications, including basic voice communications.

2.4.2. *Operation requirements of public safety communication networks*

The role of ICT has been recognized as integral to disaster management. Although communication technologies have different roles in different phases of a disaster (i.e. before, during and after), most of the applications have traditionally been found in response and recovery phases (during and after the disaster occurs) [KUN 13]. Focusing on a real-life scenario, a massive 9.0 magnitude earthquake struck Japan on March 11th 2011. The quake was centered 130 km to the east of the Miyagi prefecture's capital, Sendai. The earthquake, which caused a tsunami that crashed into the country's north-eastern coast, was originally reported at a magnitude of 7.9 but later it was updated to 8.9 and then 9.0. Japan was well prepared for earthquakes since many buildings remained standing afterward but it was not prepared for the subsequent tsunami. Damages also occurred in the capital of Japan, Tokyo, and many injuries in the north region where the quake was centered.

Statistics reported that the voice traffic registered over the commercial mobile networks in the areas affected by the earthquake (and the following tsunami) increased during and after the disasters [JAP 11]. Call-originating traffic was estimated to have jumped to nearly 60 times higher and call-receiving traffic to nearly 40 times higher than normal. Voice traffic in areas not directly affected by the natural disaster was also affected. Call-originating traffic was estimated to have jumped to nearly 50 times higher and call-receiving traffic to nearly 20 times higher than normal. After the disaster, unprecedented damages were found to fixed and mobile network infrastructures. A total of approximately 1.9 million fixed communication lines and 29,000 base stations were damaged in total. It is clear that a natural disaster can seriously affect the ICT infrastructures and techniques to prioritize services are required to cope with the peaks of traffic created by people on one hand and first responders on the other. There are both social and technical aspects to the considered in a disaster management. The ICT infrastructure may be damaged with just fewer nodes still available to support very high volumes of traffic, thus causing overloading conditions in both the control and user plane signaling. Therefore, it is imperative to identify the specific role of each communication technology before, during and after a disaster occurs and how they can interoperate is s crucial aspect for running public safety communication networks (see Figure 2.3).

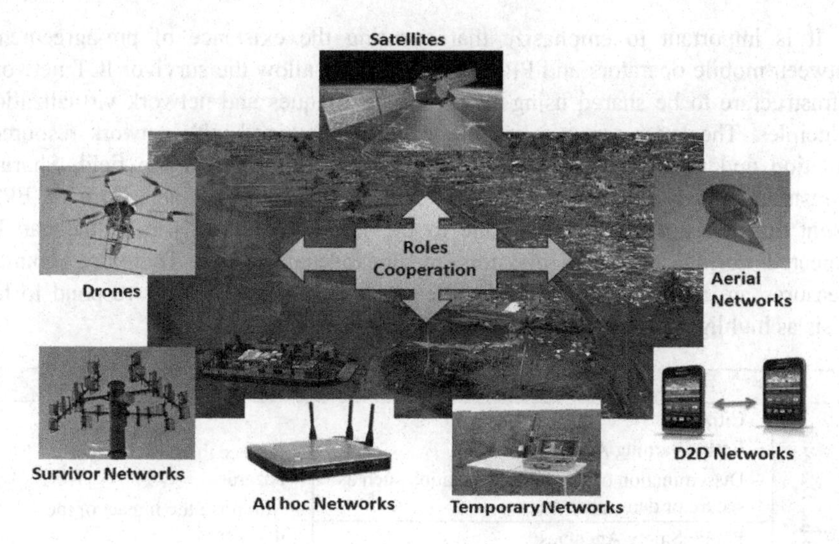

Figure 2.3. *Communications systems over disaster areas*

The types of urgent responses that are required before, during and after disasters have occurred are summarized in Table 2.2. These responses are all part of an efficient homeland security system. In the few cases in which disasters such as tsunami or earthquakes can be predicted, early alert messages and relevant information (e.g. gathering points) can be disseminated amongst the population to reduce or even prevent additional chaos caused by the citizens. Some examples of early actions using ICT technologies can include controlling train and highway traffic in order to prevent additional incidents and evacuate the people from dangerous places, schools or big assembles. PPDR organizations use ICT systems to disseminate the evacuation plans, organize rescuing activities, coordinate actions between agencies and communicate/update status of the disaster to just name some of the many activities. When a disaster occurs, communication technologies can play a vital role for saving human lives as well as for public safety agencies. For example, ICT technologies can be used for the following purposes:

1) Supporting PPDR organizations and many activities such as victims rescue, coordination, victims search, people evacuation, resource distribution and many others.

2) Identify affected areas and the level of damages in order to disseminate warning messages with the purpose of preventing people going to those areas.

3) Identify safe areas and evacuation points in order to disseminate warning messages for keeping people in safe conditions and indicating evacuation points and plans.

It is important to emphasize that ensuring the existence of pre-agreements between mobile operators and PPDR organizations allow the survivor ICT network infrastructure to be shared using scheduling techniques and network virtualization principles. The latter aspect is important when combined with network resources isolation and will play a critical role for first responders on the field. Sharing infrastructure/communication systems (eNB or BS, backhaul and Core or EPS) might anyway not be sufficient due to the large amount of traffic that can be expected during and after disasters or unexpected events. Therefore, counter measures are required for the ICT systems in order to adequately respond to the crisis as highlighted below:

	ACTIONS	OBJECTIVES
Before the Disaster	Citizens: – Early Warning Alarm System. – Dissemination of relevant information such as secure or dangerous areas.	– Reduce the probability of hazards. – Minimize the impact of the disaster. – Avoid major damages and saving lives.
	Public Safety Agencies: – Dissemination of evacuation plans. – Dissemination of plans to organize rescue and to provide first aid.	
During the Disaster	– De-prioritization and/or downgrading of normal calls and data to free resources. – Support of high prioritization (resources pre-emption) of emergency calls and relevant data. – Support of high prioritization and differentiation of users and devices identity for public safety services. – Support of resources isolation for public safety users. – Blocking incoming calls for keeping free communication resources. – Ensuring communications availability. – Optimizing use of resources. – Deploy temporary telecommunication infrastructures.	– Ensure communications. – Optimize resources. – Identify dangerous situations. – Ascertain that PPDR organizations can communicate with their personnel inside/outside the disaster area and also with the people in the disaster. – People might be severely injured or need immediate aid. – People can report back to family and friends.
After the Disaster	– Disseminating relevant information. – Affected communications nodes have to be restored. – Temporal communications nodes have to be removed.	– Enable interoperability between PPDR organizations. – Support the community and help to restore normality.

Table 2.2. *ICT considerations before, during and after disasters*

4) Design specific resource schedulers to serve the users according to service priority and real time adaptation.

5) Ascertain that high-priority calls, messages and data can be delivered in disaster conditions using pre-emption of resources.

6) Introduce time limits on the calls and differentiate between citizen and first responders.

7) Reduce the quality of telephone calls to a certain degree and use video and audio codecs tailored for low bandwidth consumption.

8) Block incoming calls to the disaster area to leverage additional free resources for emergency calls.

9) Reduce delays in the transmission of mobile phone data eliminating in parallel obsolete data which do not need to be transmitted any more.

10) Install the new temporary nodes which are easy to roll-out, self-configurable and with efficient energy consumption.

Finally, after a disaster has occurred, the dissemination of relevant information can be used to support the community and to help people to return to normal life conditions.

2.4.2.1. *Public safety and commercial users*

Notice that sharing infrastructure/communication systems immediately implies to manage different types of users connected to the network. In commercial networks, the use of the resources is done preferentially on the basis of user identity and service type. On the other hand, to fulfill the need of first responders, the way in which resources are treated by the commercial network might not be adequate to distinguish roles, scenarios and applications needed by the public safety communications.

To better exploit the resources made available by the broadband network, several methods could be used to differentiate the nature of the requests generated by PPDR organizations. Therefore, Table 2.3 shows examples of possible criteria and key factors used to differentiate resource requests by commercial and public safety communication networks. Having different criteria and key factors for managing resource requests of first responders and commercial users is a key point of differentiation that allows acquiring information such as user location, user ID and emergency status directly from the radio network.

<table>
<tr><th></th><th>Criteria</th><th>Key Factors</th><th>Examples</th></tr>
<tr><td rowspan="2">Commercial Domain</td><td>USER</td><td>Identity</td><td>Post-paid vs. Pre-paid
Limited vs. Unlimited
Businesses vs. Standard</td></tr>
<tr><td>SERVICE</td><td>Type</td><td>Voice vs. Video
Messaging vs. Email</td></tr>
<tr><td rowspan="8">Public Safety Domain</td><td rowspan="2">USER</td><td>Type</td><td>Citizens vs. Public Safety Officers</td></tr>
<tr><td>Identity</td><td>Fire vs. Police</td></tr>
<tr><td>DEVICE</td><td>Type</td><td>Portable vs. Car-mounted</td></tr>
<tr><td>SERVICE</td><td>Type</td><td>Alarming vs. Normal messages</td></tr>
<tr><td rowspan="4">USAGE</td><td>Location</td><td>Emergency area vs. safe area</td></tr>
<tr><td>Time</td><td>Before (if possible) vs. During vs. After the disasters happen</td></tr>
<tr><td>Situation</td><td>Alarm Fire vs. Terrorist Attack</td></tr>
<tr><td>Network</td><td>Normal vs. Roaming</td></tr>
</table>

Table 2.3. *Criteria and key factors used to differentiate resource requests*

2.4.3. *Public safety communications main challenges*

Embracing the approach that public safety communications could also rely on commercial standards promise to make progresses in a number of critical areas in which public safety networks still have to face several challenges [COM 14, JSA 14, PEH 07].

1) *Interoperability*: in the context of public safety, interoperability is the ability of individuals from different organizations to communicate and share information in real time. During critical situations, communications among first responders of

different PPDR organizations are hampered by interoperability problems. In Europe, incompatibility is mainly due to the lack of a harmonized approach to frequency planning and standards for public safety communications. Interoperability is a problem because decisions are made by local agencies, each of which has the flexibility to choose technology that is incompatible with that of others agencies. In this way, the possibility to reuse commercial land mobile radio systems in public safety communications is emerging as a suitable solution to solve interoperability issues. However, sharing infrastructure/communication systems is not enough to solve interoperability issues. Debates about how to improve the interoperability of PPDR organizations locally, nationally and internationally are actually needed for targeting issues such as sharing networks, data and resources in efficient way.

2) *Spectral efficiency*: public safety agencies have expressed concern that a shortage of dedicated spectrum is mainly due to the increased number of organizations installing their own network infrastructures. If public safety communication networks undergo a spectrum shortage, their communications capacity will be inadequate during emergencies. Simply allocating more spectrum resources for public safety without improving the efficiency of spectrum usage will not solve the problem. Introducing new techniques for dynamic spectrum allocation and cognitive radio can help to improve spectrum utilization efficiency of public safety communications.

3) *Dependability and fault tolerance*: critical components of public safety communication systems should rarely fail. However, some failures are inevitable in case large disasters occur, but this need not bring down the entire system. In a fault-tolerant design, other parts of the system will continue to operate, compensating for failures to some extent. This can only occur if systems are designed coherently and taking into account fault-tolerant approach. Moreover, rapid and temporary deployable public safety networks need to be adopted as vital infrastructures available to PPDR organizations. Additionally, pre-agreements between mobile operators and governments are necessary to allow agencies to use the survivor ICT network infrastructure to facilitate public safety operations during disasters.

4) *Security*: public safety networks need be designed in such a way that malicious nodes cannot easily attack or drop communications of the first responders. The greatest challenge will be in protecting inter-agency and cross-border communications, because protection must run on an "end-to-end" basis although today national and international public safety organizations adopt different technologies and policies. Therefore, efforts for harmonizing the current situation between agencies will help to solve interoperability issues and will facilitate the implementation and enforcement of end-to-end security.

5) *Cost*: the uncoordinated actions of PPDR organizations at different jurisdictional levels (local, regional or national) greatly increase costs of operation and maintenance of the communication systems. The amount of communication infrastructures deployed in a region should depend on the size of the region and its population and not on the number of local governmental entities involved. Policies for sharing infrastructure/communication systems and resources will consequently reduce the costs. However, alternative cost reduction might come from partnership between commercial telecom operators and pre-agreements between governments and operators. Tax reduction for public safety equipment need to be considered by the governments as a way of reducing deployment costs.

2.5. Conclusions

The design of public safety technologies was, is and will be greatly influenced by public safety actors and their policies. Deployment, operations and services are influenced by decisions commanded by policies with standardization determining efficiency, resilience and flexibility of public safety communication networks. This chapter delved into a comprehensive survey of the evolutionary role of telecommunication technologies in public safety, addressing several challenges still open from the deployment and operational standpoints for public safety networks. The roadmap of the evolution of communication technologies in public safety as well as the future trends were discussed providing a complete view of "where is" and "where will be" the technology to be used by the next generation of communication networks adopted by the public safety sector in the future. Special attention was paid to 4G LTE and the way it can meet current and future needs of PPDR organizations. LTE cellular technology and its advanced version LTE-A can enable the first responders with a better blend of reliability and multimedia capabilities needed for day-to-day operations. LTE will continue to develop in a backward-compatible approach and will be an important part of the 5G solution. However, 5G wireless access does not focus on specific radio access technology. It is, rather, the overall wireless-access solution addressing the demands and requirements of future mobile communication and advanced public safety communications.

2.6. Bibliography

[ABS 15] ABSOLUTE, EU FP7 Integrated Project, available at: http://www.absolute-project.eu/, 2015.

[AIR 15] AIRWAVE SOLUTIONS, Gaining strength using airwave to improve multi-agency response, available at: https://www.airwavesolutions.co.uk/newshub/assets/pdf/Interoperability-guide.pdf, 2015.

[ALC 15] ALCATEL, Mobile broadband for firstnet, available at: https://www.alcatel-lucent.com/public-safety/mobile-broadband-for-firstnet, 2015.

[APC 15] APCO, Project 25, statement of requirements, available at: https://www.apcointl.org/images/pdf/SOR-2010.pdf, 2015.

[BAL 14] BALDINI G., KARANASIOS S., ALLEN D., *et al.*, "Survey of wireless communication technologies for public safety", *Communications Surveys & Tutorials*, IEEE, vol. 16, no. 2, pp. 619–641, 2014.

[BLA 11] BLAKEY J., Case study – Japan earthquake & tsunami 11/03/11, available at: http://joeblakey.com/geography/case-study-japan-earthquake-tsunami-110311/, 2011.

[BUD 07] BUDDHIKOT M.M., "Understanding dynamic spectrum access: models, taxonomy and challenges", *Proceedings of IEEE Dyspan*, pp. 649–663, 17–21 April 2007.

[COM 14] IEEE COMMUNICATIONS MAGAZINE, 5G wireless communications systems: prospects and challenges (Part 1), May 2014.

[DOM 13] DOUMI T., DOLAN M.F., TATESH S. *et al.*, "LTE for public safety networks", *Communications Magazine*, IEEE, vol. 51, no. 2, pp.106–112, February 2013.

[ETS 01] ETSI EN 300 392-2 V2.3.2 (2001-03), Terrestrial trunked radio (TETRA), Voice plus Data (V+D), Part 2: air interface (AI), 2001.

[ETS 07] ETSI TR 102 580 V1.1.1 (2007-10), Terrestrial trunked radio (TETRA), Release 2, Designer's Guide, TETRA high-speed data (HSD), TETRA enhanced data service (TEDS), 2007.

[ETS 10] ETSI TR 102 628 V1.1.1 (2010-08), Additional spectrum requirements for future public safety and security (PSS) wireless communication systems in the UHF frequency range.

[FER 12] FERRUS R., SALLET O., GALDINI G. *et al.*, "Public safety communications: enhancement through cognitive radio and spectrum sharing principles", *Vehicular Technology Magazine*, IEEE , vol. 7, no. 2, pp. 54–61, June 2012

[FER 13a] FERRUS R., PISZ R., SALLENT O. *et al.*, "Public safety mobile broadband: a techno-economic perspective", *Vehicular Technology Magazine*, IEEE, vol. 8, no. 2, pp. 28–36, June 2013.

[FER 13b] FERRUS R., SALLET O., GALDINI G. *et al.*, "LTE: the technology driver for future public safety communications", *Communications Magazine*, IEEE, vol. 51, no. 10, pp. 154–161, October 2013.

[GCS 14] 3GPP, Technical specification group services and system aspects, Group Communication System Enablers for 4G LTE (GCSE_LTE) Release 12, 3GPP TS 23.468 v.12.2.0, September 2014.

[GOM 14] GOMEZ K., RASHEED T., REYNAUD L. *et al.*, "Enabling disaster resilient 4G mobile communication networks", *IEEE Communications Magazine*, vol. 52, no. 12, pp. 66–73, October 2014.

[GOR 13] GORATTI L., GOMEZ K., FEDRIZZI R. *et al.*, "A novel device-to-device communication protocol for public safety applications", *Proceedings of IEEE Workshop at GLOBECOM*, Atlanta, USA, pp. 629–634, December 2013.

[GOR 14] GORATTI L., STERI G., GOMEZ K. *et al.*, "Connectivity and security in a D2D communication protocol for public safety applications", *Proceedings of IEEE Wireless Communication Systems*, Barcelona, Spain, pp. 548–552, August 2014.

[GPP 12] 3GPP, Technical specification group SA, Feasibility study for proximity services (ProSe) (Release 12), TR 22.803 V1.0.0, August 2012.

[GPP 13] 3GPP, Technical specification group SA, Study on architecture enhancements to support proximity services (ProSe) (Release 12), TR 23.703 V0.4.1, June 2013.

[GPP 14] 3GPP, Technical specification group services and system aspects, Study on isolated evolved universal terrestrial radio access network (E -UTRAN) operation for public safety (Release 13), TR 22.897 V13.0.0, June 2014.

[GPP 15] 3GPP, Technical specification group services and system aspects; mission critical push to talk over 4G LTE (Release 13), TS 22.179, V13.0.1, January 2015.

[ITU 03] Report ITU-R M.2033, Radio communication objectives and requirements for public protection and disaster relief, available at: http://www.itu.int/dms_pub/itu-r/opb/rep/R-REP-M.2033-2003-PDF-E.pdf, 2003.

[JSA 14] IEEE JOURNAL ON SELECTED AREAS IN COMMUNICATIONS, Special issue on 5G wireless communications, June 2014.

[KUN 13] KUNZ A., TANAKA I., HUSAIN S.S., "Disaster response in 3GPP mobile networks", *2013 IEEE International Conference on Communications Workshops (ICC)*, pp. 1226–1231, 9–13 June 2013.

[MTS 15] MTS1, TETRA base station, available at: http://www.connectel.eu/userfiles/file/downloads/pdf/MTS1_Specification_Sheet.pdf, 2015.

[PAS 15] PAS 0001-2, TETRAPOL specifications; Part 2: radio air interface, TETRAPOL Forum, 2015.

[PEH 07] PEHA J.M., "Emerging technology and spectrum policy reform," *International ITU Workshop on Market Mechanisms for Spectrum Management*, ITU Headquarters, Geneva, Proceedings of the IEEE, vol. 97, no. 4, pp. 708–719, January 2007.

[THA 15] THALES GROUP, NEXIUM wireless, available at: https://www.thalesgroup.com/en/worldwide/security/new-broadband-lte, 2015.

[XIN 14] XINGQIN L., JEFFREY G.A., AMITABHA G. *et al.*, "An overview of 3GPP device-to-device proximity services", *Communications Magazine*, IEEE, vol. 52, no. 4, pp. 40–48, April 2014.

[YUT 12] YUTAO S., VIHRIALA J., PAPADOGIANNIS A. *et al.*, "Moving cells: a promising solution to boost performance for vehicular users", *Communications Magazine*, IEEE, vol. 51, no. 6, pp. 62–68, June 2013.

3

Next-Generation Communication Systems for PPDR: the SALUS Perspective

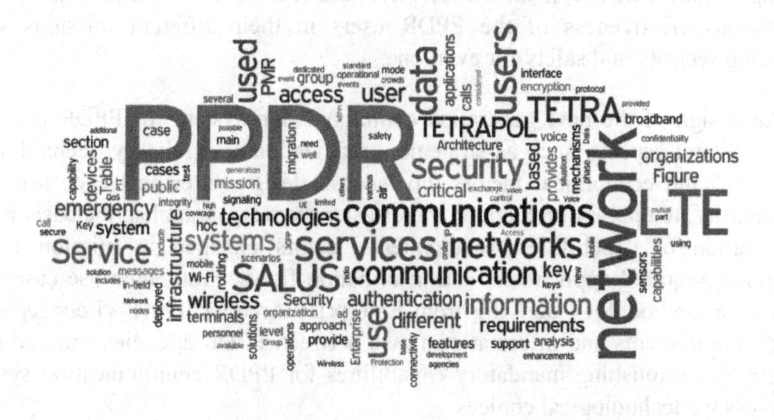

Public Protection and Disaster Relief (PPDR) agencies in European member states currently rely on digital Private Mobile Radio (PMR) networks for mission critical operations. PMR networks are based on two main standards for Europe: Terrestrial Trunked Radio (TETRA) and TETRAPOL. These networks provide secure and resilient mobile voice services, as well as basic data services. However, these traditional PMR networks show substantial limitations, when matched against modern requirements of PPDR agencies, including broadcast communications,

Chapter written by Hugo MARQUES, Luís PEREIRA, Jonathan RODRIGUEZ, Georgios MANTAS, Bruno SOUSA, Hugo FONSECA, Luís CORDEIRO, David PALMA, Konstantia BARBATSALOU, Paulo SIMÕES, Edmundo MONTEIRO, Andy NYANYO, Peter WICKSON, Bert BOUWERS, Branko KOLUNDZIJA, Dragan OLCAN, Daniel ZERBIB, Jérôme BROUET, Philippe LASSERRE, Panagiotis GALIOTOS, Theofilos CHRYSIKOS, David JELENC, Jernej KOS, Denis TRČEK, Alexandros LADAS, Nuwan WEERASINGHE, Olayinka ADIGUN, Christos POLITIS and Wilmuth MÜLLER.

dynamic secure groups, secure roaming, and emerging safety and security applications. Moreover, there are significant interoperability constraints when using multiple technologies, such as inter-technology coverage limitations, which can result in ineffective management of emergency events, both at national level as well as in cross-border regions.

This chapter discusses the evolution of PPDR communication systems toward a next generation communication network that addresses current limitations. The on-going research being conducted by the FP7 security SALUS research project (Security And InteroperabiLity in Next Generation PPDR CommUnication Infrastructures) provides significant contributions in terms of requirements and technological solutions for such communication systems. The next generation PPDR communication system enhances the current *modus operandi* by offering new services and capabilities to PPDR organizations, by leveraging on emerging broadband wireless technologies such as Long Term Evolution (LTE), in addition to existing legacy PMR systems as TETRA and TETRAPOL. This improves the operational effectiveness of the PPDR users in their different missions while enhancing security and safety for everyone.

The design of the next generation communication system for PPDR considers many different aspects, such as strategic, organizational, regulatory, technological, operational and economical factors. For such a design, three reference use cases (city security; temporary protection; and disaster recovery) establish the basis for the identification of capabilities, requirements, actors engaged in mission critical operations, required applications and information flows. From such use cases, and based on a methodology deriving from Enterprise Architecture (EA) concepts, the system requirements and information flows between PPDR agencies and end-users are refined, establishing mandatory capabilities for PPDR communication systems regardless the technological choices.

Critical mission PPDR operations demand strongly secure communications, as well as resilient, redundant and highly available infrastructures. These requirements imply support for mutual authentication of infrastructures and terminals, methods for temporarily and permanently disabling terminals and smart cards, the capability to detect and compensate for jamming at the wireless link, as well as air-interface-encryption and end-to-end-encryption of user and signaling data. Furthermore, core network elements must be physically protected against attackers. This comprehends a security analysis of technologies for PPDR communication systems, such as IEEE 802.11 (Wi-Fi), LTE, TETRA, TETRAPOL and sensor networks, such as bluetooth, IEEE 802.15.4 (Zigbee) and IEEE 802.11ah. Moreover, ICT security-risk analysis is considered for the three SALUS use cases, in order to identify the risks associated with the capabilities derived from the EA. This security-risk analysis is performed according to the NIST SP 800-30 methodology and ISO/IEC 27005 standards.

Components and interfaces for this next communication generation PPDR communication system are identified with PPDR infrastructure migration roadmaps for the evolution of legacy PPDR networks. To assist the migration decision, a Techno-Economic tool (TE) is also described. The tool takes into account not only the technological advances in terms of LTE, surveillance and their interworking with TETRA, but also dependencies among different subsystems in terms of functionality, scalability and performance. This approach allows determining and optimizing capital expenditure (CAPEX) and operational expenditure (OPEX).

The remaining of this chapter is organized as follows: section 3.1 describes each use case, as well as the associated communications requirements and validation methodologies. Section 3.2 provides an overview of the essential radio technologies used for PPDR communications. Possible migration roadmaps toward a full broadband PPDR network infrastructure are discussed in section 3.3. Section 3.4 provides a description on the security aspects related to TETRA, TETRAPOL, LTE and Wi-Fi. Section 3.5 presents the proposed technical enhancements and the final section presents the concluding remarks.

3.1. End-user validation of use cases based on operational scenarios

In order to establish the communications needs for PPDR users, leveraging the benefits of next generation technology and applications, three reference use case scenarios have been conceived. The development and validation of the use cases will be performed in collaboration with members of the end user communities across several of the European member states. This ensures that the identified requirements are capable of delivering benefits, from the end user perspective as well as validating the operational scenarios. This section summarizes each of the three use cases.

3.1.1. *SALUS use case description*

This subsection describes and characterizes the three use cases: city security, temporary protection and disaster recovery.

3.1.1.1. *Scenario 1 – City security*

The city security use case considers the development of a large-scale riot following a public protest. Increasing levels of PPDR personnel that become involved as the situation evolves and trouble caused by young people escalates. This use case describes a series of events. At first, just the police is involved at managing the protest, as a planned peaceful event. However, as the situation escalates and becomes more serious, an increasing number of PPDR personnel and agencies are brought into play. As an example, the ambulance services are brought in to treat injured PPDR personnel and members of the public; and the fire services attend to buildings set alight by arsonists. Eventually riot police are brought in to manage a full-scale riot.

Social media plays a key part in this scenario, as it is used by many of the rioting gangs to entice others to join in including inciting copycat behavior in other cities thus increasing the demands on the emergency services. During the course of the scenario, the command structure changes to reflect the escalating situation from a local police-only command and control structure to a national, multi-agency command structure.

Widely used in this case scenario are the functionalities such as group voice and video, location services and live-video streaming. The use of such features requires communication between multi-agencies and interoperability across different networks, which are also key points of this use case.

3.1.1.2. *Scenario 2 – Temporary protection*

The temporary protection use case is based on an olympic-style sporting event. As it is a high profile event with many VIP's in attendance, it includes a significant amount of planning from a public safety and security perspective. The command structure in place is typical of a large event and several resources from multiple agencies are deployed to manage public safety. The use cases include a series of events. During one of the events, a bomb threat is received from an anonymous call, forcing the evacuation of a venue holding around 50,000 spectators. The situation is aggravated by the spread of bomb threats in social media. This causes panic amongst the public who start to leave the stadium in a disorderly manner.

The emergency services utilize some of the venue's infrastructure, such as the large TV screens, to direct the public and share information. However, when a small explosion is heard, this causes widespread panic that makes controlling the public and the evacuation procedure more difficult. Furthermore, the explosion also damages some of the communications network.

In addition to the police, fire and ambulance services that are heavily involved, this use case also brings in other external factors, such as event security personnel, the bomb squad and transport services, which are used to carry spectators away from the event.

3.1.1.3. *Scenario 3 – Disaster recovery*

The disaster recovery use case relates to a flooding disaster due to a prolonged period of heavy rain extending beyond country borders. Close collaboration between emergency services of the involved countries is needed despite different communication systems being used. Floods cause devastation to areas by taking out electricity supplies that in turn take out some parts of the communications networks. Buildings collapse and, due to the high volumes of water, vehicle access is restricted in some areas and impossible in others, thus, access to stranded victims is also very difficult. Increasing the gravity of the situation, a high-speed train then derails as a result of a landslide. Several passengers are seriously injured and there are also some fatalities.

This use case presents some challenges to the emergency services, for example: interoperability problems between different countries' organizations; conveying information to the public concerned about relatives; deploying temporary emergency communications and co-ordination activities between multiple organizations.

Widely used in this use case is the functionality of air-to-ground communications, recognizing the need to utilize air support as part of the rescue operation. Also Wireless Body Area Networks (WBAN) and Wireless Personal Area Networks (WPANs) sensors are used by PPDR personnel to provide real time information of the disasters area, as well as live streaming of Closed-Circuit Television (CCTV), to the strategic tactical and operational command groups.

3.1.2. *SALUS use case requirements*

Across the European member states specific communication needs may vary depending on policy, organizational and regulatory factors of an individual country [EUR 13]. Therefore, the approach to derive the user functional requirements includes the contact of end users across a variety of public safety organizations in Europe. This allows creating a high level framework for PPDR, which is comprehensive and capable of efficiently performing cross-border PPDR operations. An iterative approach refines the functional requirements as per Table 3.1.

Phase	Description
1	Development of the use cases based on the three SALUS scenarios
2	Capture high level requirements taking into account the participating PPDR end users and possible deployable technologies
3	Refinement of requirements to a precise level of detail
4	Final refinements of requirements that will be used to implement the technical solutions and respective test

Table 3.1. *Refinement phases of use cases requirements*

Considering the uses cases requirements identification, at the highest level, PPDR functional requirements can be summarized in Table 3.2.

Requirements	Description
Voice	Users can make diverse voice call types, which include group, announcement/broadcast, emergency, individual and telephony interconnect calls. The infrastructure must ensure that emergency calls have priority over other calls, releasing capacity for an emergency call by pre-empting other users if necessary.
Video	Users can send and receive video imaging to all users, specific user groups or individuals, either from a dispatcher to the group.
Data Applications	Users can have mobile access to data applications such as messaging services and email, organization-specific databases and other data-rich applications such as location services, augmented reality and DNA/Fingerprint scanning.
Air to Ground	Seamless communications are possible between users in aircraft and users on the ground.
Ad hoc Mobile Networks	Additional network coverage or capacity can be deployed quickly and easily, for example in remote locations.
Interoperability	Communications are possible between users of different organizations whether operating on a common infrastructure or connected via different technologies or networks.
Crowd Control	The communication system can be utilized to manage large crowds of people using a combination of loud speakers, video, social networking and applications.

Table 3.2. *Use case requirements*

3.1.3. *SALUS use case validation*

Innovative features and functionality included in SALUS are to be demonstrated to end users. To facilitate this, a generic test methodology has been adopted based on the International Software and Testing Qualifications Board (ISTQB) fundamental test process [IST 15]. A test methodology is a documented and repeatable combination of test steps, types of testing and test process to support the test approach. The steps of the generic test methodology are described in Table 3.3.

Steps	Description
Control	It is present throughout the process since it is an on-going activity of comparing actual progress against the plan and reporting the status, including deviations from the plan. The control element also ensures that the Test plan is updated as the testing progress to ensure it is a true reflection of the actions taken.
Planning	Where the risk analysis takes place and involves analysis of the requirements and deliverables, for a system or software solution, to identify risk areas. The risk analysis can follow the International Organization for Standardization (ISO) 9126 [ISO 91] standard or National Institute of Standards and Technology (NIST) recommendations [NIS 12], and mainly aim to identify risks and their priority for testing effort estimation and planning.
Analysis and Design	It starts when there is sufficient detail regarding the amount of test effort required. Further analysis of the requirements leads to defined tests cases allowing test scripts to be developed. Test data is an important part of testing and will need to be representative of the data in the deployed system or software solution. A key output from this stage is the test schedule.
Implementation and Execution	The test execution phase starts once the release has been installed or setup and configured in the test environment. This is where the test scripts run and all results are recorded.
Evaluation & Reporting	This stage allows determining if the testing has been successful in the risks analysis regarding the desired quality level.
Test Closure	Collects data from the completed test activities to consolidate experience, facts and numbers.

Table 3.3. *Evaluation steps*

3.2. Emerging wireless technologies for PPDR

Nowadays, PPDR agencies are facing increasing challenges in day-to-day, planned and unplanned events. Since 1970, the number of natural catastrophes has been multiplied by four, approximately, and the number of man-made catastrophes by around three [SWI 13]. A more efficient use of communications systems and technologies with additional capabilities, such as video and data sharing, within and between PPDR agencies, is recognized as a major way to significantly enhance safety of the citizens and the PPDR personnel. In short and middle term, PMR technologies, such as TETRA and TETRAPOL, will continue to play a key role in delivering critical mission voice services, however they will soon be complemented by Commercial Off-The-Shelf (COTS) wireless technologies, such as LTE and Wi-Fi, which will enable broadband applications and services. This section provides an overview of the essential radio technologies and their additional capabilities that will support PPDR agencies in enhancing their operational effectiveness.

3.2.1. *TETRA*

TETRA is an open standard developed in the mid-1990s by the European Telecommunications Standards Institute (ETSI). The TETRA air interface is based on Time Division Multiple Access (TDMA) and Π/4 Differential Phase-Shift Keying (DPSK) modulation, over 25 kHz channels, and each radio channel carries four communications control and/or traffic channels, and can be deployed in the 400 MHz and 800 MHz frequency range.

TETRA supports mission critical voice services such as individual calls, group calls and emergency calls, among others. Voice calls are operated using Push To Talk (PTT) mode with fast call set-up times, less than 300 ms. It also supports low data rate services, such as messaging and packet data services of 4.8 kbps per slot. Communications can operate in two distinct modes: trunked when relayed by the infrastructure; Direct Mode Operation (DMO) when occurring directly between users in the case of lack of coverage or for specific mission requirements. Communications are encrypted, over the air and end-to-end if needed; and an authentication mechanism ensures that only authorized terminals can connect to a trusted infrastructure.

TETRA is used by more than 3 million users and has been deployed in most European countries with Germany, United Kingdom and Norway being the largest TETRA networks. Recently, TETRA standard has been enhanced to support higher data rate transmissions (TETRA Enhanced Data rate Services – TEDS). TEDS makes use of higher order modulation, 4-Quadrature Amplitude Modulation (QAM), 16-QAM and 64-QAM, over larger channels, 25, 50, 100 and 150 kHz, and can, in principle, increase the data rate to theoretically 540 kbps. However, practical implementations are limited to 50 kHz channel bandwidth and to less than 100 kbps. TEDS is not widely deployed today in PPDR networks.

3.2.2. *TETRAPOL*

TETRAPOL appeared in the late 1980s to provide digital voice and low data rate data mobile radio-communications. The air interface is based on Frequency Division Multiple Access (FDMA) and robust Gaussian Minimum-Shift Keying (GMSK) modulation over channels of 12.5 kHz and operates in 80 MHz or 400 MHz bands.

TETRAPOL supports mission critical voice services, such as individual calls, group calls and emergency calls, among others. The voice calls are half duplex within a group or in PTT, one talks and the others participants are listening. Groups can be configured statically or dynamically according to the mission requirements. It also supports low data rate communications of few kbps and all communications are encrypted to prevent eaves dropping. The communications may be relayed, by the network and one or more base stations, or in direct mode between terminals. TETRAPOL also supports simulcast configuration to create large coverage areas for selected groups with less radio channels.

TETRAPOL has been successfully deployed for national PPDR networks not only in European countries, such as France, Spain, Switzerland, Czech Republic and Slovakia, but also outside Europe. Around 1.85 million users use TETRAPOL today.

3.2.3. *LTE for public safety*

This section introduces features of LTE for public safety.

3.2.3.1. *LTE a global wireless standard for 4G*

LTE is the global 4G cellular system, adopted by commercial operators, with more than 331 commercial LTE networks worldwide [GSA 13]. Today, LTE is primarily used for mobile broadband applications, video and data, offering performances much superior to existing 2G and 3G systems in terms of data rates and latency. Some Mobile Network Operators (MNO) are now also offering Voice over LTE (VoLTE) services. VoLTE will be generalized in the coming years as it provides a more efficient and richer way to transmit voice with higher quality compared to 2G and 3G systems. In addition, the enhanced Multicast Broadcast Multimedia Services (eMBMS) over LTE can efficiently support new use cases, such as mobile TV, content push and file delivery.

Moreover, with the current wave of Internet of Things (IoT), LTE will also be used in the future for Machine Type Communications (MTC). The LTE standard is currently being improved by 3GPP R13 and beyond, to optimize the support of MTC, such as battery life enhancement, reduction of terminal cost and complexity and enhanced coverage. These enhancements can also benefit adjacent markets such as PPDR for connecting to environmental and chemical sensors.

3.2.3.2. *LTE enhancements for PPDR*

LTE has been selected by major PPDR end-users associations, such as Association of Public safety Communications Officials (APCO) and TETRA plus Critical Communications Association (TCCA) to be the follower technology of existing narrowband specialized voice-centric systems such as TETRA,

TETRAPOL and P25 (note that P25 is a PMR standard equivalent to TETRA and TETRAPOL mainly used in North America; there is no specific section on P25 since it is not widely deployed in Europe, however, main characteristics of P25 will be described in the technology comparison table in section 3.2.6). Nowadays, LTE can be used by PPDR agencies as a broadband data overlay, complementing narrowband mission critical voice centric systems. Indeed, the superior performance of LTE, compared to TETRA and TETRAPOL, brings additional capabilities to support new applications, such as near real-time exchange of large amount of information and high quality live video streaming of a scene. COTS LTE systems implement many advanced features [DOU 13] that are relevant for PPDR users, as summarized in Table 3.4.

Feature	Description
Excellent Radio Frequency (RF) performance	LTE air interface uses state-of-the art modulation and multiple access mechanisms (Orthogonal Frequency Division Multiple Access – OFDMA) and antenna processing (Multiple-Input and Multiple-Output – MIMO) techniques, which lead to high spectral efficiency and throughput.
Low latency	LTE is an all-Internet Protocol (IP) system with nearly "flat" architecture. Coupled with an air interface, with sub-frame duration of 1 ms, leads to end-to-end (E2E) latency as low as 10 ms, which is critical when considering applications that need minimal set-up time and transmission delays.
E2E QoS	LTE standard defines a comprehensive E2E Quality of Service (QoS) framework that allows differentiated and guaranteed delivery of services, also implementing priority and pre-emption capabilities. All the QoS attributes can be dynamically configured to guarantee that the highest priority applications and/or group of users will get access to the radio resource under any circumstances, even when congested.
Security	LTE technology includes mutual authentication mechanisms between the terminals and the infrastructure. Moreover, LTE provides over the air encryption for data, privacy; and integrity protection for the signaling information.
Resiliency	Each LTE Evolved Node B (eNB) can be multi-homed to multiple core network elements, namely the (Mobility Management Entity – MME) and Serving Gateway (SGW). This can be used for load balancing or in the event of a failure of a core network component, or to re-route traffic to/from to an alternative core network.

Table 3.4. *LTE enhancements for PPDR*

However, LTE systems cannot be considered as critical mission today. Indeed, TETRA and TETRAPOL systems include capabilities ensuring the system is operational and reliable under any circumstance. These include:

1) The capability to operate in a scalable way with very large group calls.

2) The capability for terminals to communicate directly without any support from the infrastructure, the DMO in TETRA and TETRAPOL.

3) The capability for a base station to provide services in its serving area, even if disconnected from the network; this is called Fallback Mode (FBM) in TETRA and TETRAPOL.

LTE does not provide such capabilities yet. Nevertheless, pushed by the PPDR users and suppliers, 3GPP is developing specifications for enabling similar critical mission capabilities on LTE. First enhancements for PPDR are available in 3GPP release 12 (completion planned for Q1 2015), the second wave of enhancements will be available in release 13 (completion planned for Q1 2016) and include:

– Group Communications System Enablers (GCSE) [3GP 14b]: this feature enables scalable multimedia group communications over LTE, based on eMBMS specifications augmented by a new interface to dynamically create, modify and release group communications over MBMS bearers;

– Proximity-based Services (ProSe) [3GP 14e]: this feature provides first step of DMO. However, it is limited in R12 to discovery and Device to Device (D2D) communications when under the coverage of a base station only.

Second wave of enhancements will be available in R13 and include:

1) ProSe enhancements: this will feature out of coverage D2D communications, present in Figure 3.1, as well as terminal to network relay mode for ProSe enabled terminals. GCSE will also be possible using ProSe (see Figures 3.1 and 3.2).

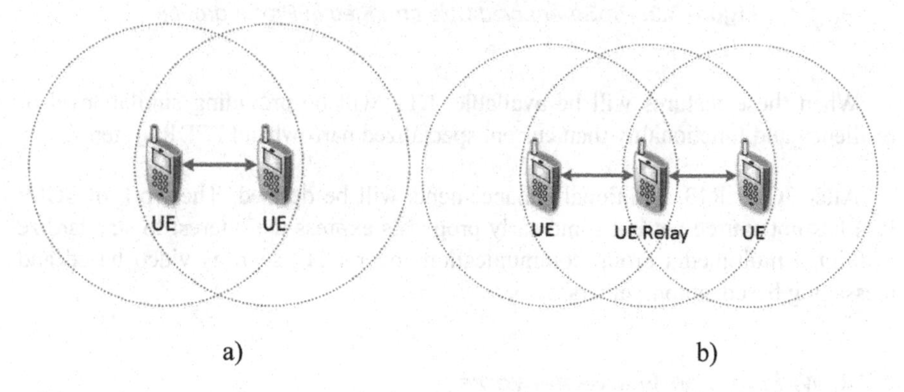

Figure 3.1 . *a) ProSe direct communication; b) User Equipment (UE) to UE relay*

2) Isolated Evolved Universal Terrestrial Radio Access Network (E-UTRAN Operation for Public Safety (IOPS) [3GP 14c]: this will mimic the existing fallback mode of TETRA and TETRAPOL, and will provide service in the coverage area of a set of eNB to PPDR users when this set of eNB is disconnected from the network.

3) Mission Critical Push to Talk (MCPTT) [3GP 14d]: PTT is a major application for PPDR users. This feature will define a standard and interoperable PTT application that will leverage current LTE standard and additional mission critical enablers, such as GCSE, ProSe and IOPS.

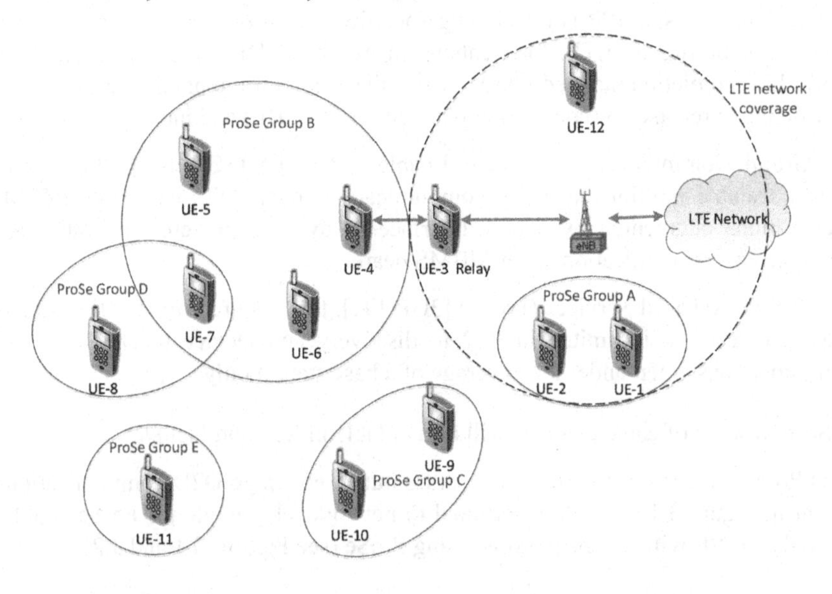

Figure 3.2. *ProSe-enabled UEs arranged in ProSe groups*

When these features will be available, LTE will be providing similar level of resiliency and functionality than current specialized narrowband PPDR systems.

After 3GPP R13, additional enhancements will be defined. The work of 3GPP R14 has not started yet but some early proposals express the interest to standardize additional multimedia group communications over LTE, such as video based and messaging based, among others.

3.2.4. *Wi-Fi and body area networks*

This section details functionalities of wireless local area networks (WLANs), WPANs and WBANs for PPDR.

3.2.4.1. *Wi-Fi and its evolution for PPDR*

The recent growth of WLAN technologies, normally designated Wi-Fi, has enabled a wide range of applications and services with a huge impact at residential and public areas. Currently, most of the devices incorporate wireless technology, due to its low-cost chipsets and support for high data transfer rates. The Wi-Fi standard is one of the most deployed wireless technologies all over the world that aims to provide simplicity, flexibility and cost effectiveness. A wireless broadband alliance market report indicates an increase of public carrier-grade Wi-Fi hotspots from 8 million in 2014 to nearly 12 million in 2018 [WBA 14].

Wi-Fi or WLAN is part of the overall solution in at least two ways: it can be part of the infrastructure or it can be used for creating *ad hoc* or mesh networks. The infrastructure solution is the ability to provide capacity extender, also referred to as WLAN off loading, which can be used for dense urban or public areas. *Ad hoc* networks are self-organizing networks, consisting of equal peers, which can communicate without the need for central nodes and without connectivity of an infrastructure. From the PPDR security perspective, they can be considered "trusted" or "untrusted" depending if they are part of the operators own system or are provided by a third party. With the purpose of enhancing the WLAN standards for supporting QoS, the 802.11e amendment [802.11e] was established to enable other type of applications rather than best effort, such as voice in mission critical PPDR operations.

WLAN can also be used to provide indoor positioning where Global Position System (GPS) fix is not available. Such capability is highly desirable for PPDR users such as fire fighters that usually need to operate in indoor environment and that need to be tracked for their own safety.

3.2.4.2. *Wireless body area networks and sensors*

In the context of PPDR, a WBAN is a wireless network that consists of sensors attached to the body of a PPDR user and a local collector, a PPDR terminal for example, to which sensors wirelessly forward their readings. The terminal then either processes the readings locally or forwards them to back-end applications, across the network for processing, visualization or storage. By providing health-related readings such as heart rate, respiration, movement or stress levels in real-time, WBAN greatly improve situational awareness of the deployed PPDR forces.

When trying to implement such a system, several issues have to be addressed. While mechanisms for sensing and forwarding readings from sensors to the local collector are usually provided by manufacturers, mechanisms for transferring readings from the PPDR terminals to the back-end applications are not.

WBAN and WPAN are considered the major solutions for short-range sensor networks in a PPDR oriented infrastructure. Medium range sensor networks are considered as an enabler technology for indoor positioning and outdoor positioning on limited areas, when conventional positioning techniques, such as GPS or Assisted Global Position System (AGPS), are not available. The most popular solution in commercial deployments is ZigBee or bluetooth. Long-range sensor networks rely on IEEE 802.11 standards for extended communication ranges and a high number of nodes.

3.2.5. *Challenges of waves propagation in crowds*

The propagation of electromagnetic (EM) waves through human crowds in the open area, out-door, plays significant role in mobile communications, peer-to-peer radio links and various others wireless communications that are established in the presence or through the human crowds. Such scenarios can happen for various reasons, for example at the large open-air sports or arts events [DOR 13], consider the example of the temporary protection use case. Estimation of radio channel parameters, in the presence of human crowds, is of interest at the stages of radio system design and planning, and at the time of the deployment.

While International Telecommunication Union (ITU) has recommendations for radio-channel attenuation in the presence of vegetation [ITU 03], there is no such recommendation for propagation through human crowds out-door.

3.2.6. *Comparison between the different wireless technologies*

Table 3.5 presents a comparison between different wireless technologies regarding the spectral efficiency (bits/second/Hz), latency, cell coverage, support for QoS and the number of users/cell.

Features	Technologies				
	TETRA	*TETRAPOL*	*P25*	*Wi-Fi (IEEE802.11.ac)*	*LTE*
Spectrum	Licensed (400 MHz/ 800 MHz)	Licensed (80/400 MHz)	Licensed (VHF/UHF/700/800/ 900 MHz)	Unlicensed (2.4 GHz / 5GHz)	Licensed (from 400 MHz to 3.5 GHz) Unlicensed (future, 5 GHz)

Channel bandwidth	25 kHz 25, 50, 100, 150 kHz with TEDS	12.5 kHz	12.5 kHz	20/40/ 80 MHz (802.11ac)	1.4/3/5/10/15/20 MHz (and carrier aggregation)
Duplex mode	FDD	FDD	FDD	TDD	FDD/TDD
Multiple Access	TDMA	FDMA	FDMA (phase 1) TDMA (phase 2)	CSMA/CA	OFDMA
PTT services	Integrated	Integrated	Integrated	Over the top	Over the top first; integrated from 3GPP release 13
Text services	Integrated	Integrated	Integrated	Over the top	Integrated
Data services	Narrowband and wideband (TEDS)	Narrowband	Narrowband	Broadband	Broadband
Video services	Low quality with TEDS	No	No	Yes	Yes
Data rates	TETRA: up to 9.6 kbps/slot TETRA 2 or TEDS: up to 538 kbps with 150 kHz	7.6 kbps	9.6 kbps	Up to 1.3 Gbps today	Up to 150 Mbps today with 10 MHz and MIMO 4x4
QoS	Yes (inc. Priority and pre-emption. TEDS has QoS negotiation features)	Yes (inc. Priority and pre-emption)	Yes (inc. Priority and pre-emption)	Yes (inc. Priority)	Yes (inc. Priority and pre-emption)

Latency[1]	250 ms	110 ms	250 ms	5 ms	10 ms
Coverage[1]	Up to 10 s of km; limited to 58 km (and 83 km with TETRA2)	Up to 10 s of km (limited to 30 km)	Up to 10 s of km (limited to 30 km)	100 m	Few km to 10 s of km (limited to 100 km)
Deployment	WAN/ Deployable	WAN/ Deployable	WAN/ Deployable	LAN/*Ad hoc*	WAN/small cells/ deployable
Resiliency	Design dependent for network FBM for base stations DMO for D2D	Design dependent for network FBM for base stations DMO for D2D	Design dependent for network FBM for base stations Talk-around for D2D	Design dependent for network; *ad hoc* modes for D2D	Design dependent for network IOPS for base stations (3GPP R13) ProSe for D2D (3GPP R13)
Security	Mutual authentication (TAA), air interface encryption (TEA) and E2EE	Mutual authentication and E2EE	E2EE (AES 256/DES)	Authentication, and air interface encryption (AES128 or AES256)	Mutual authentication, and air interface encryption (AES128 or SNOW 3G)
Terminal output power	1 W or 3 W for portable, up to 10W for mobiles	1 or 2 W for portable, Up to 10 W for mobiles	Up to 5 W for portable Up to 25 W for mobile	100 mW	200 mW for all terminals 5 W possible in 400 MHz for mobile

Table 3.5. *Comparison between different wireless technologies*

3.3. Migration roadmaps

Existing commercial LTE operators can support some functionalities of PPDR networks, nonetheless the migration to a full-mission critical LTE owned or shared

1 The values presented are indicative and may vary depending on environment conditions, network topology, frequency band used and distance between devices, among others factors.

network requires multiple phases, which include the coexistence of LTE and current PPDR broadband technologies. SALUS in the migration roadmaps considers migrations that span from simple cooperation to elaborated critical LTE owned and shared missions, providing an understanding of the future evolution of PPDR communications. This section describes the three phases of the migration roadmaps.

3.3.1. *Phase 1 – Non-mission critical cooperation with commercial LTE*

Phase 1 considers the case where multiple devices exist, one for PPDR broadband connectivity and another for LTE connectivity. Indeed, an LTE subscription is required, mainly for high-speed data access and non-mission critical communications [EUS 13, ISM 10, TCC 13]. All mission critical communications are still performed using legacy PPDR systems, such as TETRA and TETRAPOL [ITU 03, LRC 12, TOC 13]. Under this scenario PPDR organizations must have established some requirements:

1) A commercial contract with the LTE operator that would specify a dedicated Access Point Name (APN) to which the LTE terminals can connect;

2) A dedicated IP connectivity link, typically a leased line, from the LTE core network to the private PPDR network;

3) Optionally, a dedicated Group Communication Service Application Server (GCS AS) on the private PPDR network for non-mission-critical communications;

4) A separate channel for the management of the LTE subscriptions.

This concept, which already considers the enablers to support group communication services using E-UTRAN access, is illustrated in Figure 3.3.

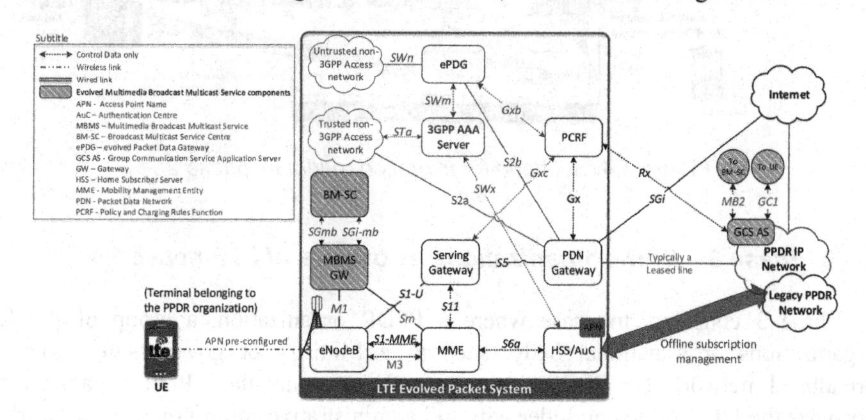

Figure 3.3. *Scenario for migration roadmap: phase 1*

3.3.2. *Phase 2 – PPDR organizations as LTE mobile virtual network operators*

Phase 2 considers the case where a PPDR organization or a group of PPDR organizations becomes a full LTE Mobile Virtual Network Operator (MVNO). Under the full LTE, MVNO the PPDR organization provides broadband wireless communications services, without owning and/or operating the wireless network infrastructure. Mission critical communications are still performed using legacy PPDR systems and LTE connectivity is available through the use of a separate device, as in the previous phase. Dual-mode terminals, LTE and TETRA in a single device, are currently not being considered due to the unclear market potential of such devices. In this phase, PPDR organizations are expected to have established a commercial agreement with an LTE operator, optional roaming agreements with other LTE operators – to ensure broad coverage and redundancy – a dedicated IP connectivity link from the LTE core network to the private PPDR network and dedicated Home Subscriber Server (HSS) server, Packet Data Network Gateway (PDN GW), Policy and Charging Rules Function (PCRF) and GCS AS on the private PPDR network. This concept is illustrated in Figure 3.4.

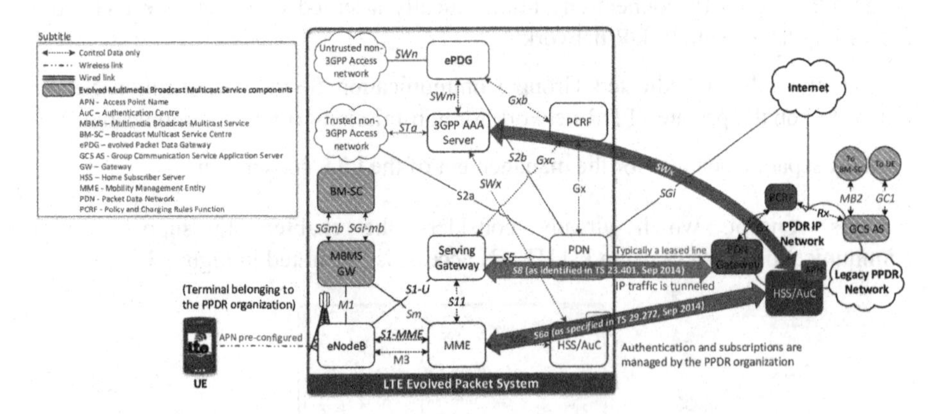

Figure 3.4. *Scenario for migration roadmap: phase 2*

3.3.3. *Phase 3 – PPDR organizations as owners of LTE networks*

Phase 3 considers the case where a PPDR organization, a group of PPDR organizations or a national body, such as a Ministry or government, owns a broadband network for PPDR operations. This means the PPDR organization becomes the LTE service provider with full administrative control over the network. It can be assumed that initially both legacy PPDR and LTE networks would coexist.

A roll-out phase toward a full LTE network with a single mobile terminal (User Equipment – UE) will then take place. Under this phase, PPDR organizations must assure that: 1) the LTE infrastructure is resilient, redundant and highly available; 2) PPDR mission-critical services, such as today's group call, PTT direct mode operation are supported by the LTE network; 3) proper interfaces are defined to interoperate with legacy PPDR networks during the roll-out phase. This concept is illustrated in Figure 3.5.

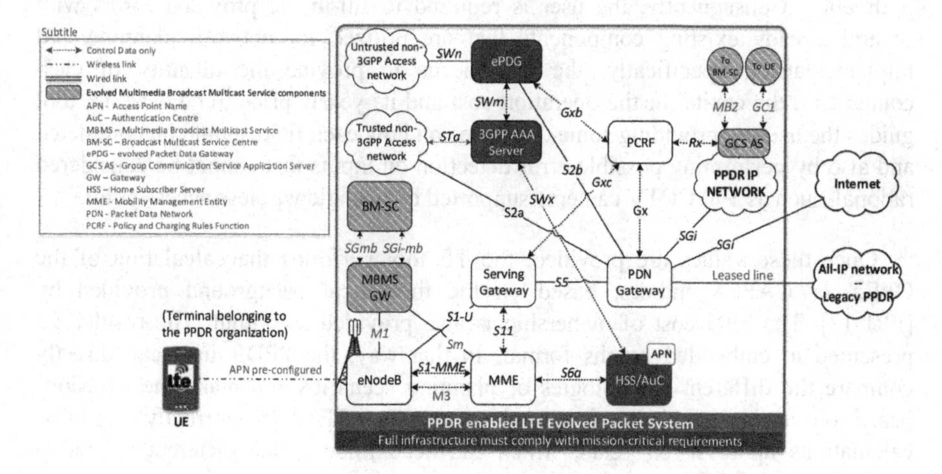

Figure 3.5. *Scenario for migration roadmap: phase 3*

3.3.4. *Techno-economic tool*

During the migration phases, the Techno-Economic (TE) tool can be used for planning of OPEX and CAPEX, which is a requirement for PPDR organizations aiming to own a full LTE infrastructure. The TE tool provides a framework that accepts entries, from the potential PPDR users and extracts results to assist the decision-making. Its architecture is divided into two components: the dimensioning part and the Techno-Economic Analysis (TEA) part. The former includes several tools that assist the user to measure the network and quantify its parameters, according to some specific requirements. Subsequently, the values that are calculated can be passed to the TEA component to proceed with the necessary financial calculations.

The dimensioning part of the TE tool, implements an LTE dimensioning process [KNO 12] for urban and suburban topologies with a carrier frequency of 700 MHz (Band 28). After the LTE dimensioning is completed, the user can proceed with the TEA, which is the major function of the TE tool. Specifically, the PPDR user is asked to select between performing a single financial analysis of one scenario or whether it is preferred to evaluate scenarios based on different technological solutions, which will be compared, based on financial data.

The next step is to decide on the existing possibilities for emergency situation scenarios that are mainly covered in the context of SALUS project, namely the city security, the temporary protection and the disaster relief. The TE tool proceeds by either calculating the OPEX, the CAPEX or both values, based again on the user preference. Consequently, the user is required to fill-in the provided forms with several already existing components that are utilized for network planning and implementation. Specifically, the user needs to provide the quantity of each component, the capital or the operation cost and its yearly price trend. The TE tool guides the user by providing some directives on how each field should be completed and also by performing possible error detection on inputs that cannot be considered rational, such as 100 CCTV cameras supported by a single wireless access point.

Once these values are provided, the TE tool performs the calculation of the OPEX or CAPEX indices, based on the theoretical background provided by [FRI 12]. The total cost of ownership is also provided as output. The results are presented in embedded graphs format. In that way, the PPDR user can directly compare the different technologies or planning scenarios and make the decisions based on a time-window of several years. The TE Tool currently performs calculations up to seven years, which can accommodate the different migration roadmaps for full LTE owned networks.

3.4. Security aspects linked to emergency communications

Mission-critical PPDR operations require highly secure communications. In the context of Information and Communication Technology (ICT), a secure communication is one that is confidential by preventing unauthorized users from understanding its content; temper resistant by promptly detecting any illegal tampering with communicated information; and highly available by preventing adversaries from disrupting communications. To elicit additional security requirements, this section provides a high-level risk analysis of a general PPDR network architecture. The section then continues with an overview of security features in PMR technologies that address identified risks.

3.4.1. *Risk analysis*

The risk analysis considers an ICT perspective of the SALUS platform. The analysis follows the asset/impact-oriented approach from NIST SP 800-30 rev. 1 [NIS 12]. First, we identify assets, threats and adverse impacts. Based on characteristics of threats, we estimate the likelihoods that threats actually lead to adverse impacts. Finally, we derive risk levels for each combination of threats and adverse impacts.

3.4.1.1. *Assets and adverse impacts*

Since the analysis is concentrated on ICT risks, the main assets represent the connectivity between various network entities and the integrity and confidentiality of exchanged information. The main impacts are, conversely, the loss of the aforementioned connectivity and the compromise of exchanged information. To illustrate various types of possible connections that have to be secured, Figure 3.6 shows a high-level architecture of the SALUS platform.

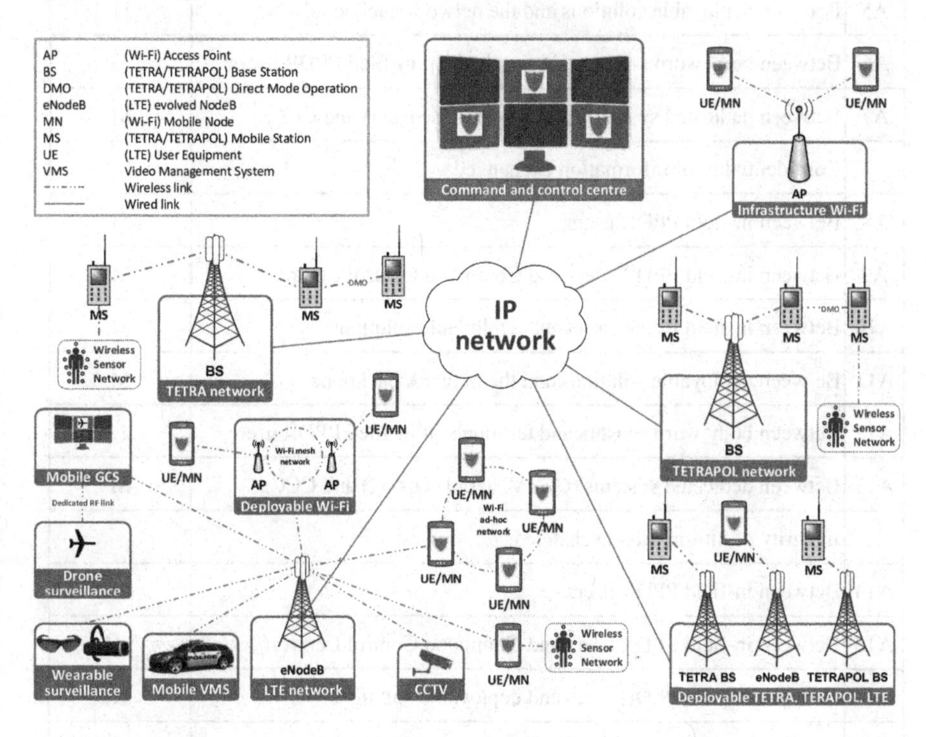

Figure 3.6. *High-level architecture of the SALUS platform*

Table 3.6 summarizes the main assets and impacts. Column impact level denotes the level of impact, if the corresponding asset is compromised. The impact level is assessed on a 5-level scale: VL, L, M, H, VH, denoting "very low", "low", "moderate", "high" and "very high", respectively. The meanings of these terms are defined in NIST SP 800-30 rev. 1 [NIS 12].

ID	Description	Impact Level
	Availability of connectivity:	
A0	Between in-field PPDR users;	
A1	Between in-field PPDR users in the same access network;	M
A2	Between in-field PPDR users in different access network;	M
A3	Between in-field PPDR users and Command Control Centre (CCC);	VH
A4	Between in-field PPDR users and deployable solutions;	H
A5	Between deployable solutions and the network backbone;	H
A6	Between body worn sensors and terminals of in-filed PPDR users;	L
A7	Between dedicated systems (CCTV, aerial sensing) and CCC.	M
	Confidentiality of information exchanged:	
A8	Between in-field PPDR users;	M
A9	Between in-field PPDR users and Command Control Centre;	M
A10	Between in-field PPDR users and deployable solutions;	M
A11	Between deployable solutions and the network backbone;	M
A12	Between body worn sensors and terminals of in-filed PPDR users;	L
A13	Between dedicated systems (CCTV, aerial sensing) and CCC.	M
	Integrity of information exchanged:	
A14	Between in-field PPDR users;	H
A15	Between in-field PPDR users and Command Control Centre;	VH
A16	Between in-field PPDR users and deployable solutions;	H
A17	Between deployable solutions and the network backbone;	H
A18	Between body worn sensors and terminals of in-filed PPDR users;	M
A19	Between dedicated systems (CCTV, aerial sensing) and CCC.	H

Table 3.6. *Main assets and impacts in emergency communications*

3.4.1.2. *Threats*

Threats can harm assets and cause adverse impact in emergency communications, as summarized in Table 3.7.

ID	Description
T1	Jamming: the adversary jams the EM spectrum and effectively blocks radio communications.
T2	Routing attack: an unauthorized malicious participates in the network and degrades its performance.
T3	Eavesdropping: a malicious user scans the network traffic and obtains sensitive information.
T4	Information spoofing: a malicious user actively modifies the network traffic.
T5	Unauthorized use of PPDR equipment for eavesdropping.
T6	Unauthorized use of PPDR equipment for misdirecting.
T7	Destruction of equipment.
T8	A malicious but authorized PPDR user causes various disarray like routing anomalies or misinforming other PPDR users.
T9	Unauthorized disclosure of information: a blackmailed PPDR user provides confidential information.

Table 3.7. *Threats in emergency communications*

3.4.1.3. *Risks*

Risks are determined by combining threats and adverse impacts and are summarized in Table 3.8. Based on the likelihood of each threat and the level of harm of the impacted asset, the risk levels are computed as described in NIST SP 800-30. The likelihoods and the risk levels are assessed with the same 5-level scale as the level of impact.

ID	Threat	Likelihood	Impacted assets	Impact level	Risk level
R1	T1	M	A1-A4, A7	H	M
	Jammers block communication among handheld devices of in-field PPDR operatives or between handheld devices and the base stations or eNodeBs.				
R2	T1	L	A6	L	L
	Small jammers, secretly attached to PPDR operatives, disrupt communication between body worn sensors and the handheld PPDR device.				
R3	T1	L	A5	H	M
	Jammers block communication between Multiple Base Stations (MBSs) and the main network.				
R4	T2	M	A1-A3	H	M
	Unauthorized devices participate in *ad hoc* network routing and impede its availability.				
R5	T3	M	A8-A13	M	M
	Eavesdropping: Unauthorized devices passively scan the wireless traffic to obtain sensitive information.				
R6	T4	M	A14-A18	H	M
	Information spoofing: Unauthorized devices actively participate in wireless communication by modifying the information to misinform PPDR personnel.				
R7	T5	VH	A8-A13	H	H
	Lost or abandoned handheld PPDR device is used to eavesdrop conversations or to accesses restricted information.				
R8	T6	H	A8-A18	H	H
	Lost or abandoned handheld PPDR device is used to misinform or misdirect other PPDR personnel.				
R9	T7	H	A2-A5, A7	H	H
	A base station, an eNodeB, a Wi-Fi access point or a deployable solution is lost to vandalism or to environmental hazards.				
R10	T6	H	A19	H	H
	Adversaries infiltrate venue information systems (screens, speakers) to mislead the crowd and cause further problems to PPDR forces.				
R11	T8	M	A1, A8-A18	H	M
	A malicious PPDR causes routing anomalies or misdirects and misinforms other PPDR personnel.				
R12	T9	M	A8-A10, A13, A14-16, A19	M	M
	Unauthorized disclosure of confidential information: unauthorized users obtain classified information from PPDR users.				
R13	T3, T4	VL	A9, A11, A13, A15, A17, A19	H	L
	Eavesdropping or information spoofing on the compromised elements of the network infrastructure.				

Table 3.8. *Risks in emergency communications*

3.4.1.4. *Recommended security controls*

The following security controls should be used to mitigate identified risks.

1) R1, R2, R3: counter jamming. Because it is practically impossible to counterpart, jamming is prohibited in most countries. But as a countermeasure, triangulation can be used to determine the location of jamming sources.

2) R4: secure the *ad hoc* routing protocol. The SALUS architecture should deploy a routing protocol for *ad hoc* and mesh networks that allow only pre-authorized devices to participate in the routing.

3) R5, R6: secure wireless communications. All wireless communication should use state-of-the-art encryption and allow only pre-authorized devices to connect to the network. The devices and the network should mutually authenticate each other.

4) R7, R8: authenticate users to devices. PPDR users should authenticate to devices and the authentication token should have a limited lifetime. This poses a challenge, because a good balance between the security and ease of the authentication method is needed. Moreover, measures that allow devices to be remotely locked or wiped should also be implemented.

5) R9: secure network access points. Losing the deployable solution, a base station, an eNB or an Access Point (AP) to either vandalism or environment may in some cases represent the single point of failure. Without the deployable solution, the connectivity between devices of different access technologies, TETRA, TETRAPOL and LTE, as well as the uplink to the backbone is lost. The deployable solution should therefore be resilient to some level of physical damage, should not be left unattended and should have a backup power supply. To deteriorate vandalism, key points of interest should be protected with surveillance mechanisms like CCTV.

6) R10: secure venue Information Systems (IS) and detect intrusions. Venue IS should employ access control mechanisms and intrusion detection mechanisms should be put in place to detect any suspicious activities.

7) R11, R12: dynamically revoke credentials. If the integrity of a PPDR user is compromised, their credentials should be remotely revoked and the compromised device closely monitored.

8) R13: end-to-end security. If a router of the internal network gets compromised or the information traverses the public Internet, the contents can be read or modified by unauthorized users. To mitigate such risks, the communication should include E2E security mechanisms.

3.4.2. *A high-level overview of security features of PMR technologies*

This section overviews security mechanisms present in current PMR technologies and exposes their drawbacks.

3.4.2.1. *Security considerations of LTE*

LTE provides several security services such as user confidentiality, anonymity, entity authentication, data confidentiality and data integrity [3SA 15]. The confidentiality is achieved mainly by the use of three mechanisms: user identity confidentiality, user location confidentiality and user non-traceability. The entity authentication is always mutual: the UE authenticates the network and the network authenticates the UE. Data confidentiality on the network access link is provided through several security features, such as cipher algorithm agreement, cipher key agreement, confidentiality of user data and confidentiality of signaling data. Data integrity on the network access link is provided by the following security features: integrity algorithm agreement, integrity key agreement and data integrity and origin authentication of signaling data.

LTE uses state-of-the-art security algorithms like Advanced Encryption Standard (AES), Secure Hash Algorithm 256 (SHA-256) and Internet Protocol Security (IPSec). Regarding the confidentiality and integrity protection of user data on the network access link, LTE makes use of strong security algorithms like snow 3G, AES or Zu Congzhi (ZUC) with keys of 128-bit length.

The Evolved Packet System (EPS) key hierarchy is triggered as soon as the identity of the mobile subscriber (i.e. Globally Unique Temporary Identity – GUTI or International Mobile Subscriber Identity – IMSI) is known by the MME. The multiple keys within the EPS Key hierarchy are obtained through a procedure known as the Authentication and Key Agreement (AKA). This procedure can be initiated by the network as often as the network operator wishes and, when triggered, the AKA procedure results in a new main key, known as KASME, that is stored both in the UE and the MME [3GP 13]. The KASME is then used to derive the Non Access Stratum (NAS), KeNB (KeNB is a key derived by user equipment and mobility management entity or by user equipment and target eNB), Radio Resource Control (RRC) and User Plane (UP) keys using a Key Derivation Function (KDF), as specified in 3GPP TS 33.220 [3GP 14a]. To increase key security, three main enhancements were considered in E-UTRAN [OLS 09]:

1) stronger key separation between networks and key-usage;

2) larger key sizes. E-UTRAN supports not only 128-bit keys but can (in future deployments) also use 256-bit keys;

3) increased protection against compromised base stations. Through the use of the forward/backward security, the air interface keys are updated each time the UE changes its point of attachment (or when the UE changes from IDLE to ACTIVE).

After a successful mutual authentication procedure, the encryption and integrity protection keys shall have been distributed according to the hierarchy depicted in Figure 3.7. The use of the "MILENAGE algorithm set" with the secret key *K* produces a Cipher Key (CK) and an Integrity Key (IK). These keys are generated in order to provide compatibility with GERAN/UTRAN networks mainly during handovers. As it can be seen, the main key in E-UTRAN is KASME. This key will be created only by running a successful authentication as explained in [3GP 13]. In case the UE does not have a valid KASME, a key set identifier, known as KSI_ASME, with value "111", will be sent by the UE to the network, which can initiate a (re)authentication procedure to get a new KASME.

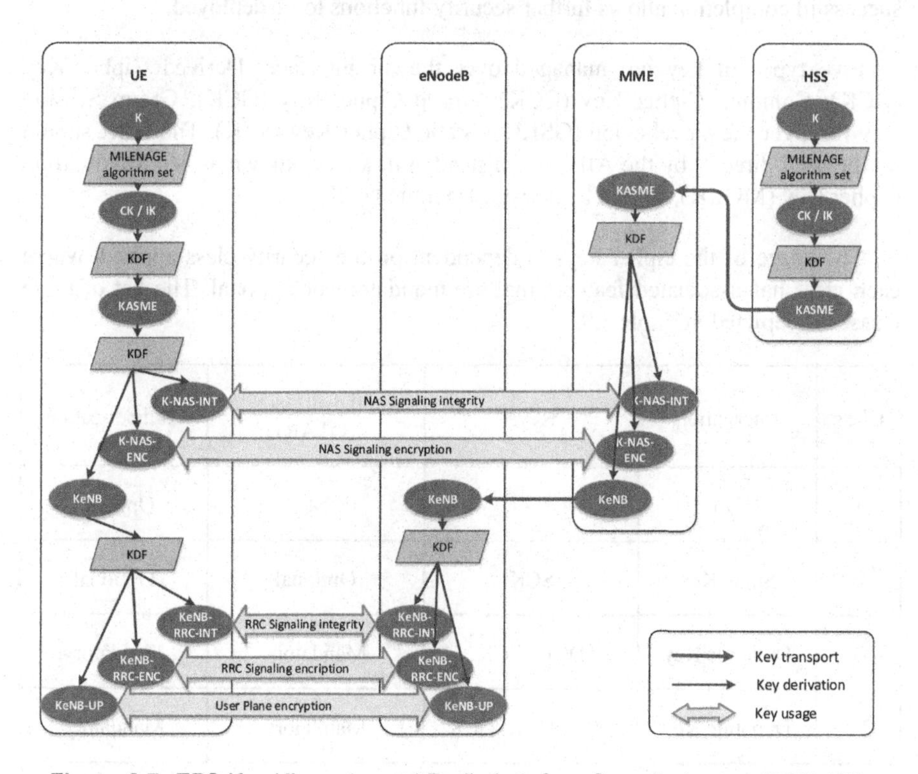

Figure 3.7. *EPS Key Hierarchy and Radio Interface Security (as per [SCH 12])*

3.4.2.2. *Security considerations of TETRA*

TETRA provides several security mechanisms: authentication to ensure trusted access from the proprietary mobile station (MS) to BS and vice versa, mutual authentication; Air Interface Encryption (AIE) to encrypt signaling, voice and data communications, which is critical to prevent eavesdropping and analysis of communication patterns; E2E encryption to additionally improve encryption of voice and data, including the transfer within the transport and core components of a TETRA network; and Enable/Disable to allow remote disabling of TETRA radios when they are lost, stolen or compromised.

Authentication in TETRA systems can be performed in three levels: authentication of the MS by the infrastructure, authentication of the infrastructure by the MS, and mutual authentication. In simple terms, the authentication takes place when a participant proves to the other that he knows the shared secret. The procedure of authentication is a two-step challenge-response protocol and its successful completion allows further security functions to be deployed.

Five types of key are managed over the air interface: Derived Cipher Key (DCK); Common Cipher Key (CCK); Group Cipher Key (GCK); Group Session Key for Over the air rekeying (GSKO); Static Cipher Key (SCK). The GCK should not be used directly by the AIE unit. Instead, a new key, known as Modified Group Cipher Key (MGCK), should be derived from the GCK.

The usage of the cipher keys is dependent on the security class applied, where each class has associated features that are mandatory or optional. The list of these classes is depicted in Table 3.9.

Class	Encryption	Keys	Over the air rekeying (OTAR)	Authentication
1	No	–	No	Optional
2	Static Key	SCK	Optional	Optional
3	Dynamic Key	DCK, CCK	Mandatory	Mandatory
3G	Dynamic Key	DCK, CCK, GCK	Mandatory	Mandatory

Table 3.9. *TETRA air interface security classes*

A summary of the TETRA air interface keys, the security classes were they are used and their usage/purpose is provided in Table 3.10.

Key	Class applied	Protection	Usage
DCK	3	Voice, data, signaling	Individual communications
CCK	3	Voice, data, signaling	Group communications
SCK	2	Voice, data, signaling	Individual / Group communications
GCK	2/3	n/a	Derivation of MGCK
MGCK	3G	Voice, data, signaling	Group communications
GSKO	2/3	n/a	Protects the distribution of SCK and GCK to groups

Table 3.10. *Summary of the TETRA air interface keys*

3.4.2.3. *Security considerations of TETRAPOL*

TETRAPOL provides several security mechanisms: end-to-end encryption, mutual authentication (to prevent masquerading intrusion, periodically, or on request, the network authenticates the terminal and the terminal authenticates the network), signaling protection (to prevent traffic analysis, a temporary identity allocated by the network is used instead of the real subscriber address), encryption diversity (multiple algorithms are permitted for end-to-end encryption, for example a national algorithm for domestic operations and a multinational for roaming), automatic rekeying (without the need of user involvement, over the air) and automatic network reconfiguration for increasing service availability (provisioning redundant critical components, having automatic fall back mode, and dynamic channel reallocation and continuous network monitoring).

TETRAPOL also provides extra security features, like remote terminal disable (temporary or permanent), subscriber identity module (Subscriber Identity Module – SIM card conforming to ISO 7810, 7811 and 7816 recommendations [ISO 10, ISO 11, ISO 16], may be used; user profile mapping is defined with a Personal Identification Number (PIN) code and authentication to the network, other configurations are possible with digital certificates) and access control (equipment is protected with smart cards or passwords; additional facilities include remote detection of physical intrusions and alarming the operators).

A simplified high-level overview of the TETRAPOL security mechanisms is depicted in Figure 3.8.

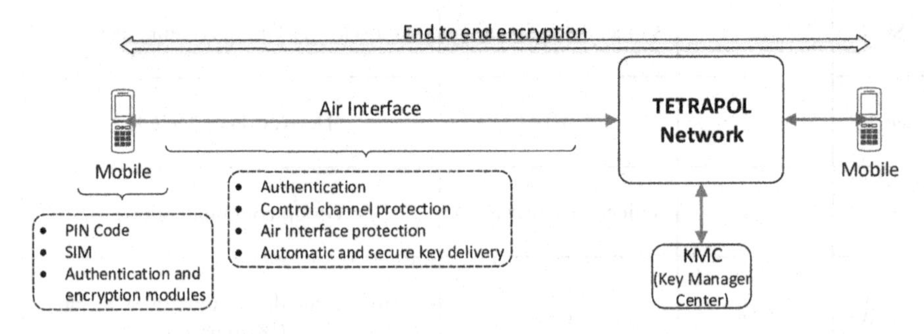

Figure 3.8. *TETRAPOL security mechanisms*

3.4.2.4. *Security considerations of Wi-Fi*

When discussing the security of Wi-Fi, it is meaningful to discuss the infrastructure and *ad hoc* or mesh modes distinctly. In infrastructure mode nowadays, the security solutions are well established and tested. And while the initial security mechanisms did exhibit some issues that were prominently showcased with several successful attacks on Wired Equivalent Privacy (WEP) protocol [FLU 01], the improvements brought forward by the Wi-Fi Protected Access (WPA) and WPA2 [802.11i] not only provide strong confidentiality and integrity, but also allow mutual authentication between terminals and access points.

However, when discussing the security of *ad hoc* mesh networking, one usually refers to the security of routing protocols. This directly affects the ability of terminals to securely connect and exchange messages: an unsecure routing protocol would allow an adversary to cause various network anomalies, such as dropping network packets, poisoning routing table entries and generally impeding the network availability. But contrary to the infrastructure mode, finding a secure and efficient ad hoc network routing protocol is still an active research topic [BOU 11]. And while some solutions have already been standardized [802.11s], finding a completely secure and efficient ad hoc network routing protocol is still an open issue. Moreover, given the wireless nature of ad hoc networking, mechanisms – similar to those presented in WPA – are needed to protect the contents of wirelessly exchanged messages.

3.5. Technical enhancements for PPDR communication systems in SALUS

PPDR organizations are confronted with a growing number of events affecting public safety and security, which require a close cooperation between the involved PPDR organizations in order to respond timely and adequately to these events. The need of cooperation requires aligned procedures and interoperable systems, which allows timely information sharing and synchronization of activities. This, in turn, requires that PPDR organizations adapt EA methodologies to develop system architectures. Also, the 3GPP organization is currently defining and specifying public safety enablers that will turn LTE system implementing enablers into a critical mission system. This section provides a simplified overview on the core concepts and their relationships in order to be used for PPDR EA development. It is also presented: the interworking of PPDR services between narrowband TETRA and TETRAPOL systems and future broadband PPDR systems LTE and Wi-Fi; and the enhancement of mobility between broadband wireless systems.

3.5.1. *SALUS enterprise architecture*

This section details the Enterprise Architecture (EA) in SALUS.

3.5.1.1 *The SALUS enterprise architecture development approach*

The Open Safety & Security Architecture Framework (OSSAF) [OSS 10] provides a framework and approaches to coordinate the perspectives of different types of stakeholders within an organization. It aims at bridging the silos in the chain of commands and on leveraging interoperability between PPDR organizations. The methodology proposed in [MUE 14] for the development of EA of PPDR organizations, in general and for SALUS specifically, uses NATO Architecture Framework (NAF) as the modeling vocabulary for describing the OSSAF perspectives and views where suitable. The NAF views are modeled with the different elements of the Unified Modeling Language (UML).

Since SALUS is addressing security and interoperability in next generation PPDR communication infrastructures and not all aspects of PPDR organizations, there is a need to tailoring the EA development and its artifacts to SALUS use cases. For SALUS only those artifacts of an EA are relevant which influence the technical development of communication infrastructures. Thus, a funding model of the PPDR organization (the "PPDR as an Enterprise") or a specific organization chart of the enterprise or a concrete product configuration used by the PPDR organization in its daily operations were not in the scope of SALUS. The EA components addressed in SALUS are the ones highlighted in Figure 3.9. The NAF views, used to describing the OSSAF views, had also to be adapted to the needs of SALUS. Figure 3.10 provides an extract from the overall model used for the development of the PPDR EA. However, for reasons of readability, not all relations, attributes, constraints, and cardinalities are shown. A detailed meta-model description as well as the description of the semantics of each concept and relationship can be found in [NAF 07].

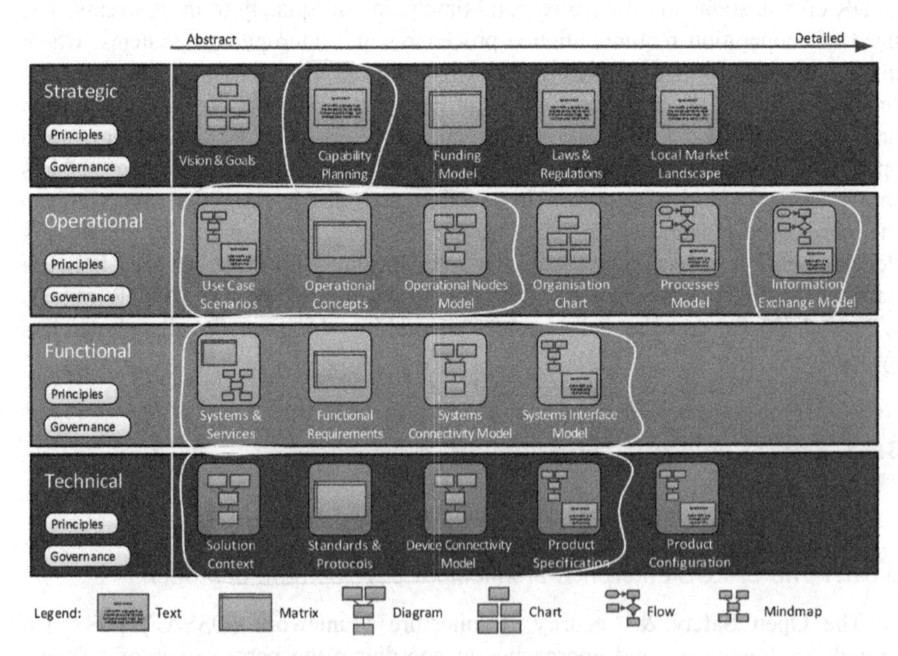

Figure 3.9. *Enterprise architecture perspectives and views and components of SALUS*

According to the approach in describing OSSAF views via single or combinations of suitable NAF-views [MUE 14], the contents of the dedicated NAF-views for use in designing the SALUS EA are described. The description contains the model elements captured in the corresponding view, which is actually a section of the overall model, proposes a suitable representation (graphical and textual) and give hints in order to support the development of the view under consideration. This is done in a way agnostic to any tool, but refers to UML modeling concepts where suitable. As depicted in Figure 3.10, the modeling is performed considering the connectivity between operational nodes and their linkage to the capabilities, in the light of NAF view NOV-2, NATO operational view, operational node connectivity description.

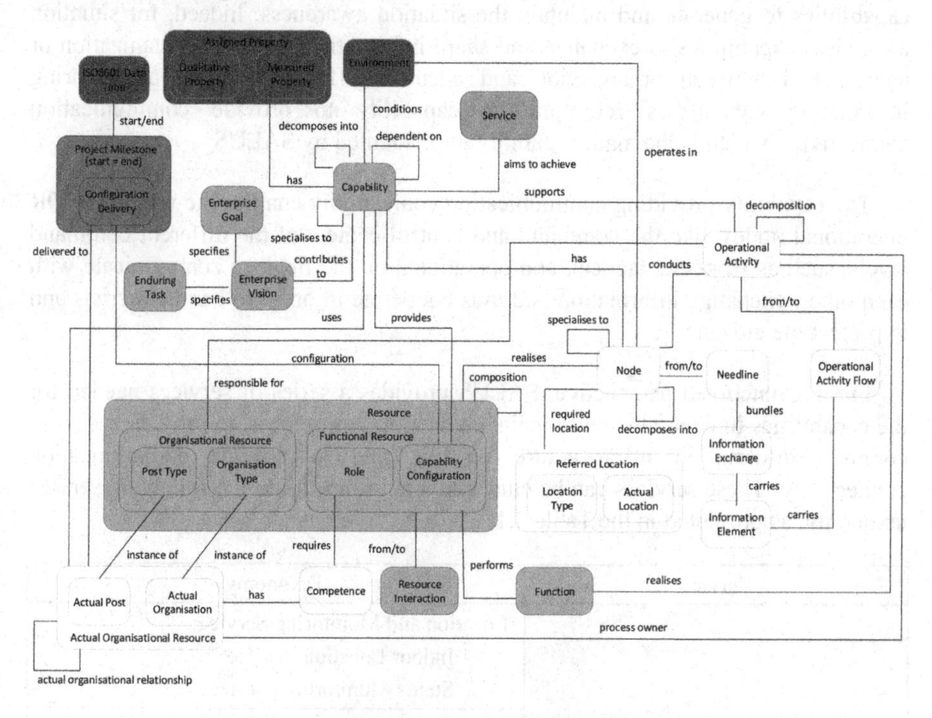

Figure 3.10. *NAF Model elements according to the strategic and operational scope*

3.5.1.2. *The SALUS enterprise architecture*

The methodology proposed by [MUE 14] is using the approach of capability based planning. One can understand a capability as [TOG 11]:

> *An ability that an organization, person, or system possesses. Capabilities are typically expressed in general and high-level terms and typically require a combination of organization, people, processes, and technology to achieve.*

Using this approach, the following SALUS EA capabilities were identified: the capability to protect the public and the citizens – public protection; the capability to conduct a mission in an integrated way – integrated mission conduction; and the capability to protect the own forces – force protection.

The SALUS capabilities also rely on others, like assessing the development of a crisis or the capability to coordinate and cooperate, which in turn depend on the capabilities to generate and maintain the situation awareness. Indeed, for situation awareness capabilities to exchange and share information within an organization or agency and between organization and agencies. The exchanging and sharing information capabilities rely on the capability to provide communication connectivity, which is the main capability implemented by SALUS.

The capability providing communication connectivity enables the various PPDR operational nodes, like the command and control centers of the different command levels, such as strategic, tactical, and operational on the field; to communicate with each other, exchange information and thus cooperate in order to handle a crisis and to protect the citizens.

From a functional perspective, SALUS provides a series of services needed for the capabilities to generate and to maintain situation awareness, to provide access to common information infrastructure services and to provide communication connectivity. These services can be clustered and grouped into the following service taxonomy, as presented in the Table 3.11.

Service	Taxonomy
Situation Awareness Service	Location and Monitoring Service: – Indoor Location Service – Status Monitoring Service
	Sensor and Tracking Service: – Sensor Data Acquisition Service – Sensor Control Service – Force Tracking Service
Information Assurance Service	Security Service: – Intrusion Detection Service – Resource Authentication Service – Policy Enforcement Service – Forensics Service
Management Service	Management Service: – User Management Service – Group Communication Management Service

	– Mobility Management Service – Policy Management Service
Network & Information Infrastructure Service	Information and Integration Service: – Message Brokering Service
	Communication Service: – Voice Communication Service, including - Push To Talk (PTT) - Group Call - One to One Call - Emergency Call - Ambience Listening – Data Communication Service - Streaming Service - Text Messaging Service
	Interaction Service: – Video Conferencing Service – Chat Service
	Network/Transport Service: – Mobility service - WiFi2LTE Mobility Service - Traffic Management Service – QoS Monitoring Service - Network QoS Monitoring Service – Communication Interworking Service - TETRA2TETRAPOL Interworking Service - TETRA2LTE Interworking Service - TETRAPOL2LTE Interworking Service

Table 3.11. *Enterprise Architecture service taxonomy*

One of the important SALUS services includes the interoperability between TETRA, TETRAPOL and LTE to allow the capability of communication activity between different PMR technologies.

3.5.2. *Wireless sensors in PPDR systems*

The use of wireless sensors carries out an important role for PPDR systems, in the SALUS, especially wearable sensing solutions and computing devices that can be carried out by emergency services, such as fire brigades, during the action. They allow the collection of information from the environment, which improves situation awareness, and can be deployed as short, middle or long range networks solutions.

As a unified solution, SALUS proposes the concept of a Message Broker (MB), which is a network element that facilitates the exchange of messages between PPDR terminals and back-end applications. MB is a connection point between terminals and back-end applications: all messages are sent to the MB and from there forwarded to the intended recipients. This allows everyone connected to the MB to asynchronously exchange messages, with the following benefits:

1) It offers a unified and readily secured transportation mechanism that can be used by any application;

2) It provides a simple discovery mechanism by which back-end applications and PPDR terminals discover and reach each other. Terminals, in particular, may often switch networks, so their IP addresses may not always be known in advance.

The MB offers two types of message exchanges: a direct one and a subscription based. In the direct message exchange, the sender explicitly denotes the recipient and sends the message to the MB, which then forwards the message to the recipient. In the subscription based message exchange, the sender tags the message with a topic, sends it to the MB, which then forwards the message to all recipients that have previously subscribed to that topic. In this case, the MB performs the application layer multicast.

The wireless sensor networks are an integral part of the SALUS project. Sensor data is usually collected by sensors attached to the bodies of in-field deployed PPDR personnel and then sent via their hand-held terminals (user equipments) to back-end applications for processing. The first type of used sensors concerns bio-signals, such as the heart rate, blood pressure, temperature and similar kinds of the user information. Other kind of personnel wireless sensor information are movement and localization, obtained from accelerometer, gyroscope and GPS sensors. The body signals are used in order to interpret current user health state, location and position allowing an analysis to search for critical events, such as heart attack and falls of users.

3.5.3. *Extensions to next generation networks*

This section includes extensions to PMR technologies developed in SALUS.

3.5.3.1. *PMR services over broadband IP*

Since narrowband and broadband PPDR systems will coexist, according to the migration roadmap presented in section 3.3, interworking of PMR services between the different wireless access technologies is a major requirement. The PMR services to be supported on all access networks can be split into four main categories: basic services, PMR supplementary services, telephony supplementary services, and security features. The basic services include the minimum feature set for a conventional PMR network. They consists of registration/deregistration, group affiliation, group calls, one-to-many communications with PTT user request to talk; individual calls, PTT or hook button based; telephony calls to/from an external telephony network, broadcast call, call from a dispatcher to all PPDR users in a group; status, such as predefined set of text messages; and generic text messaging, binary messaging, as transmit sensor information.

The PMR supplementary services are the more advanced services that are essential to PPDR users to operate safely and efficiently. They include priority calls, pre-emptive priority calls, emergency calls, late entry, dynamic regrouping, discreet and ambiance listening from the dispatcher position and location reporting. The telephony supplementary services are services related to public access telephony such as call forwarding features, when busy, or without reply; call hold, call transfer, call barring, incoming and outgoing; and call authorized by dispatcher. The security services are features that are related to the critical use of the wireless communications for PPDR users. They include mutual authentication, the ciphering on the air interface, the end-to-end encryption, the temporary and permanent disabling of a terminal.

In SALUS, the design and implementation of PMR services on top of broadband IP wireless system is loosely based on existing concepts for PMR system interconnection developed referred to as On-Demand Intelligent Network Interface (ODINI). Similar to ODINI, Critical Voice and Data Protocol (CVDP) is also based on unique Intellectual Property Rights (IPR), for fast and robust operation of PTT operations [ROH 11]. A node, such as relay, is introduced in the architecture to forward all signaling and data IP packets between the different access systems, as depicted in Figure 3.11. This approach covers the complete set of PMR features and guarantees with appropriate settings on the LTE access network parameters such as QoS Criteria Indicators (QCI) and priorities but also the parameters and restoration mechanisms of the IP transport networks to achieve the performance targets of call set-up/floor grant of less than 300 ms and 99.99% service availability. On top of the essential voice and text messages centric services, similar approach will be used to transmit over type of information such as pictures, data and video.

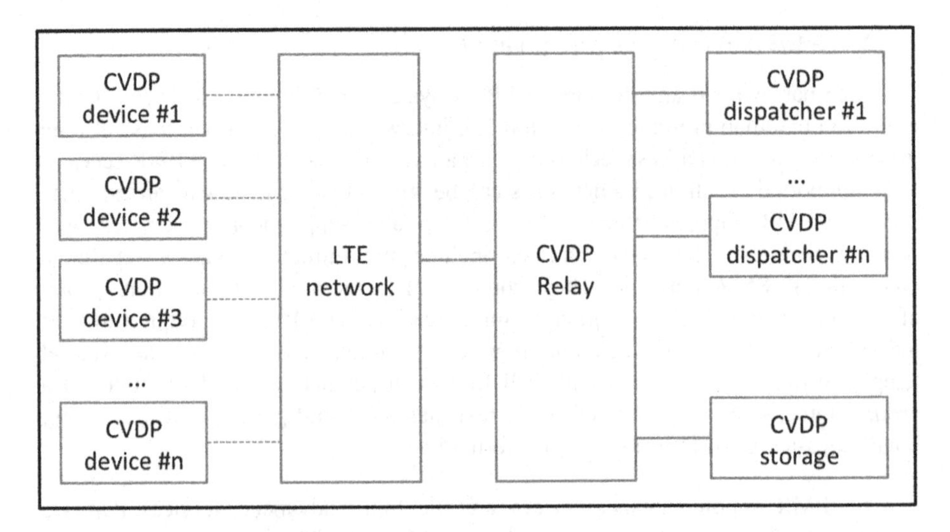

Figure 3.11. *CVDP high-level architecture*

3.5.3.2. *Seamless mobility for HetNet*

SALUS system is a system providing the combination of narrowband, TETRA and TETRAPOL; and broadband technologies, such as LTE, WLAN in infrastructure mode and WLAN in *ad hoc* mode. In SALUS, all services are being carried over IP and should be supported seamlessly when moving between systems that could be private and/or public access networks. SALUS approach is to use fast handover extension for proxy mobile IPv6 (PFMIPv6) [YOK 10]. This allows for seamless handovers between devices in the same IPv6 domain using either predictive or reactive mechanisms. Performing predictive handover requires signaling exchanges between the mobile node and the previous and new mobile access gateways, respectively the previous Mobility Access Gateway (pMAG) and next MAG (nMAG). In SALUS this signaling exchange is performed using Media Independent Handover (MIH) protocol, which is specified in IEEE 802.21 [MIH 09]. Figure 3.12 gives an example of this complementary approach.

The IP mobility mechanism is being extended in SALUS to support mobility for multicast traffic which is a new area in IP mobility and a key requirement for group communications to save bandwidth in delivering the same content to a large number of PPDR users.

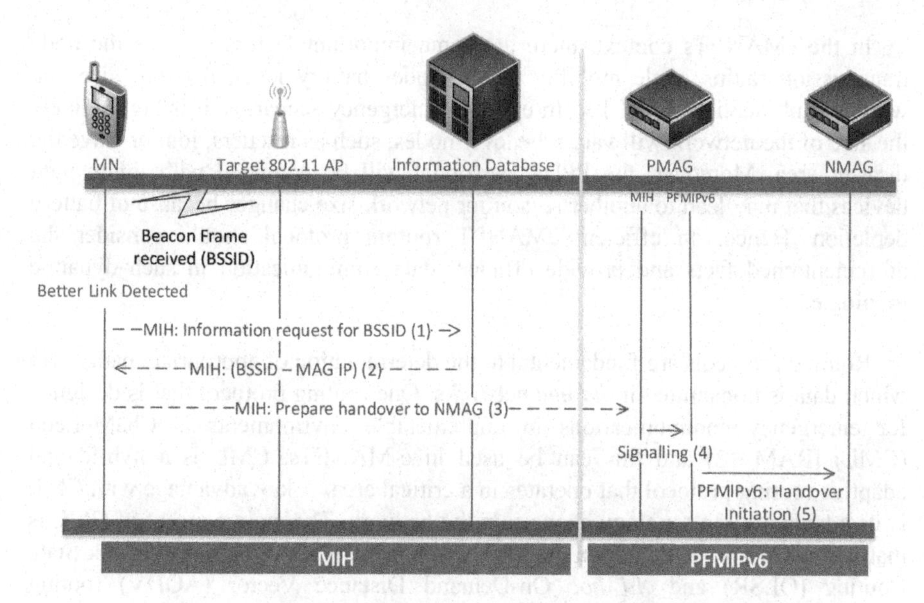

Figure 3.12. *Example of MIH signaling for PFMIPv6 handover*

3.5.4. *Mobile ad hoc networks support for emergency communications*

Mobile *ad hoc* NET works (MANETs) consists of a collection of autonomous and collaborative nodes, devices, which can communicate directly with each other in a decentralized manner. The role of each node is to send, receive and forward messages. MANETs can provide a purely and instantaneous distributed Peer-to-Peer (P2P) wireless communication architecture in the realm of wireless communication where lightweight devices are widely used. Considering the next generation networks, MANET devices that are IP based can be used for providing *ad hoc* and lightweight communication solutions for the PPDR users in various emergency scenarios. The term "emergency scenarios" includes a plethora of extreme disaster scenarios e.g. earthquakes, major floods, terrorist attacks, tsunamis and forest fires [MIL 10]. In catastrophic events like these, there is always the likelihood that the day-to-day communication infrastructure such as LTE, TETRA and TETRAPOL is damaged. Repairing the aforementioned infrastructure is time consuming, and also the emergency personnel require functional communications by the time they arrive on the scene, in order to communicate. MANETs have the ability of instantaneous setup, hence makes them an ideal solution for use by the first responders on scene. Typically, a MANET deployed in such context, strictly for the purpose of first responders' communication is termed as emergency MANET (eMANET).

In the eMANETs context, there are some important factors such as the node transmission radius, node mobility levels, nodes battery level, network size and wireless link quality [RAM 10]. In extreme emergency scenarios, it is highly likely the size of the network will vary whenever nodes, such as rescuers, join or leave the disaster area. Moreover, the PPDR personnel will be equipped with lightweight devices that may lead to another reason for network size changes because of battery depletion. Hence, an efficient eMANET routing protocol should consider the aforementioned facts and provide efficient data communication in such dynamic topologies.

Routing protocols are fundamental to the determination of appropriate paths over which data is transmitted in *ad hoc* networks. One routing protocol that is designed for emergency communications in unpredictable environments is ChaMeLeon (CML) [RAM 12] and this can be used in e-MANETs. CML is a hybrid and adaptive routing protocol that operates in a critical area. A key advantage with CML is its adaptability to topology changes in the network. The main concept of CML is that it combines in its operation the intrinsic characteristics of Optimized Link State Routing (OLSR) and *Ad hoc* On-Demand Distance Vector (AODV) routing protocols by setting a network threshold. According to [RAM 12] CML performs better than AODV and OLSR for varying size networks; consequently, it is a better alternative to both protocols for such growing or shrinking networks.

Currently, the routing protocols used for MANET operations lack security implementations, since every node is considered trustworthy, cooperative and non-malicious. Thus, the exchange of messages is done under the assumption that all nodes can be trusted. This can lead to various negative consequences, such as revelation of confidential data as a result of different attack types or destroy of communication links. To achieve an effective solution in terms of security for topology discovery in MANETs, the [SAX 09] approach can be applied in PPDR *ad hoc* networks. The approach is a certificate-less identify key management framework, which utilizes bi-variant and uni-variant polynomial and ElGamal public key cryptography types.

The process of encrypting the broadcast messages in MANETs is very crucial. The MANET is initiated by a trusted dealer to be known as the squad leader. The broadcast messages encryption is achieved with the help of partial secret shared to the participating nodes, so every legitimate node can decrypt each of the broadcasted, encrypted, messages without the need of exchanging additional keys. By following this scheme, every node can calculate the decryption key, using information from the squad leader and the identity of the node that broadcasts the messages. The security procedure of this scheme is deployed based on a partial secret, which is provided by the squad leader. Hence, any possible intruders that aim to perform malicious actions are limited only to listening to network messages, without having the ability to gain access to the wide private network information. Another significant benefit of this approach is that it is not affected by nodes joining or leaving the network. The proposed security extension is scalable because the computational cost is independent of the network size.

3.5.5. *Radio-channel attenuation within crowds*

The resilience of various radio-systems used in the context of PPDR systems can be improved by knowing in advance the radio-channel attenuation in the presence of crowded areas. The PPDR radio-systems are normally used to convey sensor data from body-area networks to data hub, such as command and control centre. For that reason, a research into the radio-channel attenuation in the presence of human crowds is conveyed within the framework of SALUS project. In general, the data for the radio-channel can be obtained through numerical simulations of electromagnetic field and large measurement campaigns. Both approaches must deal with the fact that the human crowds are stochastic in nature, and hence a large amount of data must be processed in a statistical way, in order to estimate the radio-channel attenuation. Moreover, various sensors use different radio-systems to communicate, such as bluetooth, Wi-Fi, LTE, among others; and therefore frequencies for all those radio-systems should be taken into the account. Particularly, the electromagnetic simulation of human crowds presents state-of-the-art in the computational electromagnetic community. It needs the most sophisticated simulation algorithms and high-performance computers [OLC 14, VEL 15].

3.6. Concluding remarks

Nowadays, PPDR organizations rely on PMR networks for voice and data communications in emergency operations. Such networks are designed to be resilient and to cope with crisis and emergency handling. The two main standards for digital PMR networks in Europe are TETRA and TETRAPOL.

SALUS has considered three distinct use cases reflecting the need for new functionalities in real public safety scenarios: the city security includes a large scale riot and public protest with different types of PPDR agencies (police, ambulance, fire fighters); the temporary protection, which includes a big sport event with crowded stadiums; and finally, the disaster recovery includes flooding in a cross-border set-up including a derailed train. These use cases are a product of exhaustive gathering of PPDR user's requirements through user interviews.

Current PMR technologies cannot be neglected in favor of new ones that are feature-enriched, such as LTE. Indeed, TETRA and TETRAPOL standards play an important role in PPDR missions. For instance, 'security features like encryption, authentication and integrity mechanisms, as well as Push to Talk (PTT) and group call, are features that PPDR users rely on to perform their mission efficiently. Nonetheless, these technologies have accrued several limitations that impose restrictions to the exchange of information between control centers and PPDR users on the field, for instance: video calls are limited (low data rates in TETRA and TETRAPOL), restricted situation awareness, which cannot be complemented with sensor data such as the monitoring of bio signals data or detecting a man-down event.

LTE has been selected to be the follower technology of existing narrowband specialized voice-centric systems such as TETRA, TETRAPOL and P25, mainly by enabling broadband data overlay and complementing mission critical voice centric systems. SALUS proposes PPDR infrastructure migration roadmaps within three distinct phases for PPDR organizations. In particular, the coexistence of current PMR technologies with LTE will be a fact. Nonetheless, a final phase will allow PPDR organization to run LTE as a dedicated network for mission critical activities, acting as a Mobile Virtual Network Operator. Within such migration roadmaps, SALUS is also releasing a techno-economic tool that facilitates migration toward a broadband PPDR network.

The main goal of SALUS is to design, implement and evaluate a next generation communication network for PPDR organizations, supported by network operators and industry. As specified in the migration roadmap and according to the PPDR user requirements, several technologies can coexist in PMR networks. As such, one of the main outcomes of SALUS is the interoperability between different standards, such as TETRA, TETRAPOL and LTE enhancing the actual mode of operations by offering new services and capabilities to PPDR organizations. This will improve operational effectiveness of the PPDR users in their different missions, while improving safety of the involved actors. However, the definition and design of an efficient next generation communication system for PPDR is a complex and demanding task. Many different aspects such as strategic, organizational, regulatory, technical and economical aspects have to be taken into account. The EA has been employed to facilitate such task through diverse methodologies.

The outcome of the project is expected to further enhance interoperability in the next generation PPDR networks, promoting modernization and standardization in the European Union, encouraging investments in competitive access networks with innovative techno-economic solutions. Indeed, technological and economical aspects are equally important to offer an effective solution for PPDR users in terms of cost of operations and safety.

3.7. Bibliography

[3GP 14a] 3GPP TS 33.220 v12.3.0, Generic Authentication Architecture (GAA); Generic Bootstrapping Architecture (GBA), available at: www.3gpp.org/DynaReport/33220.htm, June 2014.

[3GP 14b] 3GPP TS 23.468 v12.2.0, Group Communication System Enablers for LTE (GCSE LTE), Stage 2, available at: www.3gpp.org/DynaReport/23468.htm, September 2014.

[3GP 14c] 3GPP TR 22.897 v13.0.0, Study on Isolated Evolved Universal Terrestrial Radio Access Network (E-UTRAN) Operation for Public Safety, available at: www.3gpp.org/DynaReport/22897.htm, June 2014.

[3GP 14d] 3GPP TS 22.179 v13.0.0, Mission Critical Push to Talk, available at: www.3gpp.org/DynaReport/22179.htm, September 2014.

[3GP 14e] 3GPP TS 23.303 v12.2.0, Proximity-based Services (ProSe), stage 2, available at: www.3gpp.org/DynaReport/23303.htm, September 2014.

[3GP 13f] 3GPP TS 33.401 v12.10.0, 3GPP System Architecture Evolution (SAE); Security Architecture, available at: www.3gpp.org/DynaReport/33401.htm, December 2013.

[BOU 11] BOUKERCHE A., TURGUT B., AYDIN N. *et al.*, "Routing protocols in ad hoc networks: a survey", *Computer Networks*, 2011.

[DOR 13] DORTMUND S., "Impact of human crowds on the radio wave propagation in large concert halls", *Antennas and Propagation (EuCAP), 7th European Conference*, pp. 8–12 and 3030–3033, April, 2013.

[DOU 13] DOUMI T., DOLAN M.F., TATESH S. *et al.*, "LTE for public safety networks", *IEEE Comm. Magazine*, vol. 51, no. 2, pp.106–112, February 2013.

[EUR 13] ELECTRONIC COMMUNICATIONS COMMITTEE (ECC), User Requirements and Spectrum Needs for the Future European Broadband PPDR System (Wide Area Network), 2013.

[FLU 01] FLUHRER S., MANTIN I., SHAMIR A., "Weaknesses in the key scheduling algorithm of RC4", in VAUDENAY S., YOUSSEF A.M., *Selected Areas in Cryptography*, Springer Berlin Heidelberg, 2001.

[FRI 12] FRIAS Z., PEREZ J., "Techno-economic analysis of femtocell deployment in long-term evolution networks", *EURASIP Journal on Wireless Communications and Networking*, 2012.

[GSA 13] GLOBAL MOBILE SUPPLIERS ASSOCIATION, Status of the LTE Ecosystem, available at: www.gsacom., com/gsm_3g/info_papers.php4, October 2014.

[IEE 05] IEEE 802.11E STANDARD, Local and metropolitan area networks-specific requirements Part 11: wireless LAN medium access control (MAC) and physical layer (PHY) specifications amendment 8: medium access control (MAC) quality of service enhancements, available at: ieeexplore.ieee.org, 2005.

[IEE 04] IEEE 802.11I STANDARD, Local and metropolitan area networks-specific requirements Part 11: wireless LAN medium access control (MAC) and physical layer (PHY) specifications amendment 6: medium access control (MAC) security enhancements, available at: ieeexplore.ieee.org, 2004.

[IEE 09] IEEE Std 802.21-2008, IEEE standard for local and metropolitan area networks – media independent handover services, available at: www.ieeexplore.ieee.org, January 2009.

[IEE 11] IEEE 802.11S STANDARD, Local and metropolitan area networks-specific requirements Part 11: wireless LAN medium access control (MAC) and physical layer (PHY) specifications amendment 10: mesh networking, available at: ieeexplore.ieee.org, 2011.

[ISM 10] INDUSTRIEANLAGEN-BETRIEBSGESELLSCHAFT MBH, Medium-term and long-term capacity requirements of PPDR services in wireless communications, May 2010.

[ISO 03] ISO/IEC 7810, Identification cards – physical characteristics, available at: www.iso.org, 2003.

[ISO 14] ISO/IEC 7811-1, Identification cards – recording technique – part 1: embossing, available at: www.iso.org, 2014.

[ISO 11] ISO/IEC 7816-1, Identification cards – integrated circuit cards – part 1: cards with contacts – physical characteristics, available at: www.iso.org, 2011.

[ISO 91] ISO/IEC 9126, Information technology – software product evaluation – quality characteristics and guideline for their use, available at: www.iso.org, 1991.

[IST 15] INTERNATIONAL SOFTWARE AND TESTING QUALIFICATIONS BOARD (ISTQB), available at http://www.istqb.org/, 2015.

[ITU 03] ITU-R P.833-4, Recommendation ITU-R P.833-4 attenuation in vegetation, available at: www.itu.int, 2003

[KNO 12] KNOLL T.M., "Techno-economic modelling of LTE networks", *ITG Fachtagung Mobilkommunikation*, Osnabrück, 2012.

[LRC 12] LAW ENFORCEMENT WORKING PARTY (LEWP) – Radio Communication Expert Group (RCEG), Public safety applications matrix Excel table, 2012.

[MIL 10] MILLAR G.P., PANAOUSIS E.A. and POLITIS C., "ROBUST: reliable overlay based utilisation of services and topology for emergency MANETs", *Future Network & Mobile Summit*, Florence, Italy, pp.16–18, June 2010.

[MUE 14] MÜLLER W., REINERT F., "A methodology for development of enterprise architecture of PPDR Organisations", *Proceedings SERP 2014, the 2014 International Conference on Software Engineering Research & Practice*, pp. 259–263, July 2014.

[NAF 07] NATO ARCHITECTURE FRAMEWORK (NAF), Version 3 Annex 3 to AC/322(SC/1-WG/1)N(2007)0004, 2007.

[NIS 12] NIST SP 800 30 Rev 1, "Guide for conducting risk assessment", *National Institute for Standards and Technology*, 2012.

[OLC 14] OLCAN D.I., KRNETA A.J., KOLUNDZIJA B.M, "Modeling of human bodies for analysis of wireless body area networks in crowds", *Proceedings of 2014 IEEE ISAP and USNC-URSI Radio Science Meeting*, Tennessee, pp. 406–407, July 2014.

[OLS 09] OLSSON M., SULTANA S., ROMMER S. *et al.*, *SAE and the Evolved Packet Core*, Elsevier, 2009.

[OSS 10] OPEN SAFETY & SECURITY ARCHITECTURE FRAMEWORK (OSSAF), available at https://sites.google.com/site/openssaf/, 2010.

[RAM 10] RAMREKHA T.A., POLITIS C., "A hybrid adaptive routing protocol for extreme emergency ad hoc communication", *Proceedings of ICCCN*, August 2010.

[RAM 12] RAMREKHA T.A, TALOOKI V.N., RODRIGUEZ J. *et al.*, "Energy efficient and scalable routing protocol for extreme emergency" *Ad Hoc Communications. Mob. Netw*, April 2012.

[ROH 11] ROHILL TECHNOLOGIES, "Fast Inter System Push to Talk Operation", Patent no. US 20110305118 A1, December 2011.

[SAX 09] SAXENA N., TSUDIK G., JEONG HYUN Yi, "Efficient node admission and certificateless secure communication in short-lived MANETs", *Parallel and Distributed Systems, IEEE Transactions*, vol. 20, no. 2, February 2009.

[SCH 12] SCHNEIDER P., How to Secure an LTE-Network: Just Applying the 3GPP Security Standards and That's It?, available at: www.tropers.deg, 2012.

[SWI 13] SWISS RE, Natural catastrophes and man-made disasters in 2012: a year of extreme weather events in the US, available at: www.swissre.com, Zurich, March 2013.

[TCC 13] TETRA CRITICAL COMMUNICATIONS ASSOCIATION (TCCA), Mission Critical Mobile Broadband: Practical Standardization & Roadmap Considerations, White Paper, February 2013.

[TOG 11] TOGAF, available at http://www.opengroup.org/togaf/, 2011.

[VEL 15] VELJOVIC M.J., OLCAN D.I., KOLUNDZIJA B.M., "Full-wave simulation of propagation in human crowds", *2015 IEEE ISAP and USNC-URSI Radio Science Meeting*, Vancouver, Canada, 2015.

[WBA 14] WIRELESS BROADBAND ALLIANCE, Carrier Wi-Fi: state of the market, available at: www.wballiance.com, December 2014.

[YOK 10] YOKOTA H., CHOWDHURY K., KOODLI R. *et al.*, "Fast handovers for proxy mobile IPv6", *RFC 5949*, 2010.

4

From DMO to D2D

Direct communications have been used for long time in PSNs. Today, due to the expectations of the transmitted traffic through the network and the popularity of proximity services, direct communications may also be present in the future generation of public cellular networks.

One of the most important requirements for PSN's is the ability to establish direct communication links between terminals without the presence of any form of infrastructure. The link must be direct between terminals. This communication mode is called direct mode operation (DMO).

With the intuition that new attractive services can emerge for users who are close to each other, a global framework has been proposed by the public wireless ecosystem composed of manufactures, operators and standardization bodies. Since

---

Chapter written by Xavier PONS-MASBERNAT, Eric GEORGEAUX, Christophe GRUET, François MONTAIGNE, Jean-Christophe SCHIEL, Guy PHILIPPE and Lirida NAVINER.

2011, the third generation partnership project (3GPP) has decided to start, through multiple endeavors, studies on services (ProSe – proximity-based services) and on associated radio aspects (D2D – device-to-device communication).

4.1. Direct mode operation communication in current PSN

4.1.1. *DMO overview*

4.1.1.1. *DMO introduction*

The walkie-talkie was the predecessor of direct communications using hand-held devices, and it was used in World War II. Walkie-talkie terminals also have a direct broadcast communication mode, also used in PSN that can be activated by pushing the push-to-talk (PTT) button. With the deployment of PSN's for emergency services, industries, oil rigs, etc., direct communications have been widely used for PMR users. Even for certain user groups such as the French intelligence agency, this is the most widely used mode.

However, until now, commercial operators have seen this communication mode as a threat, because they were afraid of losing the control of their own frequencies and, more importantly, have had trouble billing the users for this service.

With current DMO for PSNs, the network operator would not have any way of controlling direct communication between two terminals. Therefore, there would not be any exchange with the billing server. Another fear of the operators is that a multi-hop *ad hoc* network could be established, avoiding the billing server for all communications.

4.1.1.2. *TMO versus DMO*

DMO is the operation mode performed without any infrastructure, where different mobile terminals communicate using radio frequencies which are outside of control of the network. The mode that uses the base station (BS) and the network infrastructure to establish a communication between two or more terminals or other type of networks, like typical public radio mobile networks, is called trunked mode operation (TMO).

4.1.1.3. *PTT and Group Call*

The group calls in PSN's can be done in two different ways. On the one hand, there are the group calls using the infrastructure. In this case, the group is defined beforehand and only the users belonging to this group can receive the messages sent by other terminals of this specific group. This mode can be compared to multicast, because all the users belonging to the group will receive the message through their serving BS, whether they are attached to the same cell of the transmitter terminal or another one. When the PTT button is pressed, the group call starts. The message is sent through the BS which will forward the message to all the members belonging to the same transmitter device group. If there are other members of this group inside other geographical area covered by another BS, the message is forwarded to upper layers which will send the message to these BS where other group members are attached.

However, we have the DMO communications when a terminal broadcasts a message without using any infrastructure. Only the terminals under the radio coverage of the transmitter terminal and monitoring the right frequency can receive the message. Usually, a channel is assigned beforehand to a specific group of users in order to establish DMO communications. For this purpose, different frequency bands or channels are explicitly allocated. This is the typical DMO used to communicate with multiple radio terminals which are geographically close to the transmitter terminal. The main problem of this solution is that all the group members use the same channel which could cause collisions when two or more terminals access the channel at the same time. In current PSNDMO, any type of collision avoidance mechanism does not exist.

4.1.2. *DMO in TETRA*

TETRA is a Time Division Multiple Access (TDMA) system operated in FDD mode. Four timeslots of 85/6 ms (~14.167 ms) each compose the TDMA frame transmitted on the 25 kHz TETRA channel (Figure 4.1). Symbols are π/4 DQPSK (differential quadrature phase-shift keying) modulated at 18 ksymb/s. The resulting raw timeslot bit rate is 9 kbps.

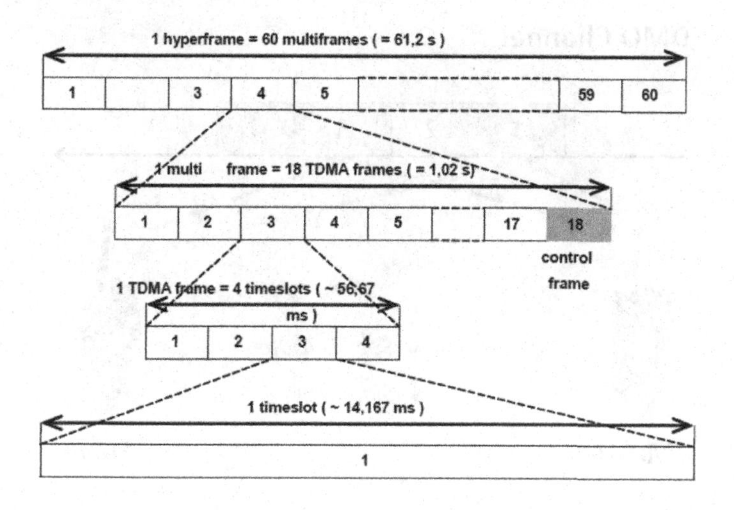

Figure 4.1. *TETRA TDMA structure*

A full TETRA channel is devoted to DMO between at least two users. Two of the four timeslots are used to provide one communication in each direction. The two corresponding time slots are separated by one timeslot in order to facilitate the Tx/Rx switching. The transmitting DMO user is considered as "master" while the others are "slave" and only authorized to preempt the DMO transmission (Figure 4.2). As only two timeslots per frame are used (TS#1 & TS#3), a second DMO communication can take place on the two remaining timeslots (TS#2 & TS#4) improving by this way the DMO capacity.

DMO is used basically to facilitate voice communications using the TETRA 4.566 kbps algebraic code-excited linear prediction (ACELP) speech coder, strongly associated with a push-to-talk (PTT) mechanism, considering either individual calls (point-to-point) or group calls (point-to-multipoint).

Regarding data services, the TETRA DMO standard proposes the following possibilities:

1) Circuit mode data bearer: 2.4 kbps, 4.8 kbps or 7.2 kbps selecting different protection levels;

2) Short data: 16 bits, 32 bits, 64 bits or up to 2047 bits;

3) Status messages: 16 bits allowing up to 65536 predefined messages.

While in a point-to-point communication checking or acknowledgment mechanisms can be operated between the two end-to-end entities, these facilities are not available in a point-to-multipoint communication.

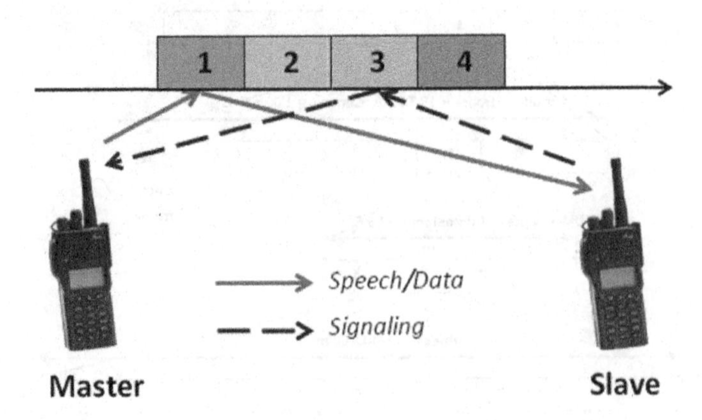

Figure 4.2. *DMO TETRA channel usage*

The term "back-to-back'" operational mode refers to DMO allowing standard terminal to terminal communications. But DMO can also be used to extend either DMO or TMO coverage (Figure 4.3), we speak about DMO Repeater or DMO Gateway respectively. To match the different scenarios, different repeater modes are proposed in the standard: Type 1A, Type 1B and Type 2.

Type 1A and 1B repeaters support only one call at the same time while Type 2 repeaters are able to manage two simultaneous calls. One unique frequency channel is required for Type 1A repeater while both Type 1B and Type 2 require two frequency channels.

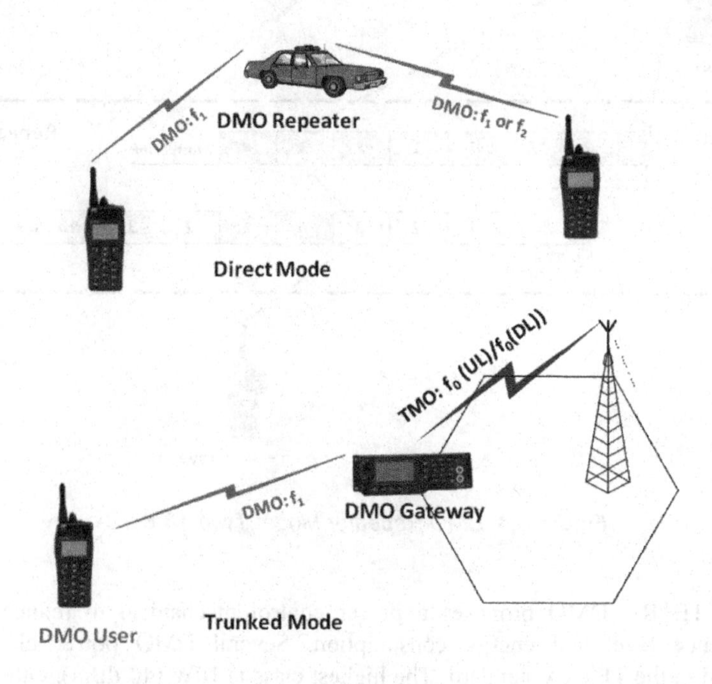

Figure 4.3. *DMO &TMO coverage (UL:Uplink / DL: Downlink)*

In mode Type 1A, all the timeslots of the DMO channel are used: TS#1 and TS#3 to manage the direct communication between the master terminal and the repeater, acting as slave, and TS#4 and TS#2 for the second direct communication between the repeater, acting as master, and the slave terminal. From the slave terminal point of view, slots 4 and 2 are respectively seen as slots 1 and 3 as in the basic mode of operation (Figure 4.4).

Type 2B mode is identical to Type 1A except for the frequency which is different for the master and slave link.

Type 2 DMO Repeaters also support dual carrier operation but they are able to offer two simultaneous calls. The delay between master and slave link is four timeslots. Consequently, two DMO channels can be used: the first one operated on TS#1 and TS#3 the second one on TS#2 and TS#4.

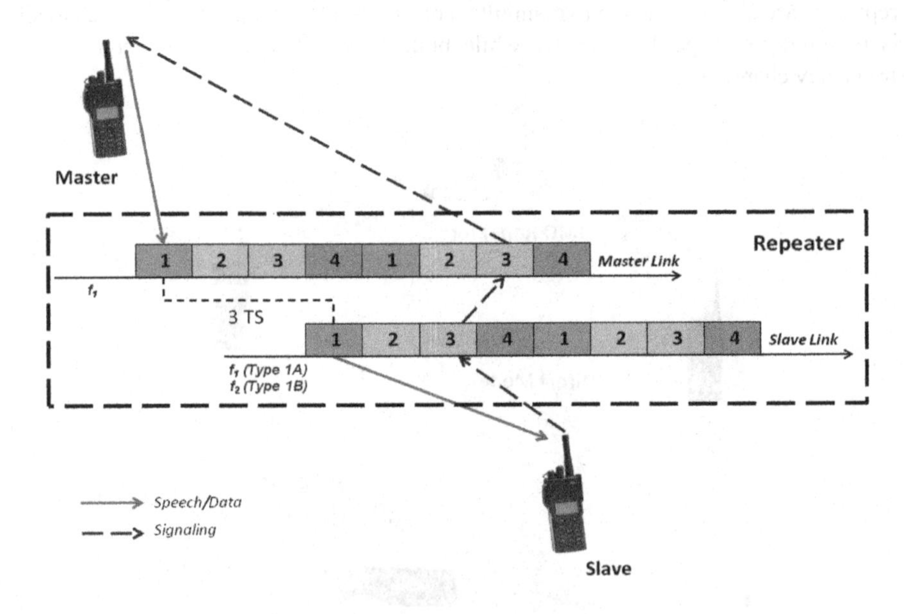

Figure 4.4. *DMO Repeater Mode: Type 1A & 1B*

The TETRA DMO proposes a power control mechanism to reduce overall interference level and energy consumption. Several DMO power classes are proposed in the TETRA standard. The highest class is 10W (40 dBm). Other power levels are obtained by decreasing the peak power by step of 2.5 dB down to 22.5 dBm (0.18 W).

The DMO signaling timeslot (TS#3 in "back-to-back" mode) is used to manage preemption mechanisms when one slave terminal requires the DMO channel. But it is also used for channel maintenance purpose (for example radio link quality reported to the master during a call) or late entry events. Late entry refers to users joining a DMO group during an ongoing call, just after switching on the radio, after leaving another DMO communication channel, after switching from TMO to DMO, or after changing the coverage area.

4.1.3. *DMO in TETRAPOL*

Main competitor to TETRA, TETRAPOL is a FDMA system operated in FDD mode and using 10 and 12.5 kHz channels. Symbols are GMSK modulated at 8 ksymb/s. Consequently the raw air interface throughput is 8 kbps. Frames of 20 ms are continuously transmitted according to a 4s multiframe structure (Figure 4.5). Each frame contains 160 bits composed of one 8 bits synchronization word and 152 bits of protected data.

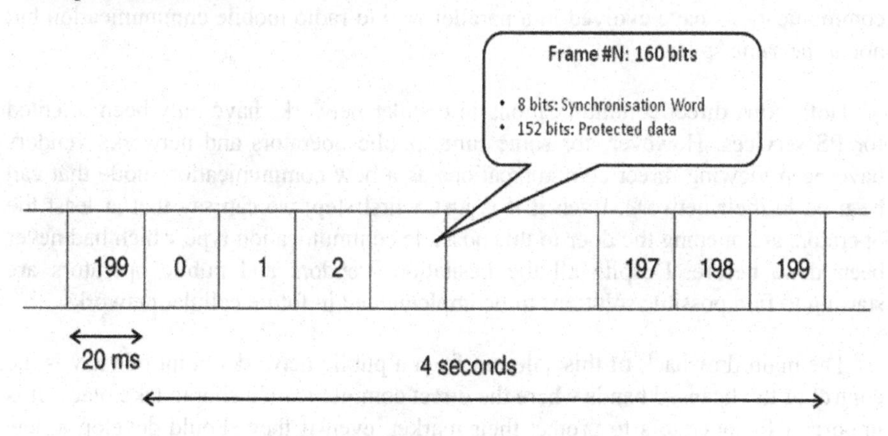

Figure 4.5. *TETRAPOL transmission scheme*

Voice services managed in a PTT way are proposed using the 6.0 kbps RPCELP speech coder. Data services are also available. A basic circuit mode at 3.3 and 4.6 kbps is proposed as well as short and status messages.

Standalone DMO and DMO repeater schemes are of course present in the TETRAPOL standard. The FDMA structure offers less possibility than TETRA. TETRAPOL repeaters allowing DMO and TMO coverage extension require two frequency channels.

No power control mechanisms are proposed for DMO, Consequently the TETRAPOL terminals are using their maximum power being 10 W (40 dBm) for mobile station mounted in vehicles and 2 W (33 dBm) for handheld.

Moreover, the FDMA structure does not offer opportunity to slave terminals to feedback information to the master. Therefore no preemption is possible. To avoid DMO users keeping the channel indefinitely, a specific activity timer (typically 10 s) is managed by the master terminal, switching off the transmission when reached.

4.2. D2D solutions for future LTE PSNs

4.2.1. *D2D overview*

DMO, as previously said, is the operation mode performed without any infrastructure where different mobile terminals communicate using radio frequencies which are outside of control of the network. The first direct communication using mobile terminals was with walkie-talkies during the World War II. Since then, direct communications have evolved in a parallel way to radio mobile communication but not at the same speed.

Until now, direct communications in cellular networks have only been oriented for PS services. However, for some time, public operators and networks vendors have been viewing direct communications as a new communication mode that can be used in their network. Even if it is just a first step, we can say that at least the operators are opening the door to this possible communication type which had never been done before. Despite all the hesitation, vendors and public operators are starting to find possible solutions to be implemented in future cellular networks.

The main drawback of this solution from a public networks point of view is the control of the licensed bands where the direct communication should take place. It is important for operators to protect their market, even if they should develop a new business model to establish direct communications in public cellular networks. On the other hand, the potential that direct communication has is huge. Direct communications could improve data rates, spectral efficiency, the consumed power on both the network and terminal sides, and could be interesting for offloading in dense scenarios.

With the interest of operators, the arrival of the internet of things (IoTs), the adoption of LTE for the new PSN's direct communications have been renamed device-to-device (D2D). Within this new name, the aim was to include all types of direct communication, from typical DMO to typical machine-to-machine (M2M) communications. In this section, we present a classification of D2D communications and some of the works and standardization efforts already made to support this kind of communications in future PSN.

4.2.2. *Classification of future D2D solutions*

Over the last few years, several solutions have been developed to find a model to establish direct communications in cellular networks such as LTE. In [ARA 14], a D2D communication mode classification is proposed depending on the spectrum band used. Basically, the D2D communications can be divided in two main groups, *Inband D2D* and *Outband D2D*, as can be seen in Figure 4.6.

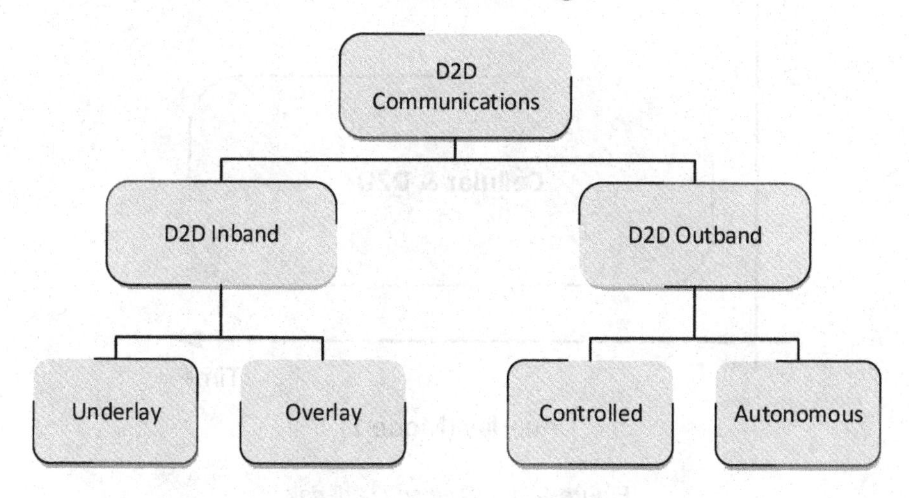

Figure 4.6. *D2D classification*

4.2.2.1. *Inband D2D*

In the first group *Inband D2D*, the spectrum bands used for cellular and D2D communications are exactly the same. In this case the operator can have control over all its cellular spectrum, and can manage the quality of service (QoS), and even in some cases, due to the interferences between cellular and D2D communications, the interference level can be controlled.

Within the Inband D2D group, two additional subcategories can be defined as follows:

1) The first is known as underlay and the same resources are used either for cellular or D2D communications. The same resources can be simultaneously used by different terminals only if they do not interfere with each other (see Figure 4.7).

2) The second subcategory is called overlay. Communications in overlay mode have a dedicated band or a certain amount of resources only for D2D communications. In this case there is no interference between D2D and cellular communications, as was the case in underlay mode (see Figure 4.8).

Both categories are also defined in 3GPP radio access networks (RAN) working group (WG) 1. Mode 1 refers to shared resources between cellular and D2D communications, underlay. Mode 2, refers to different resources for each communication mode, overlay.

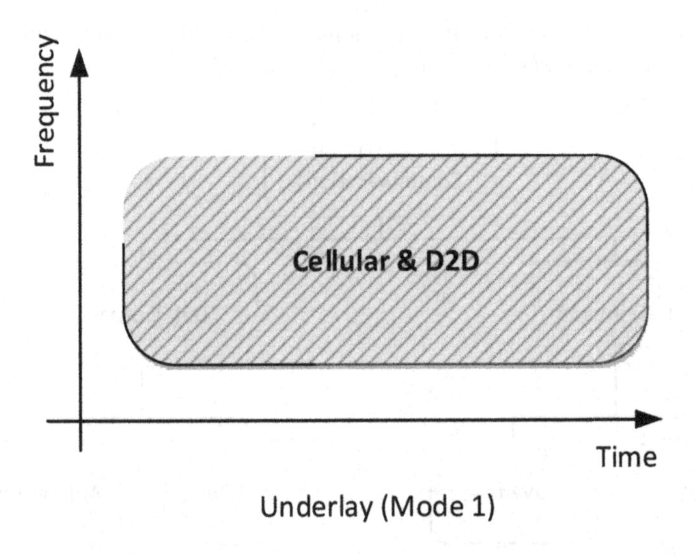

Figure 4.7. *D2D inband underlay*

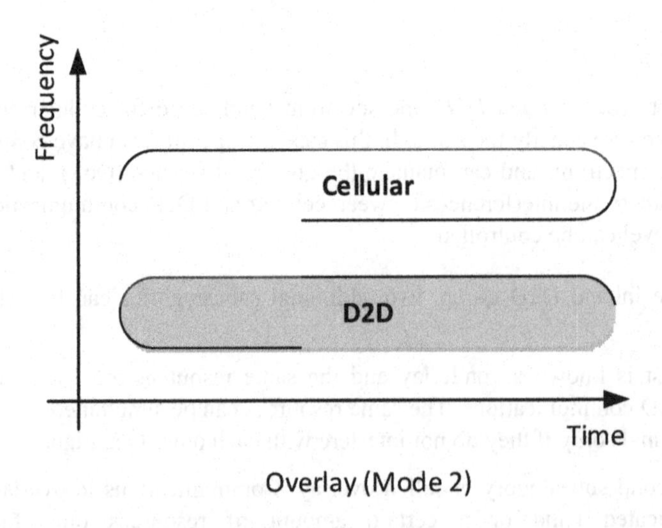

Figure 4.8. *D2D inband overlay*

4.2.2.1.1. Underlay

The underlay mode is that which uses the same resources either for D2D communications or cellular communications.

The most important advantages of using an underlay solution are the control over all the cellular spectra allowing a QoS management and the possibility to improve the spectral efficiency due to the spatial diversity, for instance, when the same resources are used for D2D and cellular communications simultaneously with different users.

However, this mode has some drawbacks compared to overlay mode and out-of-band solutions. One of them is the impossibility of having simultaneous D2D and cellular transmissions even in different resources. This can be done using an out-of-band system for D2D communications. Furthermore, when the underlay mode is used the interference level can increase considerably, unless it is managed through a resource allocation algorithm. Hence, one of the key points for a system using D2D Inband underlay communications is the resource allocation algorithm which could have a very high complexity.

For these reasons, this mode is one of the most targeted and developed by the research community; not only for the spectral efficiency improvement, but also for finding an interference management solution with a low complexity and computational cost.

The D2D Inband underlay category is the most widespread option for D2D communications. Ongoing communications, such as cellular, Wi-Fi, WiMax or others should not be degraded due to the D2D links. Therefore, within this category several solutions addressing the interference problem can be found. Basically, two main strategies are adopted: power control and resource allocation mechanisms.

For example, in [HON 10] the authors study the impact of limiting the transmission power by using a centralized power control mechanism. The goal is to manage the interference and reuse the same resource for D2D and cellular communications simultaneously.

In [FOD 11], the authors propose a distributed power control algorithm to minimize the transmission power in both transmission modes: cellular and D2D. An analysis on optimum resource allocation and power control between cellular and D2D is also provided. Resource re-utilization in D2D underlay system is implemented and evaluated for different scenarios.

Another interference management scheme was proposed in [WAN 12]. This time the proposed mechanism is oriented to D2D multicast groups under the control of a BS. First of all the enhanced NodeB (eNB) sets an upper power transmission bound to the transmitter terminal, and then it allocates the most appropriate resources to establish the D2D communication. Following this scheme, the interference between typical cellular transmissions and D2D multicast receivers is reduced.

Another strategy adopted to reduce the interference problem is the resource allocation mechanisms. In [JAN 09] a new interference aware resource allocation algorithm is presented. This algorithm tries to reduce the maximum received power in D2D pairs, and thus allow the reuse of these resources by other users.

Another way to manage the resource allocation is by defining an interference limited area in order to control the interference between cellular and D2D transmission. Therefore, the authors in [MIN 11] and [CHE 12] consider that an eNB cannot allocate the same uplink resources used in D2D communication to a cellular communication if the cellular user equipment (UE) is inside the defined area; otherwise the signal-to-interference noise ratio (SINR) at the receiver side will exceed a defined threshold. A similar strategy is used in [BAO 12], where an interference limited ring is defined to guarantee a D2D outage probability.

In [KAU 13] the authors present a new distributed resource allocation protocol, where D2D users opportunistically select the resources entailing an SINR degradation, but still allowing a spectrum reuse.

4.2.2.1.2. Overlay

In the Inband D2D overlay mode, two different bands and/or resources are defined and each one dedicated to different purposes. One is only used for typical cellular communications, while the other is used for D2D communications.

The Inband D2D overlay category is less popular than the underlay. Probably one of the reasons is that having a dedicated band to establish D2D communications reduces the challenges that exist in the underlay.

One of the most important advantages of using an overlay solution is the reduced complexity because due to the non-existence of interference between cellular and D2D communications. On the other hand, one of the major drawbacks is the reduced spectral efficiency. Using an overlay solution, there is less flexibility and the spectral efficiency cannot improve at the underlay levels.

Moreover, within the public domain, the fact of having a D2D dedicated band does not satisfy public operators because the solution is much less flexible than the previous one. For example, in [FOD 12] a comparison between underlay and overlay solutions is made. The authors conclude that network controlled solutions are more efficient and with better performance. Since the BS can control all the terminals under its coverage, the synchronization, the discovery and the communications procedures can be established and controlled easily reducing the interference between terminals. The authors assume that without the network control, the D2D terminals should send beacons to become synchronized and find each other, and sense various channels.

Despite some of the pros and cons presented above, some solutions addressing overlay problems can still be found. For example in [ZHO 13], the authors study the number of optimal relays to send multicast messages to a cluster improving the data rate transmission. Their mechanism is under the control of a BS that provides orthogonal resource between D2D and cell communications.

In [LI 12] another relay mode is proposed for forwarding a D2D message, improving the total throughput and providing more reliable D2D communications. The BS is capable of listening to D2D message and it will forward the messages when the D2D link is in outage.

4.2.2.2. *Outband D2D*

In the second group of D2D communication, called *Outband D2D*, two different frequency bands are used for each single transmission mode. Thus, two different technologies are used: one for typical cellular communications and another one for D2D communications. In this case the most used technologies for direct communications are typical short range communications: Wi-Fi, Zigbee and Bluetooth.

The advantages of using other frequencies for D2D communications is the non-existence of interferences between typical cellular and D2D communications, and the possibility of establishing two simultaneous communications: one in cellular mode, and another one in direct mode.

However, this solution entails hard requirements on the terminal side. Outband capable terminals should have the capability to receive and/or transmit simultaneously in two different technologies, thus having at least two radio interfaces. An additional problem to take in consideration is that the out-of-band communications are often based on technologies using the Industrial, Scientific and Medical (ISM) radio bands. These bands are used by several technologies causing unmanaged interference problems. These problems are difficult to control because there does not exist any kind of master that has the overall control of all the terminal transmissions. Even if the transmitted power is limited, interference is the big problem to overcome for Outband D2D types of solutions.

The Outband D2D solution has to face up other problems. For instance, if a terminal wants to relay the message received from one interface to another one, it needs to decode and recode the message before sending it through the other interface. This entails a waste of time and energy to the terminals which already have limited battery capacities.

In PSNs, security and latency are one of the most important requirements. Thus, PSNs use specific frequency bands where only their own users can get access to it.

In Outband D2D solutions that use ISM bands like Wi-Fi or Bluetooth, multiple collisions can happen in dense scenarios, involving high latency and security problems.

As in Inband D2D, the Outband D2D communications can also be divided in two different groups: *Controlled* and *Autonomous*. In the controlled mode the cellular network takes the control of the other radio interfaces used for D2D communications. By giving the control of the D2D interface to the cellular network, some improvements can be made in terms of throughput, energy efficiency (EE) and fairness. However, having control over the D2D interface through the network increases the overhead. To solve the overhead problem, some direct communication solutions allow the terminals themselves to control the D2D interfaces. This mode is called *Autonomous Outband D2D*.

4.2.2.3. *Controlled*

Controlled D2D Outband solutions are implemented to improve the efficiency and reliability of D2D communications. The main difference between controlled and autonomous solutions is that in controlled category the cellular technologies manage the interface or technology used to establish D2D links. Within this category just a small number of solutions can be found. Most of them want to improve the system performance, mainly the throughput and the EE.

For instance in [ARA 13] and [AR2 13] Wi-Fi is used to establish D2D communications improving the throughput and the energy efficiency of cellular networks. The aim of these solutions is to create cluster between the terminal that are under Wi-Fi coverage, and select one of them as the cluster head (CH). This CH will be the responsible to establish communication with the BS and relay the information to the BS or to the UE that belong to its cluster. To create the clusters the authors use a game theory approach.

4.2.2.4. *Autonomous*

The aim of Autonomous D2D Outband solutions is to reduce the overhead due to D2D communication in cellular networks. Very few solutions can be found in this category. For example in [WAN 13] the authors present a user-initiated BS-transparent traffic spreading which strengths the user-to-user communication to increase BS scheduling flexibility. The proposed scheme can reduce the file transfer delays and it reduces the consumed power compared to performance-centric algorithms.

Most of the reported works on D2D are only oriented for under coverage scenarios. Considering the work done in [FOD 12] and [ARA 14], a cluster mode for *ad hoc* networks seems to be a good option to provide seamless D2D communications between terminals in all possible scenarios cases: under cell coverage, out of cell coverage, and in a mixed scenario where some terminals are in under coverage situation while the others are in out of coverage.

Nevertheless, the main problem for PSN oriented solutions is the battery life time. The terminal which is acting as a CH will consume a lot of energy to control all the communications. Hence, it seems that the most appropriate way to allow D2D communications in a seamless manner is by creating a traditional *ad hoc* network without any CH, allowing a seamless transition between out of cell and under cell coverage.

4.2.3. *Standardization/3GPP efforts*

4.2.3.1. *Introduction*

Up to now, there have been two separate technology families for commercial cellular networks (based on standards such as GSM, UMTS-3G, IS-95, etc.) and dedicated public safety systems (based on standards such as TETRA, TETRAPOL, P25, etc.).

The national public safety telecommunications council (NPSTC) and other US organizations were the first to recognize the benefits of a single interoperable standard able to support requirements from both commercial operators and public safety organizations and offering broadband capabilities. In 2009, three major public safety organizations in the US (APCO: Association of Public-Safety Communications Officials, NENA: the National Emergency Number Association and NPSTC) endorsed LTE as the converged technology standard.

In Europe, the TETRA and Critical Communications Association (TCCA) sent a liaison statement to the 3GPP confirming that LTE had most of the features required for public safety users. However, direct communication between devices in out-of-network coverage was listed as one of the key requirements for public safety users that was not in the scope of the 3GPP work plan.

As a consequence, 3GPP approved in September 2011 a first study item on Proximity Services (ProSe) based on direct communication in which the specific public safety use cases were to be considered.

The outcome of the study item is a technical report (TR 22.803) approved in December 2012 in which 12 public safety use cases were identified. Those use cases require two main direct communication mechanisms:

1) Discovery: this mechanism is needed for use cases requiring detection of the proximity of a public safety user without network assistance in either in-coverage or out-of-coverage scenarios.

2) Direct communication: this mechanism is needed for use cases requiring transmission of voice or data between public safety users, supporting one-to-one, one-to-many, broadcast and relay modes.

At this point, several work items were initiated at 3GPP in working groups involved in service and system aspects (SA working groups) and radio access network definitions (RAN working groups), with a common target of having the first support of proximity services using direct communication in Release 12 (frozen 13th March 2015). The scope of the Release 12 content is described in Table 4.1. Discovery mechanisms primarily focused on commercial application for in-coverage devices. Direct communication can be used in commercial networks to off-load the network but the main target was public safety use cases, to allow communication within, in partial or even outside network coverage. These two latter scenarios were identified to be specific to public safety.

	Within network coverage	Outside network coverage
Discovery	Non public safety & public safety requirements	Public safety only
Direct Communication	At least public safety requirements	Public safety only

Table 4.1. *LTE D2D ProSe scope in Release 12*

The 3GPP reference architecture was updated to support ProSe as shown on Figure 4.9. The main new elements are:

– interface PC5, the direct interface between two devices supporting ProSe;

– the "ProSe function", in charge of:

- providing to the UE the parameters needed for ProSe discovery and communication, including the parameters to be used when the UE is not served by E-UTRAN (out-of-coverage scenario for public safety),

- allocating and storing the application identifiers and filters for the Discovery services,

- communicating with the Home Subscriber Server (HSS) to check whether the UE is authorized to use ProSe;

– the interface PC3, the interface between the devices and the above ProSe function.

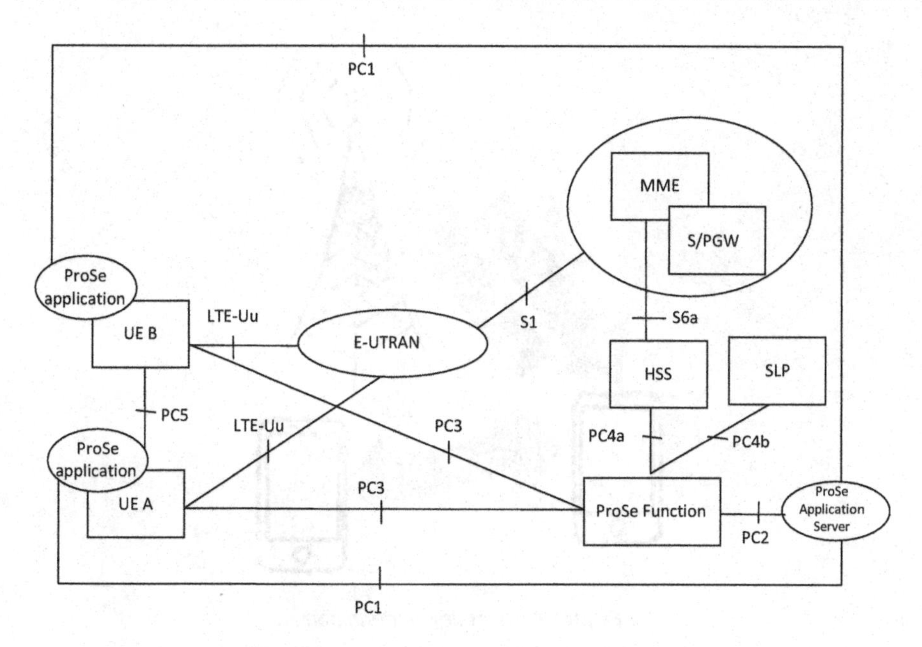

Figure 4.9. *ProSe reference architecture*

In current communication standards supporting D2D or direct mode, a portion of the spectrum is dedicated to the support of direct communication. This situation which was easily manageable for narrowband systems cannot be assumed for broadband system such as LTE. Thus, it was decided that the PC5 interface will re-use existing frequency allocation, whatever the duplex mode is (frequency division duplex (FDD) or time division duplex (TDD)).

To minimize the hardware impact on the user equipment (UE) and especially on the power amplifier, transmission of the D2D links occurs in the uplink band in the case of FDD. Similarly, the PC5 interface uses the sub-frames which are reserved for uplink transmission in TDD.

It was also decided to keep the SC-FDMA (single carrier frequency division multiple access) modulation currently used for uplink transmission. The new channels also inherit as much of the channel structure as possible applicable to the transmission of the PUSCH (physical uplink shared channel).

While “Uplink” (UL) designates the link from UE to eNodeB and “Downlink” (DL) the opposite direction, the name “Sidelink” was selected to designate the radio links over the PC5 interfaces (Figure 4.10).

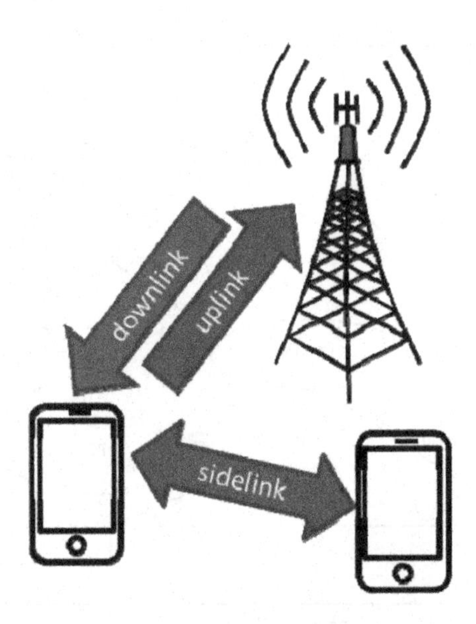

Figure 4.10. *Sidelink definition*

Figure 4.11 depicts the uplink channels (on the left-hand side) and the new sidelink channels introduced for the support of ProSe on the right-hand side.

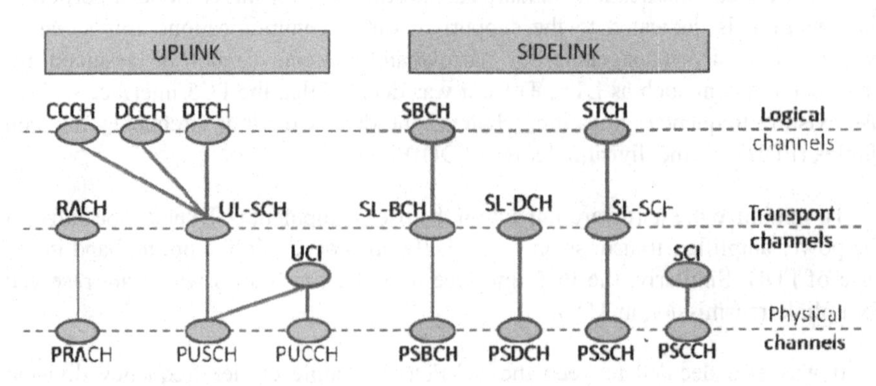

Figure 4.11. *Uplink and new Sidelink channels*

The acronyms for the sidelink channels and their main usage are described in Table 4.2.

Acronym	Channel name	Usage
SCI	Sidelink Control Information	Communication
PSCCH	Physical Sidelink Control Channel	
STCH	Sidelink Traffic Channel	
SL-SCH	Sidelink Shared Channel	
PSSCH	Physical Sidelink Shared Channel	
SL-DCH	Sidelink Discovery Channel	Discovery
PSDCH	Physical Sidelink Discovery Channel	
SBCH	Sidelink Broadcast Channel	Synchronization
SL-BCH	Sidelink Broadcast Channel	
PSBCH	Physical Sidelink Broadcast Channel	

Table 4.2. *Sidelink channels*

4.2.3.2. *Synchronization*

When the UEs are within coverage, the synchronization is derived from the downlink signals transmitted by the eNodeB. In LTE, these signals are the primary and the secondary synchronization signals (PSS and SSS). The UEs derived from these two signals the frequency and the time references so that they can access the radio resources without creating interference for other users. In addition, these two signals carry the identifier of the cell (Physical Cell Id).

In the direct mode, and especially in the out-of-coverage scenario, there is a similar need to share common frequency and time references between users. Hence at least one user will transmit some signals for that purpose. Although it was previously mentioned that the sidelink channels are designed to maximize the commonalities with the uplink channels, the sidelink synchronization signals reuse the principles of the downlink PSS and SSS: the primary and secondary sidelink synchronization signals (PSSS and SSSS) are also built from Zadoff–Chu sequences of length 62 and are transmitted on the 62 carriers in the center of the spectrum.

Instead of carrying the physical cell identifier, the PSSS and SSSS carry a "sidelink identifier" (N_{ID}^{SL} or SID). The sidelink identifier indicates whether the UE transmitting the sidelink synchronization signals has derived its synchronization from the network or not, as described in Table 4.3.

Set	$(N_{ID}^{SL}$	UE coverage
id_net	$0 \leq (N_{ID}^{SL} \leq 167$	The UE transmitting PSSS and SSSS derives its synchronization from the downlink synchronization signals received from eNodeB or from a UE transmitting PSSS/SSSS which is within coverage.
id_oon	$168 \leq (N_{ID}^{SL} \leq 335$	The UE transmitting PSSS and SSSS derives its synchronization from a UE transmitting PSSS/SSS which is out-of-coverage or from its own synchronization source (e.g. local clock, GPS signals...) if no synchronization signals are received

Table 4.3. *Relationship between SID and synchronization source*

In addition, the UE sending the sidelink synchronization signals also broadcast some information in the PSBCH. These information are transmitted in the same frequency region (the six resource blocks at the center) and within the same sub-frame, as depicted in Figure 4.12.

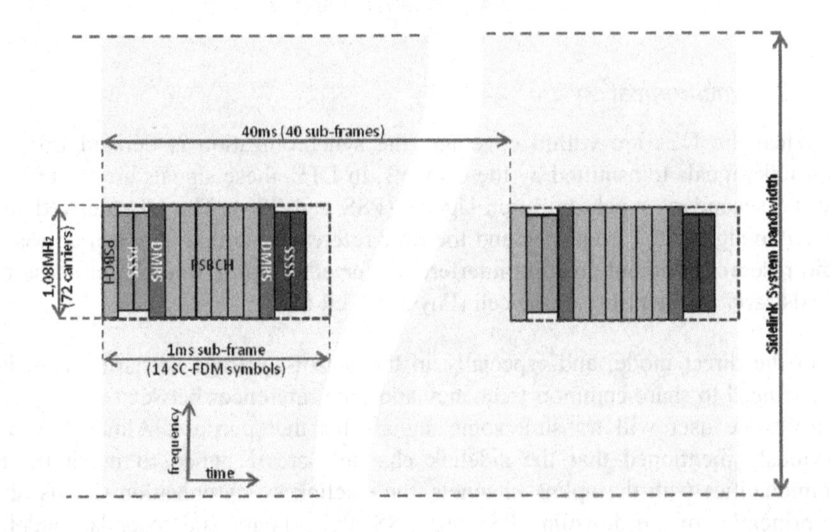

Figure 4.12. *Transmission of the sidelink synchronization signals and broadcast channel*

The SBCH contains the following information:

1) system bandwidth (1.4, 3, 5, 10, 15 or 20 MHz);

2) for TDD mode, the sub-frame configuration (which sub-frames are used for uplink and which are for downlink);

3) frame and sub-frame numbers of the SBCH, PSSS and SSSS transmission;

4) a Boolean flag indicating whether the UE is within or outside eNodeB coverage.

The conditions under which a device will transmit the synchronization signals are depicted in Figure 4.13.

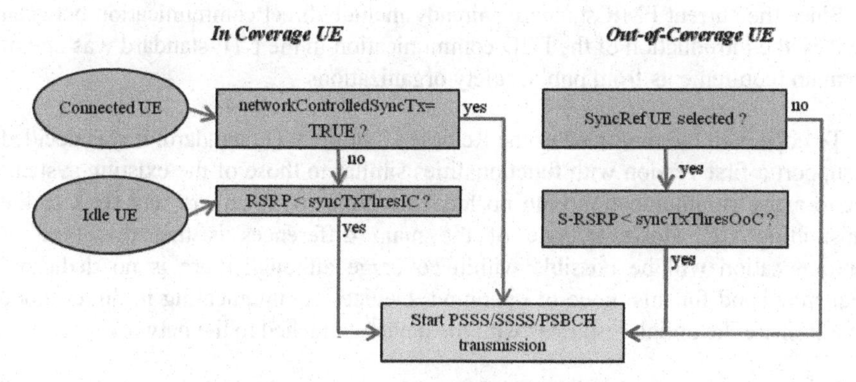

Figure 4.13. *Conditions to start synchronization signals transmission*

A device within coverage can be configured by the network to transmit PSSS/SSSS (by setting the parameter networkControlledSyncTx to true in the RRCConnectionReconfiguration message). Otherwise, it transmits the synchronization signals in case the level of the received signal from the eNodeB (RSRP: reference signal receive power) is below a threshold (syncTxThresIC). The value of the threshold is broadcasted by the eNodeB in system information block (SIB).

A similar procedure applies for out-of-coverage devices: if the device has not selected another device to derive its synchronization from or if the received signal from the selected device is below a pre-configured threshold (syncTxThresOoC), it starts transmitting the synchronization signals.

The device which is used as a synchronization reference is called SyncRef UE in 3GPP specifications. The out-of-coverage UEs periodically search candidate SyncRef UEs by detecting and measuring the level received from UE transmitting

PSSS/SSSS/PBSCH. Three levels of priority are defined amongst the candidate SyncRef UEs (in decreasing order):

1) UE transmitting PSBCH with the flag indicating it is in coverage set to true;

2) UE transmitting an identifier N_{ID}^{SL} which is in id_net set;

3) UEs not fulfilling the above conditions.

The UE with the highest level of received signal (S-RSRP) in the highest priority category will be selected as the SyncRef UE.

4.2.3.3. *Communications*

Since the current PMR standards already include direct communication between devices, the introduction of the D2D communication in the LTE standard was one of the main requirements from public safety organizations.

To cope with the timescale of the Release 12 of the LTE standard, it was decided to support a first version with functionalities similar to those of the existing system: one-to-many communication with no feedback from the receivers sent back to the transmitting UE. However, one of the main differences is that this type of communication will be possible within coverage although there is no dedicated frequency band for this mode of operation. Devices communicating in direct mode have to share the uplink resources with the devices attached to the network.

For that purpose, as mentioned in Table 4.2, two new physical channels have been introduced: PSCCH carrying the control information, and PSSCH carrying the data.

A pool of uplink resources is reserved for the PSCCH as shown on Figure 4.14. The pool is defined:

1) In frequency: by three parameters;

2) *PRBnum* that defines the frequency range in physical resource block (PRB) bandwidth units (the bandwidth of a PRB in LTE is 180 kHz) and there;

3) *PRBstart*, *PRBend* which defines the location in the frequency domain within the uplink band;

4) In the time domain: by a bitmap that indicates the 1ms sub-frames used for PSCCH transmission.

This block of resources is repeated with a period defined by a parameter *SC-Period* (expressed in sub-frame duration, i.e. 1 ms). The range of possible values for *SC-Period* is from 40 ms to 320 ms: low values have to be supported for voice transmission.

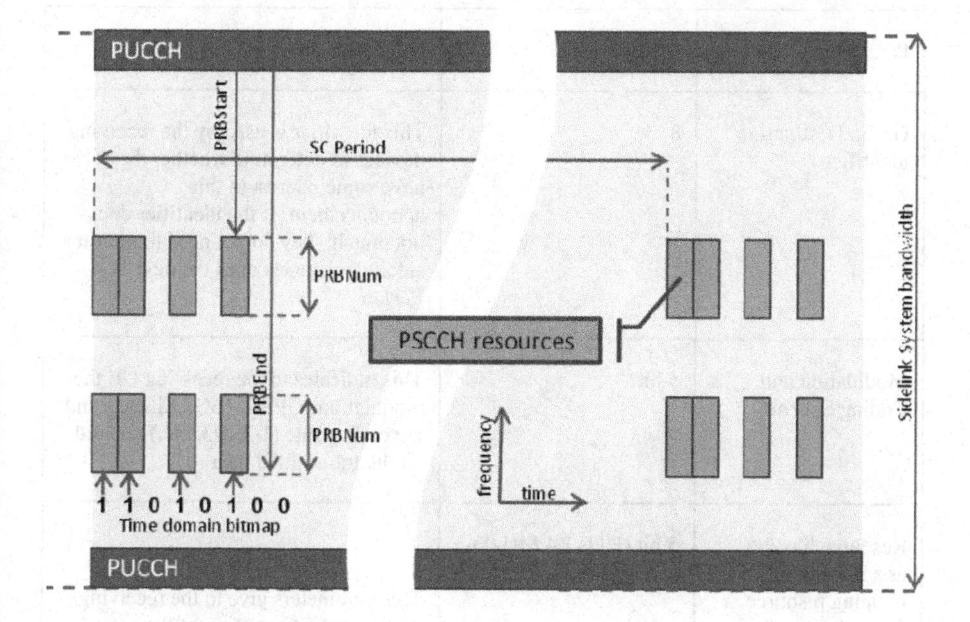

Figure 4.14. *PSCCH resource pool*

All the parameters needed to define the resource pool are broadcasted in a System Information Block (SIB) by the network. The devices which are not within coverage (and hence cannot acquire the SIB) will use some pre-configured values internally stored.

The PSCCH is used by the D2D transmitting UE to make the members of its group aware of the next data transmission that will occurs on the PSSCH: the transmitting UE sends over the PSCCH a "sidelink control information" block (SCI) that contains the information listed in Table 4.4.

Devices interested in receiving D2D services blindly scan the whole PSCCH pool to search if an SCI Format 0 matching their group identifier can be detected.

On the transmitting device side, resources to transmit the SCI format 0 information will be selected within the PSCCH pool. For the in-coverage scenario, two modes of resource allocation have been defined:

1) Mode 1, also referred as "Scheduled resource allocation";

2) Mode 2, also referred as "UE autonomous resource selection".

Parameter	Length	Usage
Group Destination identifier	8 bit	This identifier is used by the receiving devices to determine whether they have some interest in this announcement. If the identifier does not match, they do not need to monitor sidelink channels until the next *SC-Period*
Modulation and Coding Scheme	5 bit	This indicates to the receiving UE the modulation (QPSK, 16QAM, etc.) and the coding rate (1/2, 1/3, etc.) applied on the transmitted data
Resource block assignment and hopping resource allocation	5 bit (LTE 1.4 MHz) up to 13 bit (LTE 20 Mz)	This parameters give to the receiving devices the information of the resources of the PSSCH that they will decode in the frequency domain
Frequency hopping flag	1 bit	
Time Resource Pattern (T-RPT)	7 bit	This parameters give to the receiving devices the information of the resources of the PSSCH that they will decode in the time domain
Timing advance	11 bit	

Table 4.4. *SCI Format 0*

4.2.3.3.1. Scheduled resource allocation (mode 1)

In mode 1, access to the sidelink resources is driven by the eNodeB. The UE needs to be connected to transmit data:

1) The UE wishing to use direct communication feature sends an indication to the network. It will be assigned a temporary identifier sidelink radio network temporary identifier (SL-RNTI). This identifier will be used by the eNodeB to schedule the future D2D transmission.

2) When the UE has some data to transmit in D2D mode, it sends a sidelink-buffer status report (BSR) to the eNodeB which gives an indication on the amount of data to be transmitted in D2D mode. Based on this information, the eNodeB sends to the UE the allocation on both PSCCH and PSSCH for its D2D transmission. The allocation information is sent over the physical downlink control channel (PDCCH) by sending a DCI Format 5, scrambled by the SL-RNTI. The information contained in DCI Format 5 is detailed in Table 4.5. A large part of the DCI Format 5 information are directly reflected in the content of the SCI 0.

3) Based on the information received in the DCI format 5, the D2D transmitting devices sends the SCI format 0 over the resources within the PSCCH pool allocated by the eNodeB, followed by the data over the resources allocated by the eNodeB for PSSCH transmission.

Parameter	Length	Comment
Resource for PSCCH	6 bit	Provides the information of the transmitting UE of the resource to be used for SCI format 0 transmission within the PSCCH pool
TPC command for PSCCH and PSSCH	1 bit	If this bit is not set, the transmitting UE is allowed to transmit D2D signals at maximum power. Otherwise, it will comply with power control rules based on open loop (specific parameters for D2D transmission sent in the system information block)
Resource block assignment and hopping resource allocation	5 bit (LTE 1.4 MHz) up to 13 bit (LTE 20 Mz)	These parameters define the allocation for the PSSCH and have to be transmitted in SC I format 0
Frequency hopping flag	1 bit	
Time Resource Pattern (T-RPT)	7 bit	

Table 4.5. *DCI Format 5*

Thus, in mode 1, there are no pre-allocated or reserved resources for PSSCH: it is assigned "on-demand" by the eNodeB.

Moreover, since the eNodeB is responsible to give access to the resources within the PSCCH pool, collision on the PSCCH transmission can be avoided.

4.2.3.3.2. UE autonomous resource selection (mode 2)

In mode 2, the UE transmitting D2D data does not need to be connected to the eNodeB: it selects autonomously and randomly the resources within the PSCCH pool to transmit the SCI Format 0.

In addition to the PSCCH pool, there is also a PSSCH pool which defines reserved resources for PSSCH transmission. It is defined in a similar way as the PSCCH pool (PRBStart, PRBend, PRBNum in the frequency domain and a subframe bitmap in the time domain which is repeated up to the next PSCCH occurrence). The SCI Format 0 designates the portion of the pool that is used for D2D transmission. Since the transmitting UE is not necessarily connected to the eNodeB, the timing advance information may be not known and the corresponding parameter in the SCI Format 0 will be set to 0.

For out-of-coverage scenario (which was a mandatory requirement from Public Safety organizations), the "Mode 1" type of allocation is obviously not possible. D2D devices use "Mode 2" type of allocation. When in coverage, all the parameters needed for Mode 2 are read from SIB broadcasted by the eNodeB. In out-of-coverage situation, the devices will use the corresponding parameters that have been pre-configured and stored within the UE or the SIM card.

4.2.3.4. *Discovery*

ProSe direct discovery is defined as the procedure used to discover other UE(s) in its proximity, using E-UTRA direct radio signals via PC5.

In Release 12, discovery mechanisms were not the main focus for public safety and it was required that the direct communication will work without the support of discovery services.

As a consequence, discovery services are only available for devices within network coverage in Release 12.

However, it is likely that discovery services are used for public safety applications in Release 13, for instance for UE-to-network relay.

The “announcing UE” transmit a short message in order to be detectable by UEs in proximity. This short message contains:

1) a message type (which indicates open/restricted discovery and supported mode…);

2) an application code, which contains the PLMN information and some information about the application triggering the discovery service.

Devices in proximity receive the discovery message and check whether the application code matches with their discovery filter. If this is successful, the receiving devices send a matching report to the network.

As mentioned in Table 4.2, the transmission of the discovery information is done using the physical sidelink discovery channel PSDCH.

As for the reception of the PSCCH, the devices interested in receiving discovery notification will know on which resources they will search the discovery information. As for PSCCH, the eNodeB broadcast some information to define the pool of resources reserved for PSDCH. An additional parameter *NumRepetition* indicates how many times the sub-frame bitmap is repeated (Figure 4.15). The parameter *DiscPeriod* indicate the duration between discovery pool occurrences with a range from 32 ms up to 1024 ms.

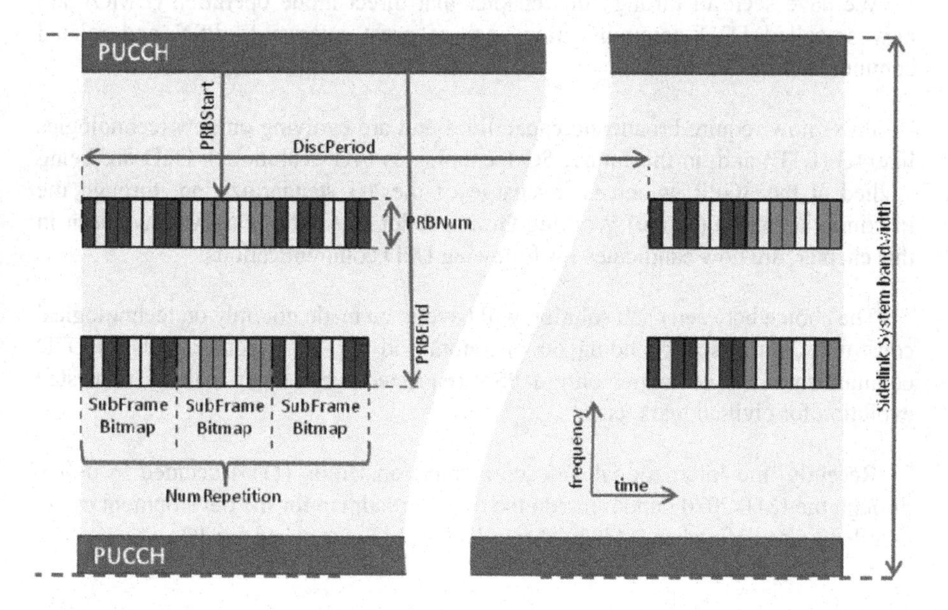

Figure 4.15. *PDSCH pool*

Similarly to communication, there are also two modes of resource allocations for discovery:

1) Type 1 or UE autonomous resource selection;

2) Type 2b or Scheduled resource allocation.

The principles are similar to those applicable for communication:

1) For Type 1, the UE randomly select resources within the pool for transmission of its discovery information (similar to Mode 2 for direct communication).

2) For Type 2b – as for communication Mode 1 – the UE will be connected to the network to request an allocation to transmit its discovery information. For discovery, the allocation information is not sent through a DCI Format on PDCCH as for direct communication Mode 1: the UE is informed through dedicated RRC signaling. While for direct communication Mode 1 the allocation is only valid for the next occurrence of PSCCH, the allocation for Type 2b discovery is semi-persistent: it remains valid until a new configuration is sent by the network or until the device enters in idle mode.

4.3. Conclusion and perspectives

We have seen all through this chapter that direct mode operation (DMO) and now so-called D2D communications are of great interest to PSN and related communications.

PSNs now require broadband capabilities and are evolving on new technologies like 4G (LTE) and, in the future, 5G technologies (4G evolutions). D2D are being studied at the 3GPP, which is in charge of the 4G standardization, through the Proximity Services (ProSe) Working Group. A lot of solutions as we have seen in this chapter, are now candidates for following D2D communications.

The choice between each solution will have to be made not only on technological constraints, but also depending on operators and end-user requirements as D2D communication is no longer only a PSN requested service but really a requested evolution for civilian markets.

Recently, the International Telecommunication Union (ITU) decided to define through the IMT-2020 denomination the overall roadmap for the development of 5G mobile (it was defined as IMT-2000 for 3G and IMT-Advanced for 4G).

The 5G challenges are mainly to offer a very high connection density and a higher data rate than in 4G, to cope with the increase of traffic volume and mobility of users while reducing the end-to-end latency issues. This has to be done taking into account spectrum, energy and cost efficiency.

Indeed, the evolution of the standard internet usage to the IoTs usage ensures that we will not have to limit the network dimensioning only to human communications.

The perspective of evolution in this market is higher than ever taking into account that now we have to also consider daily life objects and machines as end users with which we will have to interact more than ever through the same communication network.

Faced with these requirements, we can see the full advantages of D2D communications and its potential benefits. That is why D2D communications are considered as a key component of 5G networks and not solely of PSN networks.

Nevertheless, in PSNs we also have to consider this evolution of usage since in the future, we will also have to consider that we will have to communicate not only with humans, but bear in mind that security and reliability have been the key components for PSN since years.

We can see a lot of new applications that arise from PSN networks, like smart grids for critical infrastructure protection or the use of unmanned aerial vehicle/unmanned ground vehicle (UAV/UGV) swarms for surveillance and recognizance, and what is required of this level of security and reliability.

So, even if PSNs would completely benefitted from D2D evolution to improve their communication capabilities and their resiliencies, these two aspects (security and reliability) should not be neglected from a full PSN D2D evolution perspective.

The key challenge and perspective of the PSN network evolution would therefore be how to allow D2D communication evolution in PSN networks to ensure time spectrum, energy and cost efficiency but at the same time offering a very high level of security and reliability.

4.4. Bibliography

[ARA 13a] ARASH A., MANCUSO V., "On the compound impact of opportunistic scheduling and D2D communications in cellular networks", *Proceedings of the 16th ACM International Conference on Modeling, Analysis & Simulation of Wireless and Mobile Systems*, 2013.

[ARA 13b] ARASH A., MANCUSO V., "Energy efficient opportunistic uplink packet forwarding in hybrid wireless networks", ACM *Proceedings of the 4th International Conference on Future Energy Systems*, 2013.

[ARA 14] ARASH A., WANG Q., MANCUSO V., "A survey on device-to-device communication in cellular networks", *Communications Surveys & Tutorials, IEEE* 16.4, pp. 1801–1819, 2014.

[BAO 12] BAO P., YU G., "An interference management strategy for device-to-device underlaying cellular networks with partial location information", *IEEE 23rd International Symposium on Personal Indoor and Mobile Radio Communications (PIMRC)*, 2012.

[CHE 12] CHEN X. *et al.*, "Downlink resource allocation for device-to-device communication underlaying cellular networks", *IEEE 23rd International Symposium on Personal Indoor and Mobile Radio Communications (PIMRC)*, 2012.

[ETS 06] ETSI, EN 300 396-3, v1.2.1, TETRA; Technical requirements for Direct Mode Operation (DMO); Part 3: Mobile Station to Mobile Station (MS-MS) Air Interface Protocol, 2006.

[FOD 11] FODOR G., REIDER N., "A distributed power control scheme for cellular network assisted D2D communications", *Global Telecommunications Conference (GLOBECOM), IEEE*, 2011.

[FOD 12] FODOR G. *et al.*, "Design aspects of network assisted device-to-device communications", *IEEE Communications Magazine*, vol. 50, no. 3, pp. 170–177, 2012.

[HON 10] HONGNIAN X., HAKOLA S., "The investigation of power control schemes for a device-to-device communication integrated into OFDMA cellular system", *IEEE 21st International Symposium on Personal Indoor and Mobile Radio Communications (PIMRC), IEEE*, 2010.

[JAN 09] JANIS P. *et al.*, "Interference-aware resource allocation for device-to-device radio underlaying cellular networks", *IEEE 69th Vehicular Technology Conference, VTC Spring*, 2009.

[KAU 13] KAUFMAN B., LILLERBERG J., AAZHANG B., "Spectrum sharing scheme between cellular users and *ad hoc* device-to-device users", *Transactions on Wireless Communications, IEEE* 12.3, pp. 1038–1049, 2013.

[LI 12] LI J.C., LEI M., GAO F., "Device-to-device (D2D) communication in MU-MIMO cellular networks", *IEEE Global Communications Conference (GLOBECOM), IEEE*, 2012.

[MIN 11] MIN H. *et al.*, "Capacity enhancement using an interference limited area for device-to-device uplink underlaying cellular networks", *IEEE Transactions on Wireless Communications*, vol. 10.12, pp. 3995–4000, 2011.

[WAN 12] WANG D., WANG X., ZHAO Y., "An interference coordination scheme for device-to-device multicast in cellular networks", *Vehicular Technology Conference (VTC Fall), IEEE*, 2012.

[WAN 13] WANG Q., RENGARAJAN B., "Recouping opportunistic gain in dense base station layouts through energy-aware user cooperation", *IEEE 14th International Symposium and Workshops on a World of Wireless, Mobile and Multimedia Networks (WoWMoM)*, 2013.

[ZHO 13] ZHOU B. *et al.*, "Intracluster device-to-device relay algorithm with optimal resource utilization", *IEEE Transactions on Vehicular Technology*, vol. 62, no. 5, pp. 2315–2326, 2013.

5

Interoperability for Public Safety Networks

It is widely recognized that efficient communications are of paramount importance in order to have effective Public Protection and Disaster Relief (PPDR) operations. Indeed, there is great interest from governments and organizations involved in PPDR toward the evolution of existing wireless systems for critical communications to have higher communication capacity. However, an efficient communication system depends also on the capability of providing service interoperability among different networks. Interoperability barriers among the communication systems of various PPDR organizations are present both at national level (among public safety organizations of the same region or nation) and among PPDR organizations from different nations. Interoperability barriers are usually based on historical reasons: communication networks are created by each PPDR organization with a vertical structure to address the specific requirement of the organization. In some cases, interoperability barriers are also due to security reasons.

Chapter written by Federico FROSALI, Francesco GEI, Dania MARABISSI, Luigia MICCIULLO and Etienne LEZAACK.

In an effort to secure and protect the network data, cryptography mechanism and cryptography keys are different even in networks based on the same technology. Interoperability barriers are more often operational than technical. Common procedures and organizational schemes during a national disaster may be defined at national level, but rarely at international level. Cross-border operations are particularly affected by lack of interoperability because of linguistic barriers or because national organizations use different network technologies. A critical issue is the lack of roaming capability, which is available for users of commercial networks (e.g. GSM/UMTS), but it is not available for the PPDR community. As a consequence, a PPDR officer moving from one dedicated PPDR network to another will lose the communication instead of being transferred to the new network.

The situation across Europe is a good example of the issue. The strong political support for common public safety communications (Schengen Agreements) led SCHENGEN Telecommunications Working Group (later replaced by the Police Co-operation Working Group) to adopt TETRA for pan-European police communications and CEPT/ERC to designate a harmonized spectrum for police forces' usage of TETRA across Europe in 1995. Despite that, most EU countries deployed national networks based on TETRA or TETRAPOL, which are therefore almost uninteroperable. TETRA and TETRAPOL are probably the most advanced communication technologies available for public safety (PS) together with APCO P25 in USA. For TETRA technology, ETSI provided guidelines for interoperability with the TETRA Inter-System Interface (ISI), which defines the common protocol to support main services between interconnected networks. The first version of the ETSI ISI standard was released in 2000 and plans for ISI certification for individual call and ISI mobility management were completed in 2005 but, after a decade, there are few operational interoperability implementations. Similar ISI specifications are available for TETRAPOL, encompassing even TETRA–TETRAPOL interworking; they were released in 2000, but never implemented in practice.

Interoperability issues for mission critical communications are currently still the object of research projects, aimed at realizing frameworks to assess components required to implement interoperability at the system level.

The lack of interoperable communication systems has impeded real-time cooperation of PPDR forces, despite political being further reinforced by recent significant safety and security incidents. Experience shows that only a holistic approach, encompassing joint integration of procedures, technologies and legal agreements, may achieve feasible and effective solutions for PPDR interoperability in the field.

This chapter covers the main issues regarding interoperability, first analyzing the current status of European networks and giving a quick overview of some agreements and treaties active between particular countries (most of them in north

Europe), then identifying a scenario where the communication interoperability may play a particularly important role. In the final sections of the chapter, a possible framework to implement interoperability is described; the framework encompasses technical and procedural topics for networks and terminals and here is mainly illustrated in its technical aspects. The interoperability framework may represent a solution available in the short term, supported by current professional mobile radio (PMR) networks and bridging them toward an IP-based common European standard for broadband PPDR communications, shared by operators of all countries.

5.1. The role of PPDR communication system interoperability in national and international cooperation

Emergencies, natural calamities and crimes can be managed only by cooperation of multiple agencies; since they cannot be bounded inside national borders, they often require joint international actions. In the last few years, the need for cooperation during and after natural calamities (flooding, earthquakes, fire etc.), man-caused disasters (bomb attacks, aircraft crashes, chemical/nuclear alerts, etc.) or to support medical care and civil protection operations has been increasing. Crime organizations as well are nowadays globally connected, while cross-national police cooperation is often limited, especially in terms of interoperable radio communications and jointly agreed-upon procedures. International crime fighting (especially for drugs, human trafficking, smuggling, etc.) requires joint police operations as cross-border pursuit of criminals, cross-border patrols and border control, joint investigation teams, international observations, and controlled deliveries.

A transnational deployed network, sharing technologies, processes, and legal rules between international agencies, would greatly enforce security against crime and improve responsiveness to disasters.

The lack of interoperable communication systems hold back the real-time cooperation of PPDR forces, despite the strong political support driven by recent significant safety and security incidents.

Notably, uncertainty regarding costs, time scale, and functionalities is currently slowing down the integration of PPDR radio networks at the transnational level. Moreover, since national resources are limited, and time is critical in disaster relief, international cooperation will enable a greater effectiveness.

5.1.1. *State of play of public safety interoperability in Europe*

In an international framework, it is quite common to have national PS agencies employing different communication technologies (TETRA, TETRAPOL, P25,

DMR, etc.). In most cases, this is due to different requirements in terms of security, performance, traffic load, and available budget. This leads to a great challenge for national security as the lack of interoperability limits cooperation and coordination capabilities among the field deployed forces. In this context, interoperability refers to seamless operation between public safety responders using different communication technologies or vendor systems. In general, interoperability can be defined as the ability to communicate and distribute information across different wireless communications systems. For example, police, firefighting, and emergency medical services responding to an incident can cooperate if they can communicate with each other even through usually incompatible wireless communication systems. This is an open issue at national level, and even more so at a transnational one. With reference to a European context, nations share most technological standards, laws and regulation, but unfortunately not PPDR communication systems. Boundaries between countries are less evident year by year, but safety agencies, civil protection and medical organizations are often not able to coordinate their actions in the field due to uninteroperable communication networks.

The strong political support for common public safety communications, following the signing of the Schengen Agreement, led the Schengen Telecom working group (later replaced by the Police Co-operation Working Group – PCWG) to adopt ETSI TETRA for pan-European police communications [SCH 03, SCH 95]; in support, CEPT/ERC designated a 5 MHz spectrum in the frequency band 380÷400 MHz for police force usage of TETRA across Europe in 1995. Despite this, most EU countries deployed different and almost uninteroperable national networks, based on TETRA [ETS 15] and TETRA + Police (TETRAPOL) [TET 15], expected to be operative until 2025.

TETRA technology is an ETSI standard, defined with the aim of providing a unique trans-European digital professional communication system. It is currently the dominant technology used by public safety organizations, either for the deployment of a brand new PPDR network (e.g. in Poland) or replacing previous existing PMR technologies and becoming the unifying technology (e.g. in Portugal).

TETRAPOL is the other prevalent technology in Europe. Although a proprietary technology (compared with the open ETSI standard TETRA), it is a public standard and has achieved significant diffusion in the European market, mainly boosted by the French market and other countries in its area of influence.

Figure 5.1 shows the European presence of TETRA and TETRAPOL networks, while their main characteristics are summarized in Table 5.1.

Figure 5.1. *TETRA and TETRAPOL presence in Europe*

PPDR Technology	TETRA Release 1	TETRA Release 2	TETRAPOL	DMR
Frequency bands	Around 400 MHZ	Around 400 MHZ	Around 400 MHZ	UHF (406–470 MHz), VHF (137–174 MHz)
Data throughput	7.2–28.8 kbit/s	15.6–269 kbit/s	7.2 kbit/s	Up to 9.6 kbit/s
Range	Limited to 58 km Typically 8 km (urban)	Limited to 58 km Typically 8 km (urban)	Limited to 28 km Typically 6 km (urban)	Up to 40 km
Latency Delay	Around 250 ms	Around 200 ms	Around 250 ms	Below 100 ms
Protocol supported	Clear/encrypted speech, circuit mode, IP, short data service, supplementary services	As TETRA release 1 (fully compatible)	Teleservices, data services, supplementary services	All the ones defined in ETSI TR 102.361
Security capabilities	Terminal authentication with three classes of data encryption	Expansion of TETRA release 1 with AMR and MELPe	End-to-End encryption	40-bit ciphering
Location Capabilities	Location information protocol	TETRA location service	GPS-based positioning services	Supported

Table 5.1. *European communication technologies comparison*

TETRA and TETRAPOL technologies are incompatible mainly because their different radio air interfaces, but even networks using the same technology (TETRA-TETRA, TETRAPOL-TETRAPOL) may be not able to interoperate. In fact, also within the same standard, different national or organizations' networks are often not yet compatible, due to proprietary implementations of part of the networks.

In order to achieve the interoperability, the TETRA standard provides a specific protocol, called intersystem interface (ISI) [ETS 10]; this is however not yet fully available at the industrial level in multivendor network infrastructures. TETRAPOL also provides a similar intersystem interface [TET 98], including interworking with TETRA, that is even less mature compared to ETSI TETRA ISI. Thus, TETRAPOL-TETRAPOL network interconnections are achieved today just through specific gateways, but terminal migration is not yet feasible.

The interoperability between networks based on different PPDR technologies can be partially reached through solution bypassing the incompatibility issues. One option is the deployment of analogue patching, which enables terminals from one network to place basic group call communications with terminals operating on another network. In this case, only audio information passed between networks, i.e. TETRA identity information and SDS (short data service) are not available. Another option is the deployment of a digital interface (manufacturer specific) using a fixed mobile connection from one network operating within the coverage area of another network. In this case, since the connection between these networks is digital, double vocoding between networks can be avoided if they employ the same codec and thus audio quality maintained. This kind of interface is manufacturer specific, and no common solution for interconnecting terminals of different manufacturer is currently available. The infrastructure functionality available can either be more sophisticated compared to the analogue patching, like talking party identity, call priority, SDS messages, and so on, depending on the actual chosen implementation by each manufacturer.

The most flexible solution is represented by roaming interfaces, like TETRA ISI, which in principle are able to provide all native services (voice and data) to the users. In the actual practice implementation, also ISI interconnection often present some limitations, mainly related to talk group management, security etc. (see ISI IOP phases in Table 5.4, 5.5 and 5.6). TETRAPOL has defined a roaming interfaces as well, inspired to the TETRA ISI standard. These interface are specified e.g. within the ISITEP (Intersystem Interoperability for TETRA-TETRAPOL Networks) project [ISI 15], and allows voice and data roaming; due to the differences in the air interface migration is not supported.

The following tables represent the possible interconnection solutions between the several European PPDR communication technologies, and the roaming capabilities.

	TETRA1	TETRA2	TETRAPOL	DMR
TETRA1	Analogue Patch/B2B Digital Patch ISI	Analogue Patch/B2B Digital Patch ISI	Analogue Patch/B2B Digital Patch ISI*	Analogue Patch/B2B Digital Patch
TETRA2	Analogue Patch/B2B Digital Patch ISI	Analogue Patch/B2B Digital Patch ISI	Analogue Patch/B2B Digital Patch ISI*	Analogue Patch/B2B Digital Patch
TETRAPOL	Analogue Patch/B2B Digital Patch ISI*	Analogue Patch/B2B Digital Patch ISI*	Analogue Patch/B2B Digital Patch ISI*	Analogue Patch/B2B Digital Patch
DMR	Analogue Patch/B2B Digital Patch	Analogue Patch/B2B Digital Patch	Analogue Patch/B2B Digital Patch	Analogue Patch/B2B Digital Patch

Table 5.2. *Basic interoperability between European communication technologies*

	TETRA1	TETRA2	TETRAPOL	DMR
TETRA1	limited	limited	limited	no
TETRA2	limited	limited	limited	no
TETRAPOL	limited	limited	limited	no
DMR	no	no	no	yes

Table 5.3. *Roaming capabilities between European communication technologies*

Interoperability barriers among member states are caused not only by the different radio technologies, but they are also present at different levels up to the operational one. In fact, the lack of interoperable communication systems led to a reduced interest in common procedures, thus limiting the number and, therefore, the benefits of joint operations. Current procedures are region specific and limited to particular scenarios, but to enable cross-border operations, the involved countries have to define multilateral agreements and operational procedures. Such a strategy has been proving to be effective, for example between Norway and Sweden who are sharing cross-border resources to fight fire or to transport patients. Since the PPDR resources of Norway and Sweden are unlikely to be fully engaged simultaneously, the sum of their resources may be available through complete cooperation. Costs and casualties can then be limited by means of an increased number of resources: the resulting savings greatly outweigh the investments required to achieve interoperability and cooperation.

The first formal framework for cooperation on rescue services between the Denmark, Finland, Norway and Sweden is the NORDRED agreement [NOR 15], signed in 1989 with the purpose of accident prevention or contingent damage limitation across the countries, facilitating the mutual assistance in emergencies and to accelerating the deployment of auxiliary personnel and equipment.

An example of a more recent political initiative toward practical cooperation across national borders to ensure the safety and security of European citizens is the Nordic Declaration at Haga, in 2009. In this political statement, ministers responsible for societal security agreed to further develop emergency management cooperation within specific areas. Several expert working groups were established and required to report annually to the political leaders. A new declaration, Haga II, was agreed on June 4, 2013 by the Nordic ministers, with the common understanding that the Nordic cooperation should be further deepened, to withstand and manage social crises with the vision of a society without borders, where the vulnerability decreases while the ability to deal with major emergencies and crises and restore functionality is strengthened.

Unfortunately, without cross-border interoperability, the cooperation is mainly ineffective and therefore there are limited initiatives for multilateral agreements and joint operational procedure models. In practice, the lack of interoperability among national PPDR networks usually inhibits radio communications a few kilometers beyond national borders and, in turn, does not give the impetus to realize interoperability at higher levels. Moreover, the 2010 trial of cross-border communications (CBC) between Germany and Sweden highlighted that common international working methods are a prerequisite for efficient cross border communication and that international procedures have to be harmonized with the national ones.

5.1.2. *Interoperability levels*

From a technological point of view, two interoperability levels can be identified:

1) Basic interoperability (sometimes called interworking) applies both to homogeneous and heterogeneous radio networks. In this case, interoperability can be defined as the capability to communicate and distribute information across different communications systems, used by different public safety organizations without support for end-user mobility (i.e. roaming). Basic interoperability is normally achieved through analogue or digital patching between networks. Analogue patching provides basic group call communications between terminals operating in different

networks. Since the connection is principally analogue, only audio is exchanged and terminal identity information is not available. Due to the simplicity of this approach, costs are in this case minimal. On the other hand, digital patching provides basic calling without audio quality loss, avoiding intermediate decryption/encryption functionality. This solution is typically achieved using either manufacturer-specific digital interfaces between different networks or a fixed mobile connection from a network that is operating within the coverage area of another one to add reference [FAN 10]. Costs are normally slightly higher than the one of analogue patching.

2) Full interoperability (sometimes called roaming or migration) allows radio networks to support end-user mobility and seamless operation between public safety responders, using differing communication systems or products. Full interoperability is possible only when the technology provides an inter system interface supporting voice, data and mobility management services, including security. This is the case for TETRAPOL and TETRA ISI interfaces [TET 15, ETS 10], where terminals visit a foreign network using services and functionalities (group call, individual call, telephony call, short data and air interface security) similar to the ones available in its home network, albeit with some inherent limitations. Costs are significant in this case and this type of solution requires the support from EU nations for its implementation.

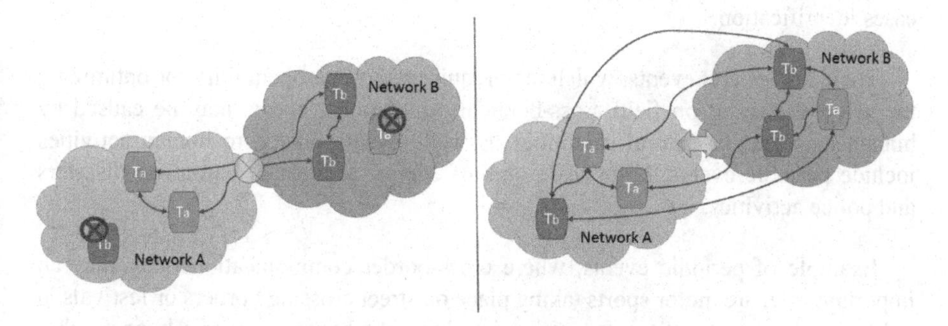

Figure 5.2. *Basic vs Full Interoperability*

5.2. Use cases

The use case identification is an important activity when defining a framework for co-operation, not only for the correct design of the framework itself, but also for preparing the users (i.e. security, rescue and protection forces operators) to prevent or afford the possible emergency situations. When using "unknown" or new resources it is, in fact, indispensable to give a preliminary training to stakeholders including, among other things, new procedures, methods and legal principles.

A good use-case scenario definition produces several benefits; some of them have been identified in the ISITEP project [ISI 15]. The procedure comprises six phases:

1) Participants and the organizers become aware of the operational and technical problems;

2) Participants from different services and countries, with their own organizational culture, know each other better and learn to better work together;

3) Participants become aware of the effectiveness of the elaborated plans and procedures, the surplus value of available resources or tools (action cards, data links, etc.), and the mono- and multi-disciplinary approach of the intervention;

4) The knowledge of the participants expands and updates skills, knowhow and behavior on the field;

5) Participants learn to deal with unforeseen circumstances;

6) Participants and the organizers evaluate the practical feasibility of the theoretical models and make adjustments, where necessary.

Specific research projects concerning interoperability and cooperation, as for example ISITEP [ISI 15] or MACICO [MAC 15], devote dedicated tasks to use cases identification.

There are several events, which may require radio interoperability for optimizing the effectiveness of on field cross-boundary operations; events may be caused by human activities or natural calamities as well. Events related to human activities include periodic events, large scale one-off events, accidents, man-made disasters and police activities.

Example of periodic events, where cross-border communications may play an important role, are motor sports taking place on street crossing borders or festivals in cities near the border of two countries, involving the aggregation of a large number of people, also moving across the border. In these cases the cooperation between international forces may be useful for the event management, as well as in case of accidents.

Example of large scale events, not periodic but happening on a specific date are concerts or festivals, prize assignment ceremonies, exhibitions, demonstrations and strikes.

Also police activities may have great benefits by the enabling their terminals for cross-border communications, e.g. for international crime fighting, wanted person research or pursuits and international investigations. Communication between police and aid operators play an important role in case of transport accidents, as well. In particular in case road/railway accidents at the border two countries, the event may block the way for rescue services to come and help may arrive from the opposite direction or by another route from a neighboring country.

The most frequent events requiring international cooperation are natural calamities, such as fires, flooding, avalanches, earthquakes or man-made disasters, such as chemical/nuclear alerts and oil spills. In these cases a quick reaction of rescue forces is essential and aid from neighboring countries assume paramount importance. Communication interoperability supports international cooperation, for example when resources are limited (e.g. aeronautical firefighting forces) or an event happening on territory between two different nations (often large rivers and big mountains border countries and flooding/avalanches may involve cross-border territories).

The time factor in all these operations is essential. Regarding public protection, potentially a few individuals may also cause extensive damage, even in a short time. Effective interoperability may greatly reduce such damages if PPDR resources can rapidly operate in foreign areas, supporting the limited local public protection resources. In disaster relief, the sooner international resources are ready to operate, the lower the costs and side effects. According to a study [WAL 91], most of the individuals involved in disasters may die within 72 h, considering injuries, outside temperature, and availability of air and water.

The following example scenario outlines the importance that interoperability and cooperation assume during operations involving territories of two or more countries.

5.2.1. *Example scenario*

An example scenario has been identified in the ISITEP project [ISI 15], based on the output of a previous three-country pilot project [COU 03]. The proposed scenario describes the situation of an armed robbery that happens near the border of three countries (The Netherlands, Belgium and Germany), therefore involving the cross-border radio communication between police forces, which can mean the connection of police forces with their own control room while they are active across the border. It can also mean the communication among the police forces of the Netherlands, Germany and Belgium including one or more control rooms.

1) Start of the event

The Dutch police receive a robbery alarm from a warehouse. The police control room orders police surveillance cars near the warehouse to take a look there and evaluate the situation.

2) Initial situation monitoring and analysis

The warehouse belongs to a security company securing high value cargo and is situated in a Dutch village, called Wijlre. The distance between Wijlre and Maastricht (NL) and Aachen (DE) is approximately 20 kilometers and the distance to Liege (BE) is 40 km. Both borders are even closer.

3) Initial National Communications

The Dutch police communicate with the C2000 radio system that is the national public safety communication system based on TETRA standard. The police are equipped with vehicular radios as well as with portable radios for communication outside the vehicle.

Two police surveillance cars respond to the alarm and start moving to the warehouse in Wijlre. While they are driving to the crime scene, they receive more information via the radio from the dispatcher of the police control room. An eyewitness told the police control room that he saw three people coming out of the warehouse carrying firearms, get into a black car and leave in the direction of the N278. The police control room also received a phone call from the security department of the warehouse. They told the police that the robbers left the building and stole a large amount of money. They even gave a detailed description of the robbers' getaway car. One of the police surveillance cars noticed the getaway car whilst driving on the N278. They communicate their position to the police control room. The getaway car is going in the direction of the Dutch–Germany border. A high speed police pursuit follows.

4) First Contact between Dutch and German police (e.g. 5 minutes from incident beginning)

The Dutch police estimate that they will not be able to catch the robbers before they arrive at the border. Therefore, they contact the police control room and ask to inform the German police.

5) Legislation issue analysis, second contact between international police forces.

The Dutch police can continue the pursuit on German territory when they fulfill the requirements laid down in the Schengen Implementation Convention. In case of the robbers being caught in the act of committing the offence, the offence must be listed in article 41.4 of the Schengen Implementation Convention [EUR 90]. The situation is urgent, the pursuing officers will, not later than when they cross the border, contact the competent German authorities and the German police are unable to reach the scene in time to take over the pursuit. There is reason to think that the robbers will try to leave the N278 and change their course to the Dutch–Belgian border or even maybe via Germany to the German–Belgian border. The getaway car has Belgian number plates and the car has been registered as stolen in the area of Liege few days before. For this reason, the Dutch police also want to inform the Belgian police. They ask the Dutch police control room to do so.

6) Cross-border operations (e.g. 10 minutes from incident beginning)

The robbers and the Dutch police surveillance cars each and pass the Dutch–German border. They inform the police control room reach. The Dutch police have a high priority to stop the pursuit quickly because of the dangerous situations during the chase. Given the seriousness of the offences there is a big incentive to catch the robbers. Due to this situation, they want to communicate with their German colleagues to make a joint operational approach.

Five kilometers beyond the border, the German police placed a roadblock to stop the robbers. The robbers are forced to stop the car and the Dutch police officers are able to arrest the three robbers. Both the Dutch and German police officers give the information about their position and the arrest of the robbers by radio to their police control room. The Dutch police control room also informs the Belgian police control room in Liege that the robbers are caught and that the Belgians can give up their stand-by position.

The described scenario highlights the importance of inter-operability in cross-border operations, for all the events happening close to state boundaries or that generally require cooperation between forces of different countries.

5.2.2. *Interoperability requirements*

An interoperability requirement depends on end-user needs and on operational scenarios, but are also strictly tied to the interoperability level provided by the underlying communication technologies. With reference to the interoperability levels defined in section 5.1.2, the following requirements can be identified:

1) Basic interoperability: communication requirements normally encompass basic group and individual calls and sometimes, depending on the radio technology involved, narrowband data too (i.e. TETRA SDS). Security features are very limited and normally provided thorough the implementation of security gateways that act as end-points for end-to-end (E2E) communications on both sides.

2) Full interoperability: in this case communication requirements encompass all the services and features available in the native network. As the plethora of features and functionality provided by a TETRA standard is quite large, and orders of magnitude higher in complexity versus commercial telephony (i.e. GSM), e.g. group management, Dynamic Group Number Assignment (DGNA), emergency,

pre-emptive priority, enable/disable, end-to-end encryption, air interface encryption with static, derived and group cipher keys, etc., TCCA has proposed an incremental approach to inter-system interoperability (ISI IOP) implementation split into four phases of increasing interoperability [TCC 15].

	Feature	Details
ISI PHASE ½	Migration	Air Interface Migration (AIM) Migration authorized or rejected by the home switching and management infrastructure (SwMI), the home SwMI may reject migration for reasons other than failure to authenticate
	Authentication	
	Individual call	Individual call between parties located in different SwMIs. Each party may be located at their home SwMI or at a visited SwMI
	Group call	Group call when located in group home SwMI Group call when not located in group home SwMI
	Status messages	Delivery of status messages to home or foreign groups and to home or foreign subscribers
	Short data service (SDS)	Delivery of SDS messages to home or foreign groups and to home or foreign subscribers
	Supplementary services	Call line identification presentation talking party identification

Table 5.4. *ISI Phases 1 and 2*

	Feature	Details
ISI PHASE 3	Mobility management	Migration authorization by the home SwMI, through pre-provisioning in foreign SwMI Group access management by the group's home SwMI, through pre-provisioning in the foreign SwMI and static group linking limited form of group attachment / detachment (opt.) individual subscriber database recovery group database recovery (GDR)
	Air Interface encryption	Derived cipher key (DCK) encryption
	Emergency call	Call priorities, including emergency and other levels pre-emptive priority
	End-to-end Encryption	Outgoing calls Incoming calls
	Telephone call	Outgoing calls Incoming calls
	Supplementary services	Late entry Air-to-ground-to-air operation

Table 5.5. *ISI Phase 3*

	Feature	Details
ISI PHASE 4	Mobility management	Group migration Restricted migration Profile update SS-profile update Over the air rekeying Complete group attachment Complete group detachment Complete group database recovery Dynamic group linking / unlinking
	Group call	Broadcast calls Call restoration within migration Withdrawing from or continuing with a call (wait/continue) Sending of profile information in ISI setup initiate Call modification DTMF procedures Critical users Resource allocation during call maintenance
	Supplementary services	Enable/disable Individual DGNA Barring of outgoing calls Barring of incoming calls Call authorized by dispatcher (CAD) Call forward
	Air interface encryption	OTAR of GCK OTAR of SCK
	SDS store and forward	
	Packet data	

Table 5.6. *ISI Phase 4*

5.3. A holistic framework for PS network interoperability

Experience shows that feasible and effective solutions for PPDR interoperability should follow a holistic approach, encompassing joint integration of procedures, technologies, and legal agreements.

The current situation can be improved by using a framework that implements interoperability at system level, allowing PPDR agencies to achieve a cross-national interoperability based on existing technologies, toward the benefits offered by emerging technologies in the long term.

An effective interoperability solution requires a framework where all the interoperability-related factors are addressed simultaneously and coherently. Effective communication interoperability in the field requires many factors to be jointly solved within a coherent framework, such as terminals, networks, as well as procedures, training and tools for logistics of communication. In international PPDR operations, visiting operators will be able to communicate with the same approach used in domestic operations. Otherwise, the time required to become operative may increase unexpectedly. Visiting PPDR operators need to cooperate rapidly with domestic resources according to joint agreed-upon procedures. This may be critical in disaster recovery, where advance planning may be limited, or in public protection, where unexpected outcomes are frequent. A common framework is the key for making this rapid cooperation possible, enabling communication interoperability in the field despite technologies, workflows and human factors (including language barriers as well).

A framework responding to the identified needs has been outlined in the ISITEP project [ISI 15]. The framework aims to achieve operational interoperability among transnational PPDR operators jointly addressing regulative, organizational, operational, and technical levels. The framework includes four components:

1) A mission-oriented structure, containing a standardized model of operational procedures and associated functional radio model;

2) A transnational ISI network, based on cloud technology and integrating the PPDR national infrastructures to allow roaming services within secure protocols; the cloud is intended to be a private network where shared applications/services are provided by distributed components within the network itself;

3) Enhanced user terminals, integrating communication technologies (e.g., TETRA, TETRAPOL) into a novel terminal architecture based on smart devices (tablets and smartphones);

4) Interoperability-enabling tools, including infrastructures dimensioning, training, business model assessment, and services for safety operations.

The cooperation framework is strictly related to the end-user model derived from [WAL 91], as shown in Figure 5.3.

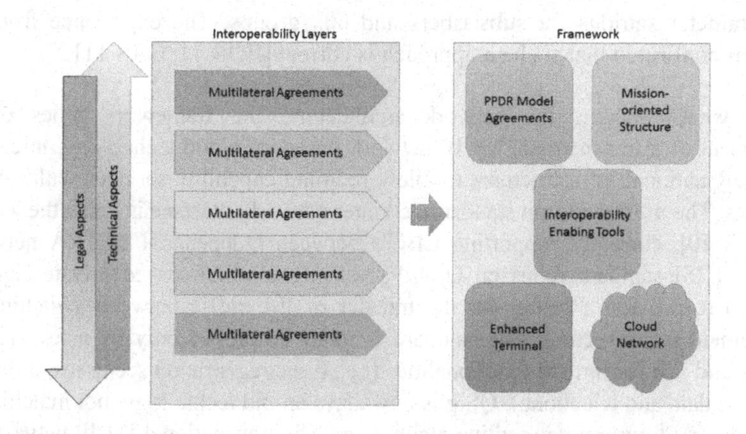

Figure 5.3. *Interoperability Layer and the components of the framework*

The interoperability is achieved at different levels:

1) By the capability to communicate among teams from different countries. This implies technical interoperability at the radio and network levels. This aspect is solved in the proposed framework by enhanced terminals and cloud network components;

2) By the sharing a set of common services, typical of PPDR operations, enabling operational interoperability; this means that without these services interoperability is not effective in international operations from the end-user point of view;

3) By common procedures, including technical, organizational, and legal aspects. Procedures will specify the essential rules of communications. Within the framework a functional model is defined to specify group interactions, and, therefore, a precise set of operational procedures are outlined, allowing cooperation in any PPDR scenario. Since a national legal framework has to guarantee rights and obligations for visiting teams in foreign areas, operational procedures have to be supported by multilateral agreements (treaties) among countries. Such agreements have to be finalized simultaneously to the adaptation of such a framework for international cooperation, remaining in line with international legislation (treaties, agreements, etc.).

All these points merge in a standardized operational model, with the associated functional radio model and common set of procedures, specifically defined by PPDR end users. Common international working methods covered in the operational model are in fact a prerequisite for efficient CBCs. Common talk groups, common languages, and radio communication rules have to be agreed by the member states using the framework. As stated in [BOR 11], regarding functional models and procedures, the template agreement has to identify the need of order routines, connection routines, guidelines, and procedures for fleet models in PPDR networks, and parameter settings for subscribers and talk groups. The experience from end users has confirmed that such an approach is correct [BOR 11, CAS 11] .

For what concerns the network architecture, the framework relies on the deployment of a transnational ISI IP network based on cloud technology, integrating the PPDR national infrastructures to allow roaming capability services within secure protocols. The most common standardized inter network interface is today the TETRA ISI [ETS 10], currently supporting CBC's between independent TETRA networks. The ETSI ISI standard relies on QSig/PSS1 (i.e., an ISDN-based private signaling standard) to provide a bearer for the transfer of ISI traffic between switching and management infrastructure. The standard is today employed only by a few TETRA vendors and just for limited functionalities (i.e., basic registration scenarios, individual call, short data and telephone). QSig is nowadays an old technology, not matching the effectiveness of current networking technology. The transnational ISI IP network will use the ETSI TETRA ISI standard as a building block to implement an IP-based protocol for network interconnection. Nevertheless, a modern framework will integrate two more elements, in order to achieve full interoperability. The first is the extension of ISI from TETRA to generic heterogeneous networks interconnection, to assure the whole range of interconnecting capabilities. The second is the adoption of a protocol stack designed to run on top of IP connectivity.

The current de facto standard for voice over IP (VoIP) communications is the session initiation protocol (SIP) [IET 02], which may be conveniently used for ISI datagram transport. In the framework proposed by ISTEP Project, the layer 3 additional network functions (ANFs) of the ISI protocol remain mainly unchanged, while the lower layers migrate to IP, as indicated in the two models, as shown in Figure 5.4.

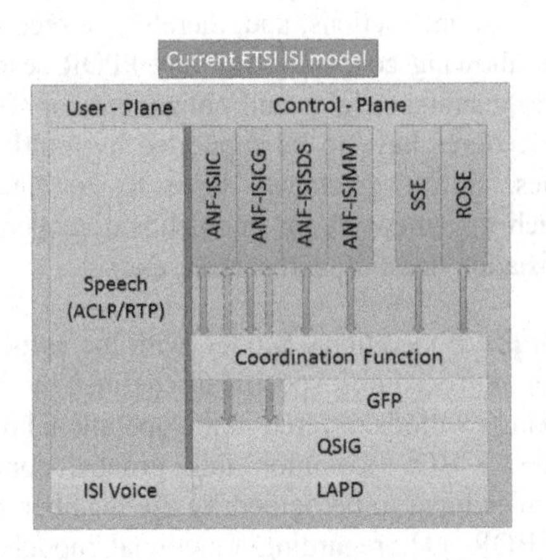

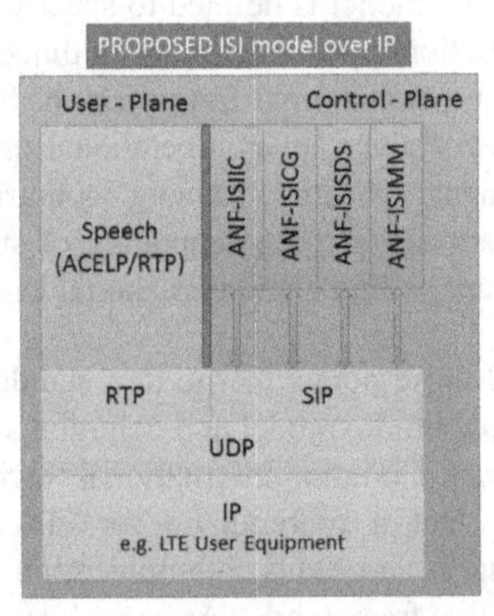

Figure 5.4. *Current ETSI ISI model and the one proposed in ISITEP project, based on IP protocol*

In this context, two main options exist for the interconnection between PMR networks deploying IP-based interfaces:

1) The use of a packet-switched intervening network, such as an intranet, to interconnect PMR networks, as identified in [ETS 05];

2) The direct connection between two PMR networks using for example leased lines or virtual private network connectivity, as identified in [ISI 15].

Recent studies [FCC 11] outlined the use of a third-party network as a feasible option for the interconnection of PMR regional networks in the United States, for its benefits in terms of savings and scalability. Such a solution implies several security challenges that may be addressed through conventional IP security systems (i.e., IPSec, firewalls, IDS/IPS, security gateways) and security-assessment certifications. The adoption of IP as the basis for the evolution of an ISI protocol stack is a good basis for the easy integration of current narrowband PPDR networks with broadband IP access technologies such as WiFi and LTE, opening professional networks toward high data rate capability. The TETRA and Critical Communications Association (TCCA) is currently foreseeing TETRA-BB as a TETRA application running over an LTE access [TCC 13a], where ISI may be seen as an API for intersystem/application communication. A cloud technology approach may represent a secure dedicated platform to deliver enabling services to PPDR users, facilitating roaming and interoperability (both at the terminal and network levels) and, more generally, joint operations addressing specific operational aspects. Some interoperability-enabling cloud applications are the dynamic functional numbering, allowing the communication with PPDR resources in charge in a specific operational area; the location-assisted numbering, allowing the calls of specific PPDR resources' type in a specific area and the enhanced message exchange application, allowing orders or information exchange through short messages.

5.4. Terminals for interoperability of current generations of PMR

The described ISI model permits terminals to migrate different networks abroad, without the necessity of performing particular activity to achieve the roaming capability. However, network solutions are not enough to obtain interoperability that will also involve users' terminals. Since in most cases current PPDR technologies (e.g. TETRA and TETRAPOL) are incompatible, as a first hypothesis, operators should be equipped with more terminals, one for each technology. However, this is not a feasible and effective solution, since the use of multiple terminals limits operators' actions and introduces costs for organizations that should buy them. Moreover, even considering terminals provided with the same air interface, interoperability is *de facto* prevented by other limits imposed by security restrictions. In particular, the main incompatibility is related to cryptography mechanisms and cryptography keys that are different in networks based on the same technology.

Therefore, the more efficient solution for interoperability of current PPDR networks consists of designing enhanced terminals able to implement different air interface standards (like TETRA and TETRAPOL) and support simultaneous communications across different radio access technologies and radio spectrum bands/channels. This multimode capability is an essential feature within the development and specification of a functional architecture, intended to properly manage terminal connectivity in a heterogeneous network environment. It is addressed in different official documents:

1) IEEE 1900.4 standard [IEE 09a], which defines the architectural building blocks enabling network device distributed decision for optimized radio resource usage in heterogeneous wireless access networks.

2) IEEE 802.21 standard [IEE 09b], also known as the media-independent handover, which develops a protocol to facilitate handovers between different heterogeneous networks (e.g., IEEE 802, Third-Generation Partnership Project (3GPP), or 3GPP2-based networks).

3) 3GPP specifications [GPP 13b], which define optimized mobility solutions for terminals supporting both 3GPP and non-3GPP radio access technologies (e.g., access network discovery and selection function).

The implementation of a multimode terminal for public safety can be approached in different ways. In the commercial world equipment provided with multiple modems are dominant: current handheld/smartphone commonly integrates a number of OEM radio chip-sets (e.g. WiFi, Bluetooth, LTE, Near Field Communications, etc.). This “hardware-based “approach is very flexible and well established but presents some drawbacks due to higher hardware design complexity, low scalability and high power consumption particularly relevant in the case of PS communications where the RF power is normally high compared to those of commercial radios. These problems can be overcome resorting to a design approach based on software-defined radio (SDR) concept. SDR allows us to obtain devices running radio applications on shared platform resources and able to install new ones even at run-time, as happens in personal computers with regular computers programs. Its reference architecture is defined in the ETSI technical report TR 102 680 [ETS 09].

The SDR standard, based on open software communications architecture, is in principle the best approach to the implementation of multi-radio terminals. Unfortunately, this requires the development of specific PPDR (TETRA–TETRAPOL, etc.) waveforms, which may be critical in the short–medium term due to business sustainability. Current standard SDR platforms are designed for military use and, therefore, hardly suited for the PPDR market in terms of performance and cost. Thus, in the short–medium term, the most promising approach is to use open programmable telecommunication devices such as Linux/Android-powered tablets or telephones, embedding a PPDR modem over a programmable platform. This approach enables cost-effective multi-technology terminals to be easily extended to various PPDR communication technologies, e.g., the next generation broadband ones (i.e. LTE). Cognitive radio is also a promising approach to optimize and share the scarce resource of the frequency spectrum in the PPDR context the reference has been substituted.

Despite the specific implementation approach (hardware-based or software-based) we can define the general functional architecture of a multi-technology terminal for public safety that exploits two or more radio modems. This architecture is independent from physical layer implementation and is therefore applicable to terminals either equipped with multiple different modems or based on SDR technology. Figure 5.5 shows the logical blocks of an enhanced terminal that integrates two different technologies, with particular reference to the European TETRA-TETRAPOL case. The reference architecture is composed of the following functional modules:

1) Physical radio, representing the radio interface and consisting in in PPDR modems (e.g. TETRA/TETRAPOL).

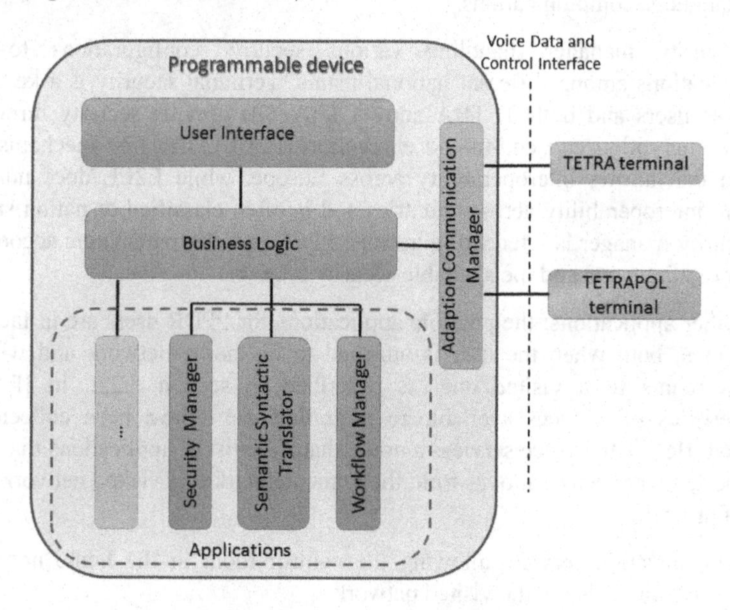

Figure 5.5. *Enhanced Terminal Architecture*

2) Adaptation communication manager, decoupling the higher stack levels from physical layer offering an abstract interface to the business logic that provides the basic PPDR services (voice and data). It implements the functionalities required to allow the communications between modems, the user interface and business logic

and implements vertical handover between different technologies when the users move across organizations or nations.

3) Business logic, realizing interaction and service logics.

4) User interface adaptor (man machine interface (MMI)), which is the set of functionalities devoted to the personalization of a human–machine interface according to specific user and operational needs, available network coverage etc.

In addition, in the architecture some applications providing interesting functions that facilitate interoperability are observed, e.g.:

1) Workflow manager, supporting the PPDR resources deployed in the field in the accomplishment of essential procedures and operational tasks of the assigned mission.

2) Semantic syntactic translator, reducing language barriers, in particular it has the task of defining and implementing the means necessary to semantically and syntactically translate the human interface on terminals facilitating cross border/language communications.

3) Security manager, handling various security configurations to allow communications among different national teams. Terminal security is a key aspect for PPDR users and both TETRA and TETRAPOL provide security through air interface encryption and end-to-end encryption (E2EE). The first mechanism is a standard that allows interoperability across Europe, while E2EE does not allow complete interoperability across countries and is often classified or nation specific. The security manager is intended to manage E2EE security parameters according to the user requirements and the available security features.

4) Other applications: the possible applications for PPDR users are in fact many and various, both when the user is attached to his home network and when the terminal roams to a visited one, as described in section 5.2.2. In [ETS 05] characteristics of services available to migrating users have been collected and described. Beside the voice services a non-exhaustive list of applications that should be supported when a user moves from the home network to a visited network are (in order of priority):

i) Localization service, allowing the control room in the home network to locate the terminal also in the visited network.

ii) Access to remote databases located in the home network.

iii) Bidirectional picture transfer.

iv) Slow video.

v) File transfer.

5.5. Interoperability on the next generations of PMR: transition between the present state of the art systems and the next generation

In previous sections of this chapter, we described a possible framework to interconnect current professional mobile radio technologies (in particular TETRA and TETRAPOL) realizing a complete inter-system interoperability. However, the communication system must also provide adequate capabilities able to respond to the needs of the PPDR operators that are changing and evolving, hence, the current PMR technologies will no longer be able to meet the desired requirements in the long term. It is reasonable to assume that the PMR technologies will evolve in the future, and the interoperability problem must be newly analyzed.

This section briefly recalls the main issues of the PMR evolution, showing that worldwide there is a strong convergence toward a unique technological solution, facilitating the system interoperability. In particular, under the assumption of adopting a unique standard based on LTE, interoperability issues a move to the application level. An example could be a video solution involving a vehicle based camera, a central server and radio terminals which are able to view the video. The goal is that the PPDR user can select different manufacturers for all these components, but this means that they implemented the same protocols over the IP level.

Another important aspect that can affect the interoperability of the PMR systems is the lack of harmonized spectrum allocation for critical services as discussed in what follows. Finally, in analyzing the evolution of the professional systems toward more efficient broadband solutions, it is necessary to take into account a transitory phase, during which current PMR systems will coexist and interoperate with new broadband systems, even developed for commercial purposes. Hence, interoperability among different generation's systems is also another important issue.

5.5.1. *Next Generation PMR requirements*

Current mobile radio systems are voice-centric technologies, whereas public service operators require more data-centric services to support applications such as [TCC 13b]:

1) improved situational awareness to/from mobile devices including video, picture;

2) GPS location, mapping;

3) streaming of live video to control rooms and back to groups of mobile users;

4) facial recognition and instant access to back office databases;

5) data collection and sharing;

6) data exchange between automatic control systems;

Indeed, a larger amount of heterogeneous data and the availability of multiple applications allow us to increase the situation awareness and decision making efficiency, avoiding corrupted, missing or late information that could result in wrong decisions causing, in the worst case, loss of life. Therefore, even if voice remains could an essential application in PPDR systems, data are becoming a mission critical service. To this end, it becomes indispensable to introduce new technologies, able to support high data rate communications while maintaining all the peculiarities of professional systems.

Worldwide governments and organizations dealing with critical communications show interest in the development of broadband professional mobile systems. In USA, the national public safety telecommunications council (NPSTC) together with other organizations such as the Association of Public-Safety Communications Officials (APCO) and the National Emergency Number Association (NENA) highlighted the need to have a national interoperable standard for next generation of public safety networks with broadband characteristics. In June 2009, NPSTC indicated LTE as the technology chosen for this evolution and contacted 3GPP, the body in charge of the development of the LTE standard, to discuss changes to be applied to the standard in order to meet the needs of public safety applications. The purpose of NPSTC is to integrate the greatest number of features in future versions of the LTE standard making it fully responsive to the needs of PRM and at the same time attractive for business and consumer applications.

In Europe, the main organizations dealing with critical communications initiated a similar process [CCB 12, CCB 13]. The definition of the new technology involves different parties, i.e. PPDR user and consulting organization together with standardization. In particular, the subjects involved in this process are:

1) TCCA: an association of users, manufacturers, application providers, integrators, operators, etc., which takes care of collecting the opinions of TETRA stakeholders, to derive requirements and operational needs in the use of these technologies [TCC 12]. TCCA also cooperates with ETSI, to create a common and interoperable solution required by the Critical Communications user community.

2) 3GPP that is in charge of integrating PMR functionalities in LTE standard.

3) Law Enforcement Working Party (LEWP): the advisory body on security and justice of the Council of the European Union composed of the police/justice and representatives of all member states. Within LEWP, the Radio Communication

Expert Group (RCEG) has the task of advising in finding solutions to address the growing need for mission critical mobile data communication. In particular, they work on the harmonization of technical solutions and frequency bands.

The LEWP-RCEG developed a set of requirements that must be met by the candidate high-speed data technology for PPDR application. Some of the most important that have been identified by the PPDR users are:

– high resolution video communications from wireless clip-on cameras to a vehicle mounted laptop computer, used during traffic stops or responses to incidents;

– video surveillance of security entry points, such as airports, with automatic detection based on reference images, hazardous material or other relevant parameters;

– remote monitoring of patients. The remote real time video view of the patient can demand up to 1 Mbit/s. This demand for capacity can easily be envisioned during the rescue operation following a major disaster. This may equate to a net capacity of over 100 Mbit/s;

– high resolution real time video from, and remote monitoring of, fire fighters in a burning building;

– building plans transmission to the rescue forces.

5.5.2. *ETSI reference model for a critical communication system (CCS)*

In the TR 103 269-1 [ETS 14], ETSI describes an architecture that will allow critical communications applications to operate across a general broadband IP network, e.g. a LTE wireless network, identifying the reference interfaces relevant to:

1) a network to terminal application;

2) a terminal to terminal application;

3) a network to terminal application via a relay terminal.

Figure 5.6 represents the reference model and interfaces for a system providing professional communication services to professional users, including both access networks and core networks, based on IP connection. The whole system is called a CCS, critical communication system, while the application providing professional communication services to critical communication users over broadband IP networks and to legacy professional networks is identified as critical communication application (or simply CCA). In order to deliver these services with adequate quality, the CCA will make use of prioritization and quality of service (QoS) functionalities of the underlying IP network; e.g. configuring the correct characteristics and attributes of the LTE QCI and ARP parameters.

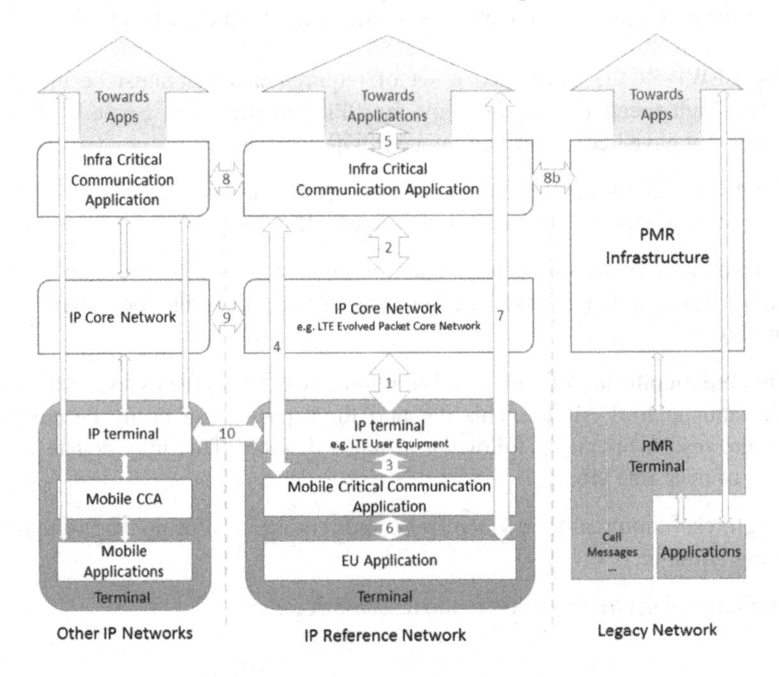

Figure 5.6. *ETSI reference model for a CCS*

ETSI identifies following main interfaces:

1) IP core network to IP terminal: following standard protocols of the underlying IP core network (e.g. the one of the 3GPP LTE UE to EPC interface).

2) Infra-CCA to IP core network: following standard protocols of the underlying IP core network, e.g. the one of the 3GPP Rx and SGi interfaces, plus the MB2 interface developed in the group communications service enabler (GCSE). The generalization of this interface allows the interworking between an Infrastructure CCA and IP Core Networks from different manufacturers, also considering that one CCA may be connected through more than one broadband IP network, which may be of the same type (e.g multiple LTE networks) or different (e.g. LTE+WiFi), obtaining a many-to-many relationship of CCA's and broadband IP access networks.

3) IP terminal to mobile CCA: this interface cannot be fully standardized, since it is dependent on the terminal and the running operating system, but will take into account the GCSE specification that will be provided by 3GPP starting from LTE Release 12.

4) Infra-CCA to mobile CCA: this interface allows interoperability between a CCA infrastructure and terminals from different manufacturers, providing functionality similar to existing digital PMR Layer 3 air interface messages (as user registration, setup and control of individual and group communications, media transfer and management and short data transport).

5) Infra-CCA to application: allowing the integration of applications in a CCA environment, independently by the CCA manufacturer. In order to support both voice calls and data, this interface consist of two components, a call interface, to provide control of sessions (control-plane) and of media transport (user-plane) within a communication, and a packet routed transport interface, for data files and signaling.

6) Mobile CCA to mobile application: allowing the integration of mobile applications in a CCA environment, with the goal of a complete portability of those applications between terminals from different manufacturers.

7) Application to application: this interface cannot be fully standardized, since it is dependent by the specific applications running on terminals and in the control room.

8) Inter CCA's or CCA to legacy PMR: objective of this interface is to allow interoperability and interworking between different CCSs or between a CCS and existing legacy PMR systems (as TETRA, TETRAPOL or P25).

9) Core network to core network: this interface is very important for interoperability and interworking, since provides support and control of mobility and roaming. It cannot be fully standardized anyway, since it is dependent on the technology of the underlying core IP network. In case the underlying network is an LTE EPC, this will be a 3GPP standard interface.

10) IP terminal to IP terminal: supporting direct communications between terminals, which may also act as repeater or network relay. This interface is strongly dependent of the used terminal technology; for LTE terminals, 3GPP is defining dedicated interfaces under the proximity services (ProSe) work item, included in LTE 3GPP activities since Release 12. For the proximity services, ETSI also identified additional dedicated interfaces (#11, 12 and 13), inter-mobile CCAs or mobile CCA to repeater CCA, infra-CCA to relay CCA and mobile CCA to relay CCA.

5.5.3. *LTE evolution toward PMR services*

Mission critical communications are characterized by more stringent requirements with respect to commercial counterpart in terms of resilience, reliability, security and availability. At present, the LTE standard is not fully compliant with professional needs. The primary criticality concerns generic support to voice calls over IP. LTE has been developed for fully IP data communications, therefore it does not include the circuit switching (CS) domain on which rely voice calls in 2G/3G technologies. Currently there are various technological options for the implementation of voice calls on LTE that include circuit switched fallback (CSFB), voice over LTE via generic access network (Volga), one voice and voice over LTE (VoLTE). CSFB consists of transferring users on a circuit switched domain provided by a pre-existing radio technology, such as UMTS and GSM for the duration of the call. This technique has two main drawbacks: LTE and legacy networks will simultaneously co-exist and call setup time is very high, causing a significant degradation of the user experience. This solution, adopted in a transitory phase on the commercial networks, is not applicable for PMR networks due to its poor performance. A solution in the medium/long term is represented by Volga, defined by the Volga forum [VOL 15]. This is based on the concept of connecting the existing mobile switching centers to the LTE network via a gateway. In this case, because the communication does not fallback to legacy networks, setup times will be lower than the CSFB. Finally, the most advanced solution consists of supporting voice calls through an IP multimedia subsystem (IMS). IMS is an IP-based connectivity and control architecture standardized by 3GPP to manage multimedia services over IP. In particular, the one voice association defined a 3GPP compliant basic profile, which contains features of network and terminal necessary to support the IMS basic voice services. Successively, the organization of operators global systems for mobile communications association (GSMA) extended this work, producing a more advanced profile called VoLTE published in [GSM 14]. At present, VoLTE is getting greater interest from the industry, but it is not the unique solution in a context of fragmented approaches to the problem. Main VoLTE features consist of:

1) set-up of the transmission path between the terminal and IMS;

2) security features for user authentication providing;

3) providing the core functionality for the establishment and termination of the call (via SIP);

4) support to call forwarding, caller ID presentation and restriction, call-waiting and multiparty conference.

However, VoLTE is not able to support some primary functions of mission critical communications, i.e. group communication, direct mode communication, push-to-talk management. In particular, it is not able to provide an efficient mechanism to distribute data, voice and video content to multiple users respecting performance and flexibility required by professional communications especially in mission critical contexts. Group communications will provide "push-to-talk" functionality, that is the ability to make a call and start talking without waiting for the response from the receiver. Requirements associated with this service are the ability of the caller to be advised if no user has received the call, or to be notified that a user joined a session, and the ability to receive a warning if the call is coming from the CS domain. Currently, 3GPP specifications for 2G/3G networks include the definition of "push to talk over cellular (PoC)" technology introduced for the first time by the open mobile alliance (OMA). The PoC has been designed to support *ad hoc* and group chat-like communications for both home users and business applications, as well as to enable individual users to dynamically create *ad hoc* groups for voice communications and chats. PoC, however, is not able to satisfy the requirements of performance and interoperability required by mission critical communications. To enable this service, an appropriate modification of the IMS framework is needed, especially to maintain a low the setup time of the call that constitutes a critical aspect in PMR contexts. Another basic functionality regards the possibility of establishing direct calls between users even outside the network coverage. This implies the need to provide the user devices with the capacity of local discovery of nearby devices without the network management control and with the ability to establish the call. To address these features, 3GPP started a number of activities performed by dedicated study and work items [GPP 15]. In particular, group communications system enablers for LTE (GCSE LTE) and proximity-based service (ProSe) are two work item description (WID) in 3GPP SA1 (System Architecture Working Group 1) that address the LTE standard support of voice/video/data group communications and direct mode for device-to-device communications. Their output will be integrated in Release 12 of the standard.

5.5.4. *Spectrum allocation of next generation PMR systems*

Suitable radio spectrum resource availability is a major issue for an effective deployment of mission critical communications. In addition, in order to have an effective interoperability of system deployed by different countries, we need to have harmonized spectrum allocations. Today professional communications, such as TETRA and TETRAPOL in Europe and APCO P25 in North America take place in reserved portions of spectrum that are located in the 380-400 MHz, 150-174 and 412–521 MHz bands. These bands are reserved exclusively for narrowband voice or low bit rate data communications (a few tens of kbps) and to TETRA wideband extension, that is the TETRA enhanced data system (TEDS), which is able to reach, at most speeds in the order of hundreds of kbps. To support broadband communications, additional resources will be located. In July 2007, the USA Federal Communications Commission assigned a portion of the 700 MHz band, released by analog TV broadcast, to safety communications [FCC 07] with the aim of establishing a nationwide interoperable communication network. FCC designated the lower portion of the spectrum (i.e. 763–768/793–798 MHz) to broadband communication and the upper half (i.e. 769–775/799–805MHz) to narrowband communications. Later, in 2012, the reserved radio spectrum was extended to 2x10 MHz in band 758 MHz–768 MHz and 788 MHz–798 MHz. In Europe, the

PPDR community expressed the need to identify a portion of spectrum below 1 GHz to be harmonized [TCC 15]. A study conducted by CEPT-ECC with the participation of the EU Council's; Law Enforcement Working Party (LEWP) published in ECC Report 199 A [CEP 13], concluded that the minimum spectrum needed for broadband data services is 2x10 MHz, in accordance with the decision made in the USA. These resources do not include voice services, air ground air, direct mode operation and *ad hoc* networks that could be allocated in additional spectrum portions. Additional studies are ongoing within the CEPT-ECC Working Group FM 49 [FM 15] with the aim of finding suitable solutions for radio spectrum harmonization. Broadly speaking, the harmonization concept implies the identification of a unique spectrum range, a tuning range, within which each administration can select the portion of spectrum that has to be used locally. In fact, not all frequencies can be available in every country, but terminals are intended to operate in the whole range. This solution presents numerous advantages in terms of increased interoperability and cooperation between different organizations and countries, and leads to higher volumes of business, which results in the possibility of obtaining wider economies of scale for terminals, infrastructures and equipment availability. Harmonized spectrum makes the emergence of competing technologies more difficult. Moreover, having a dedicated spectrum allows as to fulfill critical operation requirements of resilience, availability, coverage and security. Locating a common portion of spectrum is subject to practical constrains due to coexistence with other systems and electromagnetic compatibility other than the need to mediate different interests (public and private operators, final users, etc.) and countries' politics that reserve resources following local needs and contexts. The use of frequencies below 1 GHz is preferred due to the better signal propagation conditions with respect to higher frequencies; this implies better coverage both in outdoor and indoor environments, and allows the fulfillment of critical communication performance requirements. Currently, two frequency bands have been designated as possible candidates: 700 MHz (IMT band 694–790 MHz) and 400 MHz (subranges 410–430 MHz and 450–470 MH). The former solution is that preferred by the majority of FM 49 members. It has the advantage of allowing the deploying of the network both as a dedicated and as commercial network or a combination of theme. Inside FM 49 there are different opinions about the beneficial of using a harmonized exclusive band for PPDR: some countries take the view that there will be more changes to allocate spectra at national level if there is not competition with commercial operators. Other administrations, instead, believe that dedicated spectra will create a niche market that negatively affects economies of scale and interoperability. The 400 MHz band has the advantage of good propagation characteristics, allowing reduction of the number of sites necessary to provide the radio coverage. On the other hand, in some countries it is not possible to find enough spectrum resources in the 400 MHz band, where only 2x5 MHz bands may be available. This leads to hypothesizing about a future scenario where critical communications will be allocated in the 700 MHz band, using the 400 MHz band in some countries as complementary, taking advantage of a flexible solution with the drawback of a possible fragmentation. To make this possible and to respond to market needs, it is indispensable that manufactures will produce terminals able to operate in channels allocated in a wide range of frequencies. The results of this study will be published in the ECC 199B report and a final decision is expected to be ratified in an ITU-R resolution of the ITU-R World Radio Conference in 2015, as a consequence of agenda items 1.2 and 1.3 that deal with use of 694–790 MHz band for mobile services and PPDR needs. Special attention has to be paid to complementary applications, as they require additional spectral resources, some hypothesis can be drawn [CEP 15]. For instance, *ad hoc* networks can be used in case of unexpected big events to increase the capacity of the congested network, or where the network infrastructure is not available. The required spectrum can be obtained by installing additional temporary base stations or repeaters or in the 4940-4990 MHz band, used by PPDR in the ITU regions 2 and 3. Another complementary application is air-ground-air communications, mainly used for a monitoring video streaming from a camera mounted on a helicopter or an unmanned aerial vehicle (UAV). In this case, the use of external frequency band is demanded by national decisions subject to specific regulatory and technical conditions. Finally direct mode operation (DMO), allowing terminals to communicate directly in reduced spaces without the intervention of the network infrastructure, can be allocated in the same frequency band of the network (e.g. 700 MHz) or in other spaces to be identified in commercial IMT bands. To reserve a portion of spectrum to PPDR, on the other hand, produces a cost comparable with the value that the spectrum would have if allocated to other applications as commercial mobile communications. As an example, in [TCC 13c] this cost has been estimated, taking as reference auctions of 800 MHz bands, that took place in European countries. By comparing it with an estimate reduction of cost for disaster managing, due to the use of broadband PPDR communication, [TCC 13c] demonstrates the effectiveness of the deployment of a dedicated radio spectrum for these applications.

5.5.5. *PMR and commercial systems interoperability*

The adoption of a common standard would facilitate PPDR operators to move in different nations and to cooperate with different organizations, communicating through existing networks using their own terminals. However, another important interoperability aspect concerns the co-existence and the backward compatibility of PMR and commercial communication systems. Indeed, adopting the commercial LTE framework to answer to the PPDR operators' communication needs can open the door to new opportunities and synergies, offering advantages to both commercial and PPDR worlds. To deliver a next generation service for PPDR requires a next generation PMR network that is backward compatible with legacy communication technologies and fully converged with the 4G evolutionary wireless paradigm. The interoperability scenario will evolve over time; in particular, the TCCA produced a roadmap showing the phases of PPDR communication transition, from the existing mission critical voice networks to the upcoming broadband mission. The roadmap shows that the mission critical voice will rely on existing TETRA networks for at least another 10-15 years; it is in fact not expected that real mission critical voice group communication can be realized "over" a LTE network in the coming period. Therefore, most countries will continue operating their TETRA network, while transition to the broadband network will be gradual and will involve significant investments of cost and time. Since the transition cannot be immediate, the simplest way consists of resorting to use of commercial networks. This can be accomplished at different levels of integration and interoperability. LTE networks are composed of four main logical domains: service, evolved packet core, radio access and user equipment.

In theory, professional and commercial operators may partially or completely share one or more domains of their networks, in order to guarantee the services needed by different kinds of users. As stated before, even though LTE is provided with suitable procedures to manage different types of services with different QoS requirements, it is not able to meet the high standards in terms of reliability, resilience and security required by mission critical services. Specific PMR applications can be provided only if the service domain for professional communications remains autonomous. By means of commercial agreements, instead, it is possible to realize a co-operation between commercial and professional operators that involve other logical domains.

The simplest and least expensive solution for having available broadband multimedia applications consists of resorting to LTE commercial networks to provide services based on IP, while voice communications still remain served by legacy PMR. For IP services terminals resort to an access point name (APN) through which they become connected to the network of their company/organization. The most efficient solution would to be to use dedicated resources and leased lines that safeguard the highest level of reliability and resilience of communications. If that method is not practical and resources are shared with commercial users, some issues can arise. In particular, during disaster events, the Internet can become congested or unreachable and outages cannot be monitored and managed; if traffic prioritization, mechanisms of PPDR users are equal to the one for commercial users, the network cannot be considered reliable unless complex IP QoS solutions are used. For these reasons this solution is suitable only for non-mission critical traffic. In order to obtain multi-organization and transnational interoperability, service continuity and availability is granted only if roaming agreements are contracted between commercial operators.

A second solution is applicable when PPDR organizations own a broadband mission critical network and resort to commercial networks to cover additional areas. In this case, the PPDR operator acts as a mobile virtual network operator (MVNO). This case is partially suitable for mission critical scenarios except when users are under the coverage of commercial networks. When the two networks exist together, it could be possible to maintain an active call when passing from one to the other on the condition that co-ordination of the radio network is performed and an ISI between LTE and TETRA is defined to permit the user to have a single subscription for both networks with a single smart card.

A third scenario assumes that the PPDR operator owns a mission critical mobile broadband PPDR core network while access network is shared with a commercial mobile network operator reducing the investment and operational cost in comparison with two dedicated RANs. The spectrum can be dedicated per operator or shared. RAN sharing is defined in 3GPP specifications [GPP 13a]. This solution provides a higher support to mission critical operation that is guaranteed only if supported by shared RAN. Total support is possible only if also the access network is owned and operated by PPDR organizations in other words only if the network all network logical levels are managed by PPDR operators. In this case, transnational and trans-organization interoperability is fully supported thanks to regulations provided by LTE standards, and agreements for roaming are not influenced by commercial factors and should be modeled on real needs of PPDR users.

5.6. Acknowledgments

Concerning the ISITEP framework presented in section 5.3, the research leading to these results has received funding from the European Union Seventh Framework Programme (FP7/2007–2013) under grant agreement no. 312484.

5.7. Bibliography

[BOR 11] BORGONJEN H., European situation on mobile communication, Rescue in underground facilities, Rome, Italy, 3 March 2011.

[CAS 11] CASSIDIAN, Cross-border communications trial short summary, Report, BDBOS, MSB, EADS, 2011.

[CCB 12] CRITICAL COMMUNICATIONS BROADBAND GROUP, Mission/Vision Statement, Adopted 2012.

[CCB 13] CRITICAL COMMUNICATIONS BROADBAND GROUP, Mission critical mobile broadband: practical standardization & roadmap considerations, White Paper, 2013.

[CEP 13] CEPT ECC, User requirements and spectrum needs for future European broadband PPDR systems (Wide Area Networks), Report 199A, 2013.

[CEP 15] CEPT ECC, Harmonized conditions and spectrum bands for the implementation of future European broadband PPDR systems, Draft report 218, 2015.

[COU 03] THREE-COUNTRY PILOT, Aachen – Luik, Liège,–Maastricht, Final Report, November 2003.

[ETS 05] ETSI, Terrestrial trunked radio (TETRA), Functional requirements for the TETRA ISI derived from three-country pilot scenarios, Tech. Rep. ETSI TR 101 448 V1.1.1, 2005.

[ETS 09] ETSI, Reconfigurable radio systems (RRS), SDR reference architecture for mobile device, Tech. Rep. ETSI TR 102 680, 2009.

[ETS 10] ETSI, Interworking at the inter-system interface (ISI), European Standard EN 300 392–3, 2010.

[ETS 14] ETSI, TETRA and critical communications evolution (TCCE), Critical communications architecture, Part 1: critical communications architecture reference model, Tech. Rep. ETSI TR 103 269-1, 2014.

[ETS 15] ETSI, TETRA technology webpage on ETSI portal [Online], available at http://www.etsi.org/technologies-clusters/technologies/tetra 2015.

[EUR 90] EUROPEAN UNION, Convention Implementing the Schengen Agreement of 14 June 1985 between the Governments of the States of the Benelux Economic Union, the Federal Republic of Germany and the French Republic, on the Gradual Abolition of Checks at their Common Borders ("Schengen Implementation Agreement"), 19 June 1990.

[FCC 07] FCC, Second report and order, Tech. Rep., August 2007.

[FCC 11] FCC, Third report and order and fourth further notice of proposed rulemaking [Online], available at http://www.fcc.gov/Daily_Releases/Daily_Business/2011/db0204/FCC-11-6A1.pdf, 2011.

[FAN 10] FANTACCI R., MARABISSI D., TARCHI D., "A novel Communication Infrastructure for Emergency Management: The In. Sy. Eme. Vision", *Wireless Communications and Mobile Computing*, Wiley, vol. 10, no. 12, pp. 1672–1681, December 2010.

[FER 12] FERRUS R., SALLENT O., BALDINI G. *et al.*, "Public Safety Communications: Enhancement Through Cognitive Radio and Spectrum Sharing Principles", *IEEE Vehicular Technology Magazine*, vol. 7, no. 2, pp. 54–61, June 2012

[FM 15] CEPT Working Group FM 49 Radio spectrum for public protection and disaster relief (PPDR) web site, available at http://www.cept.org/ecc/groups/ecc/wg-fm/fm-49, 2015.

[GPP 13a] 3GPP, TS 23.251 Technical specification group services and system aspects, Network sharing, architecture and functional description (Release 10), V10.6.0, 2013.

[GPP 13b] 3GPP, 3G security, Network domain security (NDS), IP network layer security, Tech. Spec. 3GPP TS 33.210, 2013.

[GPP 15] 3GPP public safety web page, [Online] available at http://www.3gpp.org/news-events/3gpp-news/1455-Public-Safety 2015.

[GSM 14] GSMA, Permanent Reference Document IR.92, Tech.Rep. v8.0, 2014.

[IEE 09a] IEEE standard for architectural building blocks enabling network-device distributed decision making for optimized radio resource usage in heterogeneous wireless access networks, IEEE Standard 1900.4-2009, 2009.

[IEE 09b] IEEE standard for local and metropolitan area networks: media independent handover services, IEEE Standard 802.21TM-2008, 2009.

[IET 02] IETF, SIP: Session Initiation Protocol, RFC 3261, available at https://www.ietf.org/rfc/rfc3261.txt, June 2002.

[ISI 15] INTERSYSTEM interoperability for TETRA-TETRAPOL networks (ISITEP) project web site, available at http://www.isitep.eu 2015.

[MAC 15] MULTI-AGENCY COOPERATION in cross-border operations (MACICO) project website, available at: http://macico.com 2015.

[NOR 15] NORDERED, Nordic civil protection cooperation web site, available at http://www.nordred.org 2015.

[SCH 95] SCHENGEN communication requirements and the TETRA standard, SCH/I-Telecom (95) 35 (1) of 21, November 1995.

[SCH 03] EU SCHENGEN CATALOGUE vol. 4 Police co-operation recommendations and best practices, Council of the European Union General Secretariat DGH., 21 June 2003.

[TCC 10] TETRA + CRITICAL COMMUNICATIONS ASSOCIATION, CRITICAL COMMUNICATIONS, TIP phasing for ISI TF10-56-2 2 v1.2, ISI WG/SELEX Communications, 30 March 2010.

[TCC 12] TCCA, LTE broadband statement, available at www.tandcca.com/Library/Documents/LTEBoardstatement.pdf, 2012.

[TCC 13a] TETRA + CRITICAL COMMUNICATIONS ASSOCIATION, CRITICAL COMMUNICATIONS BROADBAND GROUP, Mission critical mobile broadband: practical standardization and roadmap considerations, White Paper, 2013.

[TCC 13b] TCCA, The strategic case for mission critical mobile broadband. A review of the future needs of the users of critical communications, December 2013.

[TCC 13c] TCCA WIK-CONSULT, The need for PPDR broadband spectrum in the bands below 1 GHz, Report 2013.

[TCC 15] TCCA Radio spectrum web site, available at http://www.tandcca.com/assoc/page/13043 2015.

[TET 98] TETRAPOL forum, TETRAPOL specifications part 10: inter system interface, PAS 0001-10-1/2/3 1998.

[TET 15] TETRAPOL forum portal, available at http://www.tetrapol.com 2015.

[VOL 15] VOLGA forum web site, available at: www.volgaforum.com 2015.

[WAL 91] WALKER P., International search and rescue teams: a league discussion paper, League of the Red Cross and Red Crescent Societies, Geneva, Switzerland, Paper 1375, 1991.

6

Joint Network for Disaster Relief and Search and Rescue Network Operations

6.1. Introduction

Disasters are exceptional events that are either man-made, such as terrorist attacks, or natural, such as earthquakes, wildfires and floods. Disasters create emergency situations and cause physical and social disorder. In these emergency situations, food, water, shelter, protection and medical help are needed, and the effort needed to provide these basic services to the victims must be coordinated quickly via a reliable communication network. Disaster relief operations typically involves a series of steps including establishment of communication infrastructures, performing search and rescue operations, and providing any needed first aid

---

Chapter written by Ram Gopal LAKSHMI NARAYANAN and Oliver C. IBE.

services. Disaster networks can be classified as disaster mitigation networks and disaster relief networks. A disaster recovery network is a network that is used in the pre-disaster stage to plan effective post-disaster relief operations.

A disaster recovery network, which is a part of disaster relief operation, is considered to be a life saving network that is used to provide emergency support to the disaster victims and the crew members who are helping the victims, and to provide communication infrastructure in the affected area. Sometimes the disaster relief operation involves searching for and locating the survivors, and then rescuing them. Currently this process typically involves manual searches in the disaster area, which can be hampered by the shortage of manpower in the disaster area, and it is also time consuming. To expedite this process, there must be a mechanism that enables survivors to report their locations to the Command Center, if they can, so that crew members can be directed to those locations for the rescue operation. Our interest is in the disaster recovery network and the motivating problem can be stated as follows:

– how can the survivors provide their locations quickly to the Command Center?

– what types of technologies are required, and how much time does it take for the survivors to learn to use these technologies to report their locations?

– how can these technologies speed up the rescue operation in the disaster area?

The goal of this chapter is to survey various existing solutions to these problems and their shortcomings, and propose an architecture that solves the problem. This chapter is organized as follows. Section 6.2 of this chapter provides an overview of disaster recovery network (DRN), search and rescue network (SRN) and essential network requirements. Section 6.3 discusses various existing solutions for DRN and SRN networks. Section 6.4 describes joint DRN-SRN solution called the *portable disaster recovery network* (PDRN) and its operation. Section 6.5 provides the simulation parameters for evaluating PDRN network architecture and simulation result. Finally, the chapter concludes with future activities and remarks in section 6.6.

6.2. Overview and requirements of DRN and SRN

6.2.1. *Disaster recovery systems*

As stated earlier, disasters are catastrophic events that occur unexpectedly in a random manner. They are either man-made, like terrorist attacks, or natural calamities, like earthquakes and tsunamis. The lack of infrastructure for disaster mitigation around the world and its prediction accuracy leaves civilians vulnerable to disasters. Disaster relief is an operation carried out after a disaster has occurred. A DRN is considered to be a life-saving network. The purpose of DRN is to provide emergency support to affected people and to support the crew members helping the victims. All existing networks were completely damaged in the areas that were affected by the recent incidents of Indonesia's tsunami in 2004, Hurricane Katrina in 2005, Japan's tsunami in 2011, the Haitian earthquake in 2010, and Hurricane Sandy in 2012. This rendered the crew members helpless, and many victims were trapped inside the disaster areas for a long time. There is now an increasing awareness among the government agencies for the need to implement disaster mitigation and relief systems.

Planning for disaster relief is a complicated operation that involves the use of proven technologies to coordinate, among several agencies, victims and crew members. A disaster can occur in any part of the world and no assumption must be made about any existing communication infrastructure in the disaster region. Therefore, DRN systems must be able to work autonomously, and if there is any communication infrastructure that exists before and after the disaster, then DRN must use and co-exist with such systems; this will expedite the disaster relief operation.

Since disaster relief is a life-saving operation, a DRN must be easy to deploy and operate, and it should require a short learning time by the disaster relief crew and victims in the affected area. With these basic requirements in mind, it is easy to see that wireless networks are the best choice for disaster relief operations because many of them do not require any pre-existing infrastructure to be established and are easy to operate. Several radio access technologies are currently available that can be used in cellular networks, wireless local area networks, wireless mesh networks, geographical area networks (GAN), unmanned aerial vehicles (UAV) and wireless personal area (WPAN) networks. Unfortunately, not all these technologies are directly applicable for DRN use because they are not designed for that purpose. However, several architectural solutions and protocols have been proposed and developed that use combinations of these existing technologies for partial disaster relief operations.

6.2.2. *Search and rescue systems*

Search and rescue operations are used to track individuals after a serious mishap, such as a wildfire, building collapse, earthquake, and when people are lost while hiking or trapped in mines. Data from previous disaster incidents indicate that more than 50% of deaths occurred within a few hours after the disaster event [MAJ 09]. Therefore, disaster relief and search and rescue operations (SRO) must take place within a short time after a disaster has occurred in order to increase the chances of rescuing the victims while they are still alive.

The key difference between a DRN and an SRN is that a disaster network is deployed at a particular location where a disaster has happened, and its main goal is to establish communication with the victims and then carry out the search operation. Search and rescue networks can be used in a disaster affected area as well as in wilderness scenarios where there may be no disaster and thus no impact to geographical areas. When an SRN is used in a non-disaster-affected area, the problem becomes that of searching for and rescuing people who are lost. In this case there may be no definite location or boundary defined for the search area, or it may be a well-defined large area. In a disaster-affected area, a disaster recovery network is first established followed by a search and rescue operation. Thus, a disaster leads to the use of both a disaster recovery network and a search and rescue network. In general, the area where a search and rescue network is deployed is larger than the area where a disaster recovery network is deployed.

Systems that can locate disaster victims within a short time are essential in search and rescue systems. As in the case of the World Trade Center (WTC) disaster, the nature of a disaster can be such that it can disrupt the entire communication infrastructure. In the WTC case, the rescue crew members could only use one-way and two-way radio systems to communicate. Because everyone was using the same frequency band, it was difficult to locate the victims due to signal interference.

Tracking and identifying the locations of disaster victims are two of the operations that take place in a search and rescue operation. In the past, dogs, human signs and acoustic signals were the primary means for this purpose. These techniques worked in the cases where humans and animals could access the affected areas. Unfortunately, these solutions often involve a great deal of human assistance that is typically in short supply in many disaster areas. Recent developments in sensor technology, wireless communication, robotics, computing and audio-visual technology have led to techniques that can be used to track and identify the locations of disaster victims and rescue them quickly. Some of the available sensor-based identification schemes include barcodes, biometric schemes, RFID and cell phones. These technologies are used in search and rescue network (SRN) systems, which include the global position system (GPS), wireless networks and sensor networks. These identification and tracking schemes complement each other because no one scheme can be used in all situations. For example, GPS-based solutions are applicable for outdoor situations only because they cannot provide the exact floor location inside the building. Similarly, many RFID-based schemes are useful only in indoor office applications.

6.2.3. *Key disaster recovery network and search and rescue network requirements*

Essential requirements that need to be part of a disaster recovery system are discussed here. The over-riding assumption is that DRN and SRN are wireless networks, and the requirements discussed in this chapter are specific to wireless communication networks.

6.2.3.1. *Quick response*

A disaster is an emergency situation that requires a quick turnaround for the infrastructure required to respond to the situation. Thus, a major criterion for evaluating a DRN and SRN is how quickly it can become operational and be used to meet the needs of the disaster victims.

6.2.3.2. *Life Expectancy of the Network*

Based on previous disaster incidents, the time to restore the damaged infrastructure takes on the order of months, if not years. Therefore, the disaster recovery network must be able to provide normal service until the rescue mission is completed and possibly beyond the completion of the rescue mission until normal infrastructure is restored.

6.2.3.3. *Interoperability*

Some proposed DRN's are expected to be interconnected with either the public switched telephone network (PSTN) or the Internet. Therefore, such networks must be designed to permit them to interoperate with these networks. It is known that some popular DRN's, such as police emergency networks and fire department networks, use proprietary protocols. Thus, any DRN and SRN that is required to be interconnected with these networks must have the necessary interface to enable the protocol conversion.

6.2.3.4. *Tariff-free operation*

In GSM-based networks, users are required to "pre-pay" for service before using the network by having the necessary minutes in their accounts. This is usually the case for most countries outside the USA. Thus, if a DRN supports voice-based applications that permit users to connect to the outside world, it should allow these users to use the network free of charge. This requirement is applicable to DRN system.

6.2.3.5. *Network coverage*

When a disaster occurs, the communication infrastructure may be partitioned into a number of islands. In this case, a disaster recovery system should be such that it can be quickly used to interconnect the different islands of disaster areas. If no part of the pre-existing infrastructure is available after the disaster, then it should be possible to deploy a solution that can cover the disaster area with one network or a cluster of networks that can be interconnected to permit communication across the affected area.

6.2.3.6. *Support for heterogeneous traffic types*

The ability of a DRN to support voice, data and video applications is a major concern. Some proposed solutions are voice-only solutions while others are data and/or video-only solutions. A desirable feature of a DRN is its ability to support different traffic types. This requirement is applicable to SRN systems.

6.2.3.7. *Network capacity*

Consider a situation where some or all the victims in a disaster area have devices with which they can communicate with the outside world, but the infrastructure is damaged by the disaster. Any DRN that is subsequently set up must have sufficient capacity to handle the sessions generated by both the victims and the disaster relief crewmembers. Thus, a DRN solution should have the capacity to support this traffic scenario.

6.2.3.8. *Ease of use and equipment cost*

A user terminal or mobile node is the mobile device that is used by either the disaster victims or the crewmembers to access wireless service. Such a system should be simple to use with little or no learning time due to the emergency situation. For example, if the device to be used by the victims is a voice device, then it should be as simple to use as the normal mobile phone that the victims are used to. Therefore, ease of use is a major issue in evaluating a DRN. Also, the user terminals that could be distributed to the disaster victims should not be expensive in order to ensure that every victim that needs such device gets one.

6.2.3.9. *Outdoor and indoor scenario*

The search and rescue (SAR) system must work for both outdoor and indoor scenarios. SRO applications must work seamlessly for both indoor and outdoor environments. If the SRN has two separate solutions in which one is for indoors and the other is for outdoors, then they must work together seamlessly.

6.2.3.10. *High precision for localization and search operation*

Locating the subject or a survivor is an important operation. For this reason, the system must provide the survivors' location with reasonable accuracy without any ambiguity. Using this information, the search team will be able to respond quickly to rescue the survivors.

The following are the possible scenarios that the system must consider:

1) The system must provide high-precision location information and narrow the search area for SAR operation.

2) If the system is not able to provide high-precision location information but provides or has defined target search areas, then the solution must propose a faster search mechanism to rescue the survivors.

3) If the location information is of high precision, then the search area will be very narrow. This will lead to fast search and rescue.

6.3. Previous work

6.3.1. *Disaster recovery network solutions*

Several solutions that use a combination of *ad hoc* wireless network, mesh network, satellite and wireless sensor network have been proposed as DRN solutions. Most of these proposed solutions establish a DRN for only crew member-to-crew member communication. Very few systems consider survivor-to-crew member communication or expect the survivor to be holding a tracking device.

6.3.1.1. *Ad hoc networks*

An *ad hoc* network is a network that is constructed on the fly without any pre-existing infrastructure. A mobile *ad hoc* network uses a wireless medium for communication and user terminals act as relay stations. A wireless mesh network (WMN) is a form of *ad hoc* network that is designed to provide more than one path between nodes in the network [AKY 05]. SKYMESH is an example of a wireless mesh network where the complete infrastructure is constructed using Wi-Fi access points [SUZ 06]. SKYMESH uses commercial off-the-shelf components to construct a wireless architecture in the disaster-affected areas using balloons in the sky. The goal of SKYMESH is to provide accessibility and connectivity in the disaster affected areas with no assumption of any pre-existing network infrastructure. It creates WLAN network by using a huge helium-filled balloon which floats in the air. The payload of the balloon consists of WLAN hardware that creates wireless coverage. A group of WLANs connects to the balloon in a line-of-sight manner. One balloon can be chosen to have connectivity to satellite ground station that in turn uses stationary satellite to connect to the Internet. The focus of SKYMESH is to have rapid deployment of a network and the network is assumed to be temporary. This is a short-lived network and is not suitable for all types of terrain. Note that such a network is used to provide crew member-to-crew member communication. The search operation performed to locate the survivors is manual, and will not scale when impact of the disaster is large.

Multimedia WMN for disaster network architecture uses standard WMN protocols for DRN [KAN 07]. This architecture uses PDAs and laptops to form a WMN access mesh network that connects to a Command Center via a satellite backbone. The goal of this solution is to rapidly deploy the mesh networks among disaster-affected regions, and use the devices with Wi-Fi interface for communication. It is used to share pictures and videos using peer-to-peer technologies. This architecture uses standard MANET Optimized Link State Routing Protocol in the mesh networks and is established by crew members so that they can share the pictures with Command Center and among themselves. One or more of the laptops are equipped with satellite interfaces so that they can establish communication with satellite backbone networks. This solution is used to enable communication among the crew members only. The search operation performed to locate the survivors is manual, and will not scale when impact of the disaster is large.

6.3.1.2. *Hybrid networks*

Hybrid networks are heterogeneous networks formed using cellular networks and 802.11x WLAN networks. Unfortunately, cellular networks require huge infrastructure to operate and, therefore, are not suitable for disaster recovery operations. When one or more cellular base stations is damaged due to disaster, mechanisms must be in place to quickly divert the traffic by deploying a mobile base station and provide coverage, or by quickly deploying an *ad hoc* network that will act as a relay to cover the affected area. There is an increasing interest in recent years in hybrid networks, and 3G and 4G standardization efforts are in progress. Most of the hybrid network protocols require the user terminal to have multiple interfaces in order to freely roam between a cellular network and WLAN network.

The unified cellular and *ad hoc* network (UCAN) [LUO 07], the integrated cellular and *ad hoc* relay system (iCAR) architecture [WU 01], the cellular aided mobile *ad hoc* network (CAMA) [BHA 04], and the enhanced communication scheme combining centralized and *ad hoc* networks (ECCA) architecture [FUJ 05] are based on the hybrid network design. Each of these solutions focuses on one or more specific aspects of the solution, such as to increase throughput by doing handoff from the cellular network to the WLAN, or to have uniform load across all APs, or to provide QoS guarantees within the *ad hoc* network. Because all these systems are built around existing cellular infrastructure, their applicability to DRN operation is limited. For instance, ECCA is similar to iCAR, but its focus is on having accessibility and reachability as opposed to performance improvement. ECCA assumes that only a portion of the network will be impacted by disaster and makes such solutions less applicable to disaster operation.

6.3.1.3. *Satellite networks*

Satellite networks provide robust, global coverage, complementing disaster relief efforts. Satellite images are used to identify the effects of natural disasters and are used in aiding long-term recovery efforts. Using satellite phones, one can communicate from any part of the world. A satellite phone is like any other phone and is portable.

However, due to limited adoption by the civilians around the world (there are currently an estimated 1 million satellite phones worldwide) when compared to 2G/3G phones (there are currently about 3 billion phones), satellite phones are not going to be an effective and complete solution for disaster recovery operation. Moreover, satellite phones are expensive and not easy to use. The satellite phone finds its use only among crew members that use it to assess the damage at a disaster site before embarking on the rescue operation.

The wireless infrastructure over satellite for emergency (WISECOM) uses the satellite network as backbone and connects existing wireless networks to landline systems [BER 07]. WISECOM is composed of three components, namely WISECOM Access Terminal (WAT), WISECOM Transport and WISECOM Servers (WS). WISECOM network provides both crew member-to-crew member communication and survivor-to-survivor communication. It demands huge infrastructure to operate and manage the network.

Geosynchronous Earth Orbit (GEO) satellites are at 35,786 km above sea level at an equatorial orbit and they orbit the Earth at same rate as the Earth turns on its axis. This allows the satellite to remain in a relatively fixed position and allows communication from Earth with simplified infrastructure. The drawback of a GEO satellite is that it has a very high round trip time and is not suitable for latency driven applications. The Medium Earth Orbit (MEO) satellite operates at 8,000 km and enables applications that demand low latency [BLU 13].

6.3.1.4. *Amateur radio*

Amateur radio (or HAM radio) is one of the oldest P2P technologies that are operated by certified professionals [SIL 04]. It was shown to work during recent disaster incidents, such as Hurricane Katrina and the Indonesian tsunami. It mainly provides voice and Morse code communication and Hurricane is expensive to deploy and operate. HAM radio consists of transmitter, receiver and antenna sub-systems. A separate frequency band is allocated for HAM radio communication and HAM radio operators tune to this band and use Morse code for communication. In 1997, the Internet Radio Linking Project (IRLP) started to interconnect HAM radio and the Internet using voice over IP protocols [IRL 14]. Since HAM radio is used by certified professionals, its applicability and usage are so far restricted to crew members or first responders; also, it supports only voice and no data. There is no support for crew member-to-survivor communication. The search operation performed by crew members is manual, and will not scale if the impact of the disaster is large.

6.3.1.5. *Wireless sensor networks*

Sensor systems with wireless capabilities are becoming increasingly important as they are used to collect data for different applications. Wireless sensor networks

(WSNs) [AKY 02, MAC 08, YIC 08] are a part of disaster warning systems (DWS) that perform two operations: disaster mitigation and rescue operation after disaster. Some of the useful applications of WNS are detection of floods, earthquakes, wildfires and other environmental disasters. They are also useful in finding victims who are trapped in disaster areas, and for searching for people who are lost while mountain climbing or hiking. When used in this environment a wireless sensor network is not a total solution; it complements the previously discussed disaster recovery network architectures. Sensor network design must consider scalability, fault tolerance, topology, power consumption and transmission power. A typical sensor consists of a sensing unit, a power unit, a transceiver unit and a processing unit. Sensors are designed in such a way that twice the energy is consumed in transmitting the sensed information than receiving the information and their energy consumption is higher in communicating than computing the actual sensed information. Sensors can be deployed from helicopters and are unmanaged. Sensor network routing nodes, which are unattended for most of the time, may fail for various reasons, and routing protocols must compensate for these issues in their design. Most of the sensor routing techniques use location-based information. Many of the existing routing protocols suffer from topology changes and lack scalability.

Sensor for disaster relief operations (SENDROM) [CAY 07, SAH 07] is a management system primarily used for rescue operations after large disasters. The problem with any sensor-based approach is that it demands infrastructure to collect and transmit information. Sensors are battery operated and unmanaged and are usually not deployed uniformly. The SENDROM architecture demands that the sensors be kept in every home and office for accuracy. After a disaster has occurred, it is very likely that these sensor locations will be disturbed and may not be accessible due to obstacles getting in the way. Thus, installing sensors in homes and offices prior to a disaster requires huge deployment cost.

6.3.1.6. *Unmanned aerial vehicle*

Unmanned aerial vehicles (UAV) are a class of aircrafts that can fly without the onboard presence of pilots [WAT 12]. Unmanned aircraft systems consist of the aircraft component, sensor payloads and a ground control station. They can be controlled by onboard electronic equipments or via control equipment from the ground. When it is remotely controlled from ground it is called RPV (Remotely Piloted Vehicle) and requires reliable wireless communication for control. Dedicated control systems may be devoted to large UAVs, and can be mounted aboard vehicles or in trailers to enable close proximity to UAVs that are limited by range or communication capabilities.

UAVs are used for observation and tactical planning. This technology is now available for use in the emergency response field to assist the crew members.

UAVs are classified based on the altitude range, endurance and weight, and support a wide range of applications including military and commercial applications. The smallest categories of UAVs are often accompanied by ground-control stations consisting of laptop computers and other components that are small enough to be carried easily with the aircraft in small vehicles, aboard boats or in backpacks. UAVs that are fitted with high precision cameras can navigate around the disaster area, take pictures and allow the crew members to perform image and structural analysis. As UAV operations require onsite personnel, it will be helpful for onsite crew members to access the disaster area first before entering the disaster affected area. UAVs that are suitable for outdoor operation and can fly at reasonable altitude are used for disaster impact analysis. The important aspect of such UAVs is that the initial assessment gives a clear disaster planning direction. After the survivors are detected via image analysis, crew members can then try to make contact with the survivors and perform quick rescue operations. Nano UAVs can be used in-built and combined with robots capabilities and can be a very useful in detecting structural damages to buildings and detect survivors trapped inside debris.

In recent years, increasing research efforts and developments are improving UAV for various application and reliability. UAV is still in experimental stages at the moment. Also, a shortage of skilled onsite crew member is a bigger problem. [PRA 06] highlights that a minimum of three staff members is required to operate a UAV.

6.3.1.7. *Device to device communication*

Public safety agencies and organizations have started network evolution planning for LTE-based public safety solutions. LTE supports a wide variety of services from high bandwidth data services to real-time communication services – all in common IP-based networks. Mission critical communication in demanding conditions like natural disaster sets strict requirements, which are not necessarily supported by regular commercial mobile networks. Globally, the USA is the main driving market, where FirstNet government agency has a mandate to build nationwide public safety network using 700 MHz bands (band 14). In Europe, the driving market is the UK, where the UK Home Office has established a program with a target to build new emergency service networks that will provide mobile services for the three emergency services (police, fire and ambulance). However, such deployment models will be expensive.

3rd Generation Partnership Project (3GPP) standardization bodies are responsible for defining standards for wireless cellular networks [GGP 13]. Currently, to develop and assist public safety network systems, standardization is in progress to develop direct device-to-device communication capabilities from one mobile device to another mobile device at a distance of up to 500 m. Figure 6.1

describes various possible communication mechanisms for the user to communicate to other users with and without the support of network infrastructure. These features are currently in the specification phase and are expected to be made available in coming years. One of the key features of Release 13 is to enable proximity services and communication. This useful feature allows one mobile user to communicate directly or through other intermediate mobile device that acts as relay without needing cellular infrastructure. Such infrastructure less communication is critical for disaster relief operation wherein one survivor can communicate directly with other survivors. At this moment, standardization is still in progress and aspects related to device discovery, interference free communication, privacy and other services are still under development.

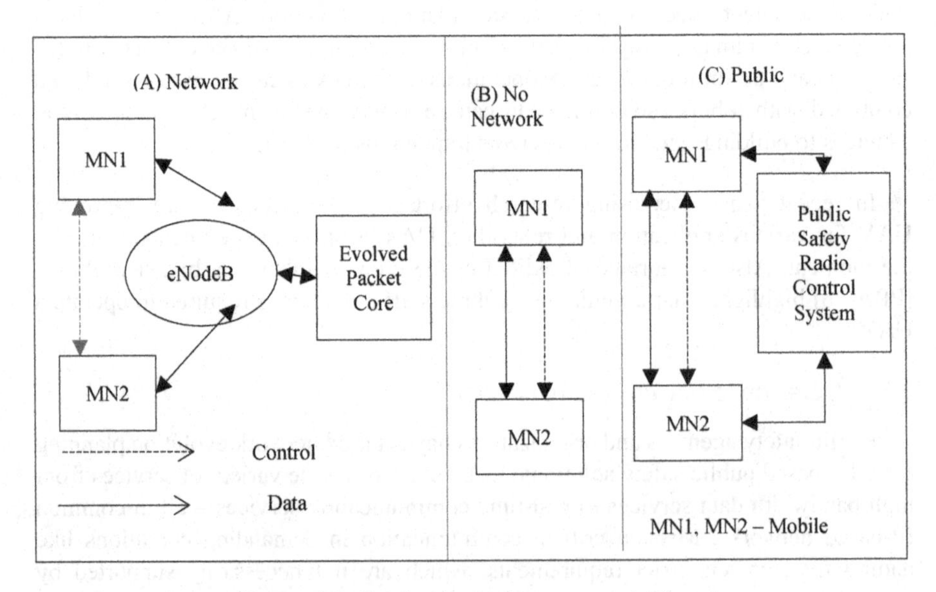

Figure 6.1. *Device-to-device communications*

6.3.2. *Previous search and rescue network solutions*

6.3.2.1. *Satellite-aided tracking systems*

The satellite-aided tracking systems that have been proposed include the use of satellite phones, geographical information systems (GIS), global positioning systems (GPS), location-based services (LBS), the cosmicheskaya sistema poiska avarynyh sudov (COSPAS, which is Russian for Space System for Search of Vessels in Distress), the search and rescue satellite-aided tracking (SARSAT), global navigation satellite systems (GNSS) and the slobalnaya navigatsionnaya sputnikovaya sistema (GLONASS, the Russian equivalent for GNSS).

COSPAS and SARSAT are satellite-based search and rescue systems that are used to detect and locate emergency beacons carried by aircraft, marines and individuals. These systems were developed by Russia and the United States, respectively. The architecture is composed of satellites, ground stations, local user terminals (LUT), mission control centers (MCC) and rescue coordination centers (RCC). The MCC and RCC are distributed geographically around Europe and US. Their main function is to coordinate with their rescue teams. The user terminals include the emergency position indicating radio beacons (EPIRB), the personal locator beacon (PLB) and the emergency locator transmitter (ELT). These are the three devices that are used by marines, individuals and aircrafts, respectively [ASC 09]. The user terminals generate beacon distress signals in emergency situations. The distress signal is received by the satellite systems, and the information is relayed to the LUT, which computes the location information. The computed location information is then forwarded to the MCC, which generates an alert message to RCC along with the received location information for rescue efforts. PLB systems are expensive and not affordable by civilians in many developing countries. Adoption of such systems is not common among civilians and requires everyone to carry these devices.

6.3.2.2. *Robot-assisted emergency and search and rescue systems*

When victims are trapped inside debris disaster areas, it is usually difficult to rescue them quickly. Mobile robots are useful gadgets that can go inside the building and detect if any victim is present and then signal the crew members for recovery. The purpose of using mobile robots is to track both the victims and to track the rescue crew who are helping the victims. The size of robots ranges from a small device to a moving crane. Robots can also be used in hazardous locations, such as chemical plants and nuclear reactors, where humans cannot reach. Due to advances in wireless technologies, robot-based solutions have been proposed for SRN operations [ASC 08, CAM 02].

A robot typically consists of electromechanical parts, processing units, sensing units and communication units. Sensors in robots fall into two categories, namely sensors used to control the movement of robots and sensors used to identify the victims or track their locations. A robot collects its location and movement from the environment and either computes the location locally or forwards it to a control station via a wireless link. Five basic requirements for robots are localization, environmental mapping, path planning, motor control and communications [KO 09].

Recently, robots, along with Nano UAVs, are getting more attention in the research community. For indoor navigation, GPSs cannot be used to track the location of the users trapped inside the building. A wireless mobile robot tracking system architecture is composed of mobile robots tied with blind nodes and reference nodes. Reference nodes are typically deployed at different locations of the building. A blind node, which is installed in a mobile robot, is allowed to move freely inside the building. ZigBee interface protocols are being used for communication between reference nodes and blind nodes. Reference nodes are static nodes and when a blind node sends a request, it responds with the location information. A blind node collects all the location information from several reference nodes and, using the received signal strength indication (RSSI) values, it computes its own location. All the signals received from various reference nodes are computed at a blind node using a centralized algorithm. The calculated location value is sent to the control station for processing and in this way the amount of information that needs to be communicated is small.

A drawback of the system is that it assumes that reference nodes are distributed uniformly prior to the disaster. Also, it uses RSSI values as a wireless signature to determine its location, which may not be accurate due to interference from other sources.

Even though the robot-based SAR operations are getting more attention and importance, the human–robot interaction is complicated and requires skilled personnel to operate the systems [YAN 02]. Some of the proposed systems demand infrastructure to be in place in indoor buildings before the disaster and such requirement limits the applicability of the solution.

6.3.2.3. *Emergency service using cellular phone*

Currently nearly three billion people around the world use cell phones. Emergency service is provided via 911 in USA, or 112 in European countries. In order for this service to work, users must have access to a phone service, and must be in a cellular coverage area. There are two types of cell phone systems commonly used: those with inbuilt GPS receivers and those without. A phone system with a GPS receiver functions as follows. When a subject is in need of emergency, he/she dials 911 (or 112 in Europe). The GPS inside the phone sends out the signal and computes the location information of the mobile phone. This information is transmitted to the mobile switching center and then appropriately routed to the nearest rescue station. The GPS performs the determination of the location using a satellite module in these systems. For the systems that do not have the satellite connectivity, such as GSM-based phones, the network determines the location of the coordinates (latitude and longitude). In these systems when a user dials out 911, the serving base station and its neighboring base station exchange messages with the phone and determine the distance offset. The network then performs the hyperbolic lateration process to determine the location of the phone.

Hyperbolic lateration is a location tracking method that uses the difference between the signal arrival times from a device to three or more reference points. The accuracy of this system varies from 100 to 300 meters. As in the GSM-based system, accuracy is limited to latitude and longitude, and it does not work for multiple floors. In order to compute the location of mobile phone with reasonable accuracy, these systems need a minimum of three base station signals, and these base stations must be non-collinear with respect to the mobile phone; otherwise the location accuracy will be very poor. Also, this solution is expensive and works only when cellular infrastructure is available.

6.3.2.4. *Sensor-based networks wilderness scenario*

The Connectionless Sensor-Based Tracking System Using Witnesses (CENWITS) [JIN 07], Yushan Nation Park (YushanNET) [JAR 97] and SenSearch [CAS 03] architectures use a wireless sensor-based scheme for SRO. They use GPS and sensors to track the location of hikers in wilderness scenarios. They are based on a connectionless architecture in which the nodes in the network use the store-and-forward mechanism to communicate with the base station and the nodes need not be connected to the network all the time. The CENWITS architecture is designed for specific scenarios, such as search and rescue operations where civilians are lost while hiking. The network consists of wireless sensors (WS), GPS transceivers, location points (LP) and access points (AP). The operation in the network can be explained as follows:

1) Each sensor is assigned a unique identification (ID), and a person is required to carry a GPS receiver and WS. The GPS receivers are used to give the exact coordinates, and the information is transferred via the WS.

2) While mountain climbing, if an individual, say Jack, comes across another person, for example Jane, moving in the opposite direction, the sensor carried by each person detects the other sensor and they exchange location information that is recorded in their local sensor storage.

3) Later, when Jane meets another person, for example John, the information about Jack's location is also exchanged with John. The process of meeting people and exchanging the stored location information of all their previous interactions with other people is called "witnessing".

The drawback of the sensor-based approach is that it demands infrastructure and expects the person to carry the devices. This makes the solution less applicable.

6.3.3. *Shortfall of the existing solution*

The existing solutions for DRN and SRN assume the infrastructure to be in place, and that survivors are already equipped with cellular or mobile device to communicate with the rest of the world or with the crew members. A disaster can occur in any part of the world and there may not be any cellular infrastructure prior to disaster in such places. Thus, survivor-to-survivor and survivor-to-crew member communication will not be possible. Past events have shown that even when a survivor has these mobile devices, the deployment of a DRN takes a long time, and by the time the DRN becomes operational the batteries of survivors were drained and this results in no communication from survivors. Also, since it is not a common practice for everyone to have their phone in hand (or a near-by place) while sleeping, if the disaster takes place at night, any solution that requires the potential victims to have disaster-related devices with them will not work. Similarly, in a wilderness scenario, the battery might die or the phone could be lost. So, in reality, it is not a valid assumption that the civilians will be holding their mobile phone before and after the disaster. The previously proposed solutions have several shortcomings and their rescue efforts will take time and may result in loss of lives due to the long deployment time.

The architectures and process flows for the existing solutions were described earlier. After a disaster incident, the DRN is deployed for the crew members only. Communication is established (or allowed) between crew members, and then the crewmembers have to search for the victims and rescue them one by one. The subsequent rescue operation often involves manual search for survivors in disaster areas, which means that it will take time before the actual rescue operation can take place. A solution is required that permits the survivors to quickly report their locations to the crew members or to an onsite Command Center so that they can be rescued on time. It is essential that a survivor has a phone to establish his/her presence quickly to the Command Center, and this will enable the SAR team to rescue him/her quickly. There is a need to introduce concurrency in operation to decrease the time in search operations.

To improve the disaster relief operation, new process and procedures are introduced [LAK 11]. A new procedure introduces the distribution of phones and how the survivors use those phones to report their locations automatically to the Command Center for speedy rescue operations.

The problems that need to be solved can be posed as follows:

1) Using the DRN, how can a survivor communicate with SAR team and reveal his/her location information?

2) Using DRN, how can the SAR team know the survivor's location and find an optimal path for rescue operation?

These issues are addressed in the next section.

6.4. Portable disaster recovery wireless network architecture

PDRN provides a communication infrastructure that enables survivors to communicate with crew members within a disaster area [LAK 11, NAR 12]. The PDRN architecture consists of one or more access points (APs), a gateway (GW) node, PDRN phones and an on-site command centre. In the PDRN, when a disaster occurs, inexpensive PDRN phones are randomly dispersed over the disaster area by, for example, being dropped off from a helicopter or by a UAV. Also, access points are deployed at the periphery of the disaster area, and some can be dropped inside the disaster area by the same mechanism that is used to drop off the phones. They are designed to communicate with a Command Center that is also located outside the disaster area. Once a PDRN phone hits the ground, it uses a built-in GPS functionality to locate its coordinates. It then attempts to communicate with the Command Center via one or more access points to register its location coordinates and thereafter begins to continuously emit a beeping sound that is designed to attract wandering survivors. Thus, because the exact location of a beeping device is known to the Command Center once a survivor reaches any such device, his/her location is completely known at the Command Center from where a rescue team will be dispatched to rescue them. Also, it is assumed that once a survivor reaches a beeping phone and talks to the Command Center with the phone, the phone will stop beeping. The architecture of the PDRN is shown in Figure 6.2.

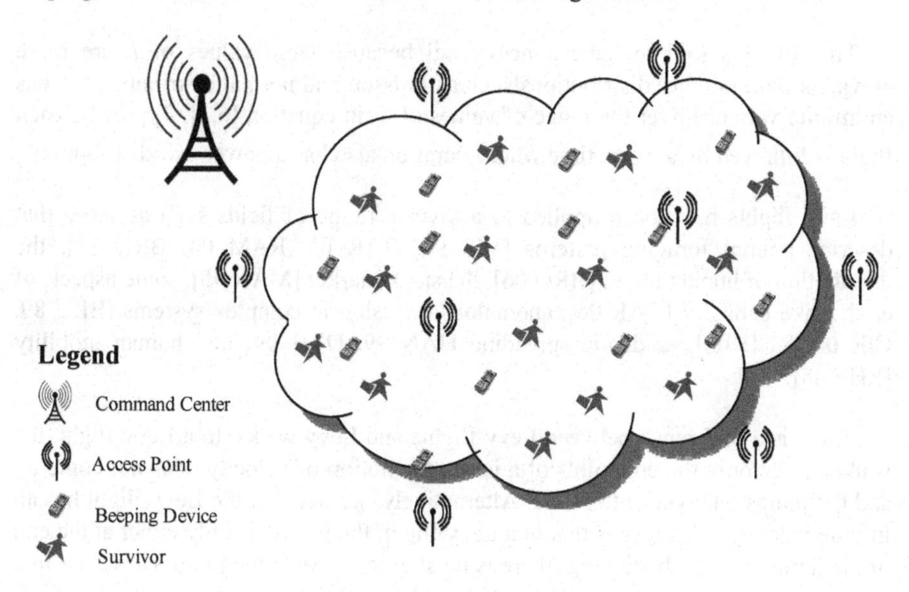

Figure 6.2. *PDRN architecture*

The effectiveness and efficiency of the solution depends on the following criteria whose analysis is provided in remainder of this section:

– What are the possible approaches that a survivor may take to reach a phone?

– How much time does it take a survivor on the average to reach a phone?

6.5. Modeling and simulation of survivor movement

There are two ways to model the movement of survivors when they are looking for beeping phones in the network. These are the random walk and the Levy walk. Different random walk models have been discussed in [LAK 11] and [NAR 12]. In this chapter we discuss Levy walk models.

A Levy flight is a mathematical description of a cluster of random short moves connected by infrequent longer ones. Thus, it consists of random walks interspersed by long travels to different regions of the walk space. Mathematically, the sequence of random movements of length L has a probability distribution function (PDF) $f_L(l)$ that obeys the power law; that is,

$$f_L(l) \propto l^{-\gamma}, \qquad 0 < \gamma \leq 3 \qquad [6.1]$$

This PDF is said to have a heavy tail because large values of L are more prevalent than in other distributions such as Poisson and normal distributions. L has an infinite variance over the range of values of γ in equation [6.1]. Typically, each flight is followed by a pause time whose duration also has a power-law distribution.

Levy flights have been applied to a diverse range of fields such as those that describe animal foraging patterns [VIS 96, BAR 05, RAM 04, BRO 35], the distribution of human travel [BRO 06], the stock market [MAN 95], some aspects of earthquake behavior [CAR 06], anomalous diffusion in complex systems [BLU 89, CIR 05, RUB 08], epidemic spreading [JAN 99, DYB 09] and human mobility [RHE 08].

There is a difference between Levy flights and Levy walks. In a Levy flight, the walker visits only the endpoints of a jump, the notion of velocity does not come up and the jumps take very little time. Alternatively, we say that the Levy flight has an infinite velocity. This means that in a Levy flight, the walker is only either at the end of the jump or at the beginning; there is no stop in between the jump. However, in a Levy walk, the walker follows a continuous trajectory from the beginning of the walk to the end and this leads to a finite time being needed to complete the walk at a finite velocity. Thus, the concept of velocity is the major difference between the two; in the case of the Levy flight, the velocity is infinite, and in the case of the Levy walk, the velocity is finite.

One of the advantages of the Levy walk over the random walk is that the probability of a Levy walker returning to a previously visited site is smaller than that in the random walk. Polya proved that a random walker on a one-dimensional or two-dimensional surface returns to the origin with probability of 1. Thus, a random walk represents a mobility model in which the walker tends to hover around its starting point. This means that random walk models have the problem that random walkers tend to return to their starting points very often. Also, the number of sites visited by n random walkers that start at the same point is much larger in the Levy walk than in the random walk. The n Levy walkers diffuse so rapidly that the competition for target sites among themselves is greatly reduced compared to the competition encountered by n random walkers. The latter typically remain close to the origin and hence close to each other.

This dispersive feature of the Levy walk is advantageous in the PDRN network where factors such as the concentration of survivors and the distribution of phones are considered. Irrespective of survivors' locations, each survivor's trajectory is likely to be different, and hence when competing to reach the dispersed phones, a Levy walk model ensures that the probability that two or more of them are heading for the same phone will be greatly reduced. Thus, with respect to the PDRN, the Levy walker will occasionally take long steps and thus is more likely to reach the vicinity of a beeping phone than a random walker.

6.5.1. *Random motion with reward*

In the PDRN, a walker (or survivor) starts out walking aimlessly (or in a random manner) until he/she reaches the vicinity of a *beeping zone*. A beeping zone is an area within which a survivor hears the beeping of a phone. Theoretically, in both the random walk and the Levy walk, the walker must complete the stipulated length of the walk process before stopping to generate the next length of the walk. Since the Levy walker occasionally takes long flights, he is more likely to "leap" over a beeping phone than a random walker. This means that when a Levy walk is used, the value of the next step is likely to fall beyond a point where a phone is located than at that position. Thus, even though the Levy walker takes short steps most of the time, the few longer steps are likely to result in his leaping over of a beeping zone. This means that a walk is not likely to come across a phone. Similarly, if a pure random walk is used, then theoretically the choice of the next direction cannot be influenced by the fact that it might lead to a location that is further from the beeping device than the current location.

To improve the performance of the system, we introduce a *reward-based* (or *rewarded*) random motion that follows one of the two models until the walker is in the vicinity of a beeping phone. Under the reward-based scheme, when the random walker enters a beeping zone, he/she switches to a biased random walk. Specifically, when the walker enters a beeping zone, he/she makes a deliberate attempt to avoid going in directions that lead to a decrease in the volume of the sound of the beeping phone. Thus, in a classical walk the walker is walking aimlessly while in a rewarded walk, he/she is attempting to walk purposely toward a beeping phone. While there are different types of biased random walk (see [LAK 11], for example, for the different types), we assume that the walker uses a symmetric random walk with a slight modification. The modification is that if at the end of the current step the intensity of the sound of the beeping device is less than what it was at the previous location, the walker returns to the previous location and will practice a non-reversing random walk that prevents him from choosing a direction that leads to the previous location. For example, if the direction that led to the decrease in the intensity of the beeping was to the right, when the walker returns to the point from where he/she took that step to the right, he will have only three choices: left, up and down. He/she will follow this strategy until he/she reaches the phone. The right becomes a forbidden direction.

6.5.2. *Levy walk models of PDRN survivor*

We consider the following types of Levy walk [IBE 13]:

1) Classical Levy walk, which involves four parameters, namely the step size, time taken during each step, the waiting time between steps and the direction for the next step. The step size is based on the Levy distribution and a survivor chooses a random location that is uniformly distributed between 0 and 360 degrees from the current position. The step time is based on Levy distribution, waiting time between steps is random and there is no correlation between the four parameters.

2) Symmetric Levy walk, which is essentially a Levy lattice walk. In this walk the survivor moves in one of the four directions: east, west, north or south. After each step, the survivor chooses the next location with equal probability, and moves in one of the four directions based on probability outcomes. The step size and step duration are based on the Levy distribution, and after each step a survivor waits a random time before the next one. The step size, step duration, direction and waiting time are independent parameters.

3) Non-reversing Levy walk [DOM 58], which is similar to the symmetric Levy walk except that here the survivor will not immediately go back to the previous location where he came from. This does not prevent him from going back to a

previously visited site, as in a traditional self-avoiding walk; but he cannot do so immediately after leaving the site, which is why we define the model as a relaxed version of the traditional scheme. In this walk, the step size and step duration are based on the Levy distribution, and after each step a survivor waits for random time. The step size, step duration, direction and waiting time are independent.

4) Alternate Levy walk, which requires a survivor to alternate between x and y directions for each step. Unlike the symmetric Levy walk where directions are chosen independently at each step, this scheme requires a movement in the x-axis to be followed by a movement in the y-axis, and vice versa. The step size in each direction and step duration are based on the Levy distribution, and after each step a survivor waits a random time. The step size, step duration and waiting time are independent.

6.5.3. *Simulation*

In a Levy walk simulation, a survivor is likely to leap over barriers, which may lead to the survivor leaping over the beeping area. To solve this problem, a hybrid model is used that utilizes the Levy walk until a survivor comes in the district of a beeping zone where he switches over to a form of the random walk. The phones that are dropped in disaster area are at discrete locations.

Because a random walk can cause a survivor to bypass a beeping phone if the step length does not terminate at the phone, we define a reward-based random walk as follows. When a survivor comes within the beeping zone of a phone, he switches to a biased random walk such that he is more likely to move in the direction that leads to an increasing loudness of the phone than in a direction with decreasing loudness. After each step, he spends a random waiting time before taking the next step. The waiting time between flights is assumed to have a Levy distribution.

The following are the different walk model considered for simulation:

1) Levy walk-to-Levy walk model (LEVY->LEVY) in which the walker always performs the classical Levy walk even when he gets into a beeping zone.

2) Levy walk-to-reward-based Levy walk (LEVY->RLEVY) in which the walker initially performs the classical Levy walk until he gets into a beeping zone when he switches to the reward-based Levy walk.

3) Symmetric Levy walk-to-symmetric Levy walk (SYM-LEVY->SYM-LEVY) in which that walker always performs the symmetric Levy walk even when he gets into a beeping zone.

4) Alternating Levy walk-to-alternating Levy walk (ALT-LEVY->ALT-LEVY) in which the walker always performs the alternating Levy walk even when he gets into a beeping zone.

5) Levy walk-to-reward-based symmetric random walk (LEVY->RSRW) in which the walker initially performs the classical Levy walk until he gets into a beeping zone when he switches to the reward-based symmetric random walk.

6) Levy walk-to-reward-based alternating random walk (LEVY->RARW) in which the walker initially performs the classical Levy walk until he gets into a beeping zone when he switches to the reward-based alternating random walk.

7) Levy walk-to-reward-based non-reversing random walk (LEVY->RNRSRW) in which the walker initially performs the classical Levy walk until he gets into a beeping zone when he switches to the reward-based relaxed self-avoiding random walk.

8) Alternating Levy walk-to-reward-based alternating Levy walk (ALT-LEVY->RALT-LEVY) in which the walker initially performs the alternating Levy walk until he when he gets into a beeping zone when he switches to the reward-based alternating Levy walk.

9) Symmetric Levy walk-to-reward-based symmetric Levy walk (SYM-LEVY->RSYM-LEVY) in which the walker initially performs the symmetric Levy walk until he when he gets into a beeping zone when he switches to the reward-based symmetric Levy walk.

10) Non-reversing Levy walk-to-non-reversing Levy walk (NR-LEVY->NR-LEVY) in which the walker performs the non-reversing Levy walk even when he gets into a beeping zone.

11) Non-reversing Levy walk-to-reward-based non-reversing Levy walk (NR-LEVY->RNR-LEVY) in which the walker initially performs the relaxed self-avoiding Levy walk until he when he gets into a beeping zone when he switches to the reward-based relaxed self-avoiding Levy walk.

6.5.4. *Simulation result*

Several configurations are possible and described in [LAK11, AKP 13]. For simulation results described in Figures 6.3–6.6, the following are the configurations of survivor and phone distributions:

– both the survivors and the phones are uniformly distributed;

– both the survivors and the phones are normally distributed;

– the survivors are normally distributed and the phones are uniformly distributed;

– the survivors are uniformly distributed and the phones are normally distributed.

The performance measures are the mean first passage time (MFPT) and the percentage of survivors who reach a beeping phone and thus are rescued. MFPT is the mean time it takes for a survivor to reach a beeping phone from the beginning of the search process.

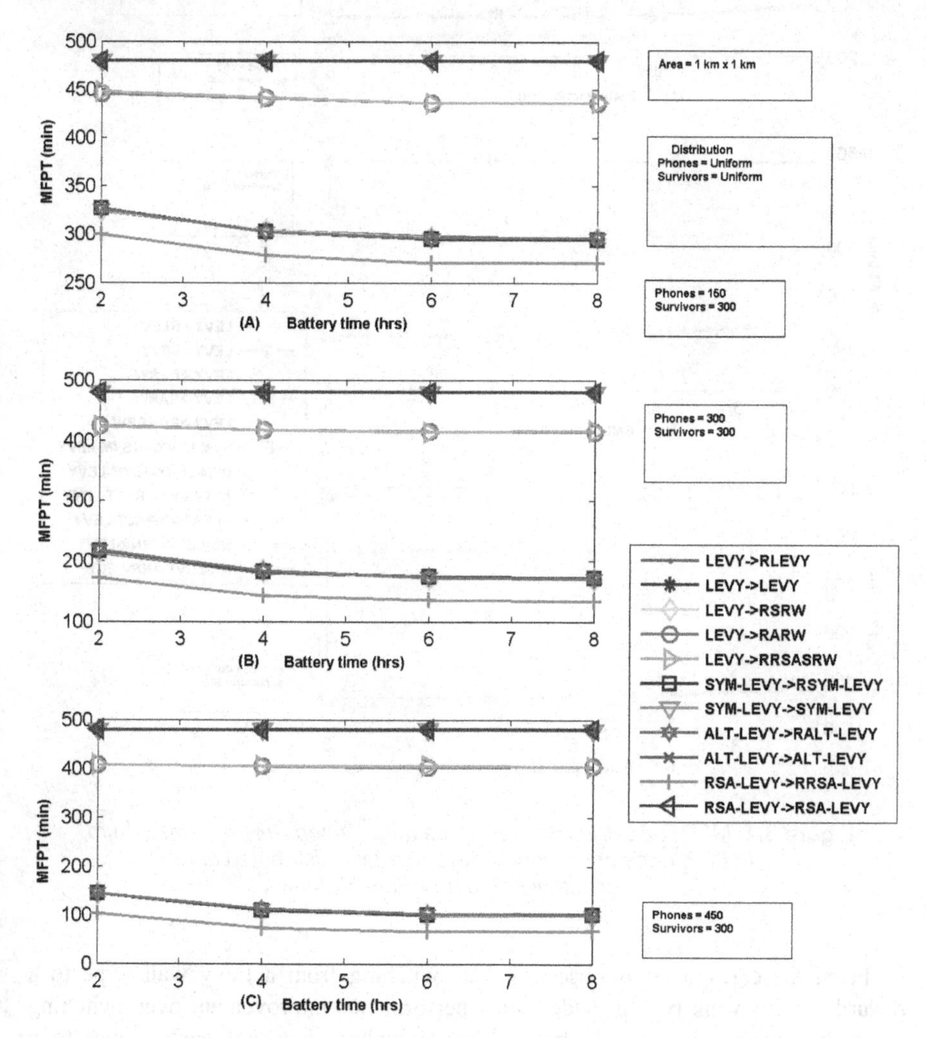

Figure 6.3. *MFPT of Survivors reaching a target when Area = 1 km x 1 km, both Phones and Survivors are uniformly distributed and constant number of Survivors*

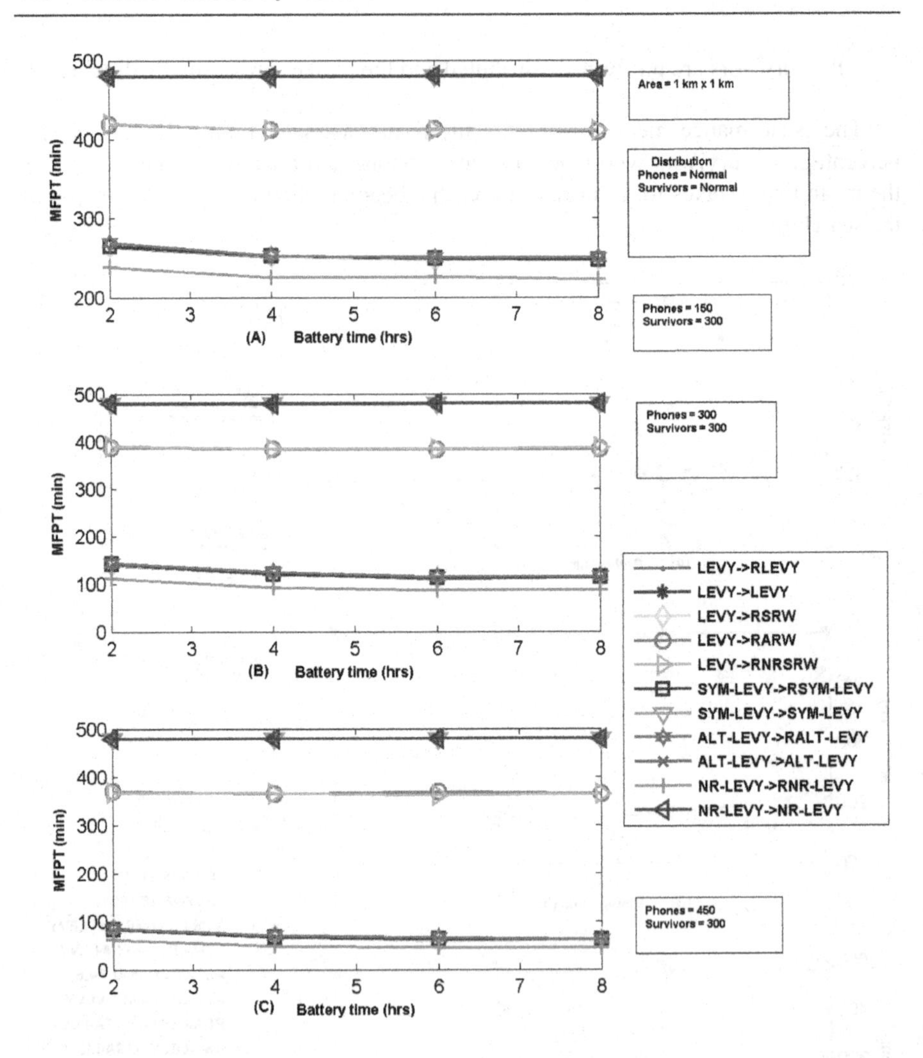

Figure 6.4. *MFPT of Survivors reaching a target when Area = 1 km x 1 km, both Phones and Survivors are normally distributed and constant number of Survivors*

From these results it is observed that switching from a Levy walk type to a rewarded Levy walk type provides some performance improvement over switching to an unrewarded Levy walk type. When a walker does not switch over to a rewarded walk in the beeping zone, he "ignores" the beeping of the phone and thus continues to wander about aimlessly until he/she "accidentally" stumbles across a phone. Thus, ignoring the beeping phone results in a poor performance, as expected. Also, models that start with a modified Levy walk type tend to perform better than those that start with classical Levy walk. It is interesting to observe that the Levy walk-to-rewarded Levy walk combination is among the worst performers, which reinforces the statement that the best performance is obtained by starting initially with a modified Levy walk type, which includes the non-reversing Levy walk, the symmetric Levy walk and the alternating Levy walk.

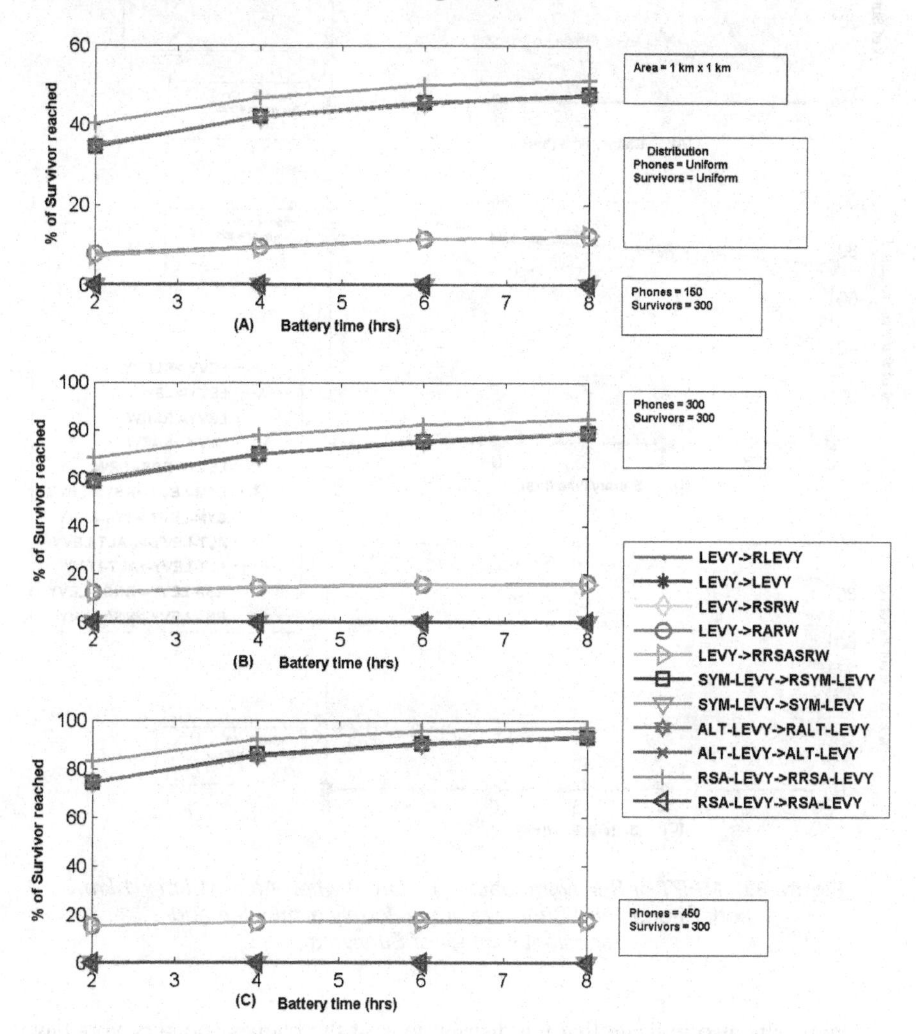

Figure 6.5. *Percentage of Survivors reaching a target when Area = 1 km x 1 km, both Phones and Survivors are uniformly distributed and constant number of Survivors*

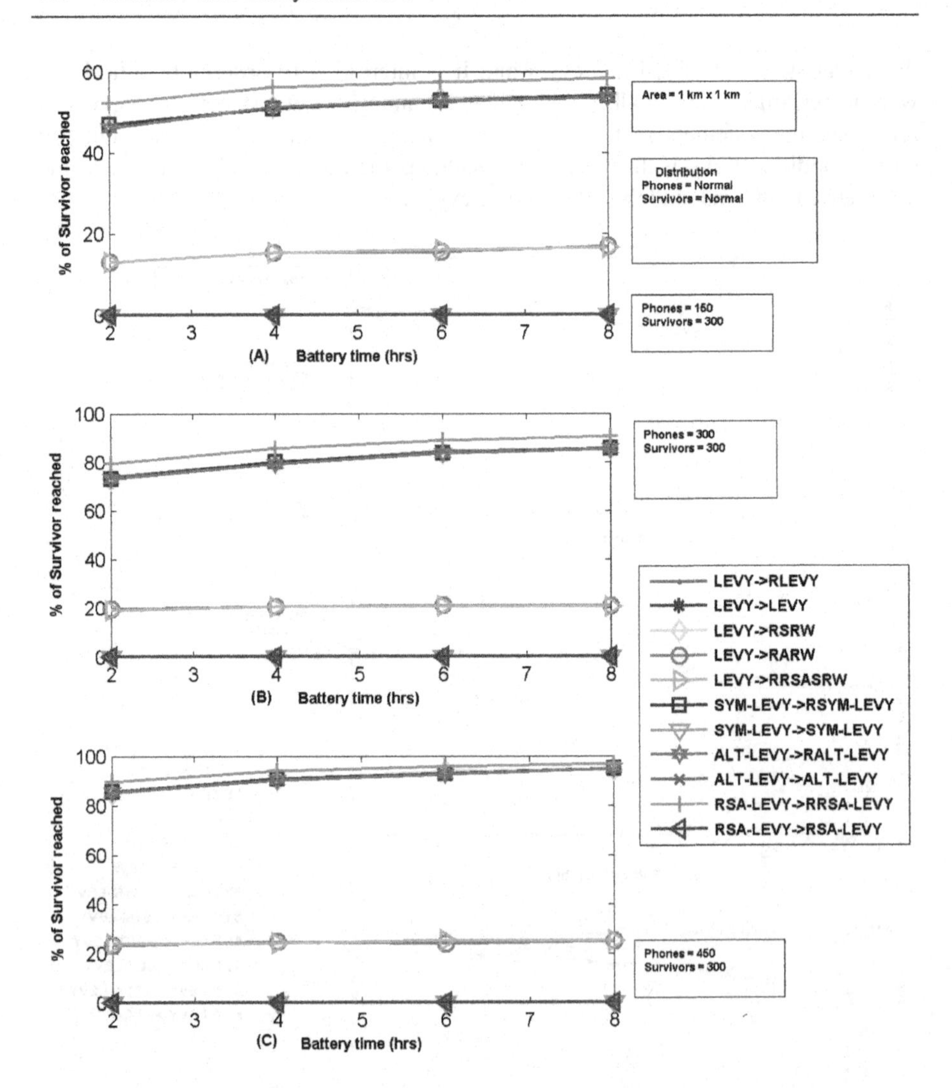

Figure 6.6. *MFPT of Survivors reaching a target when Area =1 km x 1 km, both Phones and Survivors are uniformly distributed and constant number of Survivors*

The results also indicate that the distributions of the phones and survivors have an impact on both MFPT and percentage of survivors rescued. Specifically, the best results are obtained when both the phones and survivors are normally distributed within the disaster area, followed by when both are uniformly distributed. There is not much distinction when one of them is normally distributed and the other is uniformly distributed. This observation is important for rescue operation planners because they need to drop more phones in areas of suspected concentration of victims than in other areas.

It is also observed that for a fixed number of phones, the performance of the system improves as the number of survivors in the area decreases, which is to be expected. Similarly, for a fixed number of survivors, the performance of the system improves as the number of phones dropped in the area increases, which is also to be expected.

Finally, the battery life of the phones has an impact on the performance of the system. Specifically, as the battery life increases, the percentage of rescued survivors increases. However, as the battery life increases, the MFPT first decreases and later remains constant or shows only a modest decrease.

In [LAK 11] and [AKP 13], it is shown that the results obtained for all parameter combinations. To summarize the result, it is observed that the Levy walk performs better and mimics the human mobility model. Furthermore, when battery life increases, more percentage of survivors get rescued. Also, knowledge of distribution of survivors is important when they are distributed normally; both MFPT and percentage of survivors rescued is higher.

6.6. Conclusion and future work

In this chapter we examined various network architectures that are applicable for disaster recovery operations and search-and-rescue operations. As such, each solution has its own strengths that contribute toward disaster relief operations. We also identified the gap in disaster recovery process and described PDRN solution. Using PDRN, different types of random walks are simulated to see how to quickly one can rescue trapped survivors. For future study: (1) the concept of PDRN and survivor modeling is genericthis can be applied to public safety and emergency network solutions such as UAV, LTE D2D wireless networks and need further studies; (2) also, in PDRN, the optimal path to rescue the survivor is research problem and needs to be explored further.

6.7. Bibliography

[AKP 13] AKPOYIBO S., Levy walk models of disaster recovery networks, Masters' Thesis, Department of Electrical and Computer Engineering, University of Massachusetts Lowell, May 2013.

[AKP 14] AKPOYIBO S.E., LAKSHMI-NARAYANAN R.G., IBE O.C., "Levy walk models of survivor movement in disaster area networks with barriers," *International Journal of Computer Networks & Communications*, vol. 6, pp. 1–18, 2014.

[AKY 02] AKYILDIZ I.F., SU W., SANKARASUBRAMANIAM Y. *et al.*, "Wireless sensor networks: a survey", *Computer Networks*, vol. 38, pp. 393–422, March 2002.

[AKY 05] AKYILDIZ I.F., WANGB X., WANGB W., "Wireless mesh networks: a survey", *Computer Networks*, vol. 47, pp. 445–487, 2005.

[ASC 08] ASCHENBRUCK N., GERHARDS-PADILLA E., MARTINI P., "A survey on mobility models for performance analysis in tactical mobile networks", *Journal of Telecommunication and Information Technology*, vol. 2, pp. 54–61, 2008.

[ASC 09] ASCHENBRUCK N., GERHARDS-PADILLA E., MARTINI P., "Modeling mobility in disaster area scenarios", *Performance Evaluation*, vol. 66, pp. 773–790, 2009.

[BAR 05] BARTUMEUS F., LUZ M.G.E., VISWANATHAN M.G.E. *et al.*, "Animal search strategies: a quantitative random-walk analysis", *Ecology*, vol. 86, pp. 3078–3087, 2005.

[BER 07] BERIOLI B., COURVILLE N., WERNER M., "Emergency communications over satellite: the WISECOM approach", *Proceedings of the 16th IST Mobile and Wireless Communications Summit*, pp. 1–5, July 2007.

[BHA 04] BHARGAVA B., WU X., LU Y. *et al.*, "Integrating heterogeneous wireless technologies: a cellular aided mobile ad hoc network (CAMA)", *ACM Journal on Special Topics in Mobile Networking and Applications (MONET)*, vol. 9, pp. 393–408, 2004.

[BLU 89] BLUMEN A., ZUMOFEN G., KLAFTER J., "Transport aspects in anomalous diffusion: Levy walks", *Physical Review A*, vol. 40, pp. 3964–3973, 1989.

[BLU 13] BLUMENTHAL S.H., "Medium Earth orbit Ka band satellite communications system", *Military Communications Conference, IEEE MILCOM*, pp. 273–277, November 2013.

[BRO 06] BROCKMANN D., HUFNAGEL L., GEISEL T., "The scaling laws of human travel", *Nature*, vol. 439, pp. 462–465, 2006.

[BRO 07] BROWN C.T., LIEBOVITCH L.S., GLENDON R., "Levy fights due to Dobe Juhoansi foraging patterns", *Human Ecology*, vol. 35, pp. 129–138, 2007.

[CAS 03] CASPER J., MURPHY R.R., "Human–robot interactions during the robot-assisted urban search and rescue response at the World Trade Center", *IEEE Transactions on Systems, Man, and Cybernetics – Part B: Cybernetics*, vol. 33, no. 3, pp. 367–385, June 2003.

[CAY 07] CAYIRCI E., COPLU T., "SENDROM: sensor networks for disaster relief operations management", *Wireless Networks*, vol. 13, pp. 409–423, 2007.

[CHE 08] CHECHKIN A.V., METZLER R., KLAFTER J. *et al.*, "Introduction to the theory of Levy Flights," in KLAGES R., RADONS G., SOKOLOV I.M. (eds), *Anomalous Transport: Foundations and Applications*, John Wiley & Sons, 2008.

[CIR 05] CIPRIANI P., DENISOV S., POLITI A., "From anomalous energy diffusion to levy walks and heat conductivity in one-dimensional systems", *Physical Review Letters*, vol. 94, pp. 244–301, 2005.

[COR 06] CORRAL A., "Universal earthquake-occurrence jumps, correlations with time and anomalous diffusion", *Physical Review Letters*, vol. 97, pp. 178501-1–178501-4, 2006.

[DOM 58] DOMB C., FISHER M.E., "On random walks with restricted reversals," *Proceedings of the Cambridge Philosophical Society*, vol. 54, pp. 48–59, 1958.

[DYB 09] DYBIEC B., KLECZKOWSKI A., GILLIGAN C.A., "Modeling control of epidemics spreading by long-range interactions", *Journal of the Royal Society Interface*, vol. 6, pp. 941–950, 2009.

[FUJ 05] FUJIWARAA T., WATANABEB T., "An ad-hoc networking scheme in cellular networks for disaster communications," *Ad Hoc Networks*, vol. 3, pp. 607–620, September 2005.

[GPP 13] 3GPP TR 22.803 V12.2.0 Feasibility study for Proximity Services (ProSe), Release 12, available at: www.3gpp.org/DynaReport/22803.htm, 2013.

[IBE 13] IBE O.C., *Elements of Random Walk and Diffusion Processes*, John Wiley & Sons, 2013.

[IRL 14] THE INTERNET RADIO LINKING PROJECT, available at http://www.irlp.net, 2014.

[JAN 99] JANSSEN H.K., OERDING K., VAN F.W. *et al.*, "Levy-flight spreading of epidemic processes leading to percolating clusters", *The European Physical Journal B*, vol. 7, pp. 137–145, 1999.

[JAR 97] JARVIS R., "An autonomous heavy duty outdoor robotic tracked vehicle", *Proceedings of the 1997 IEEE Intelligent Robots and Systems*, vol. 1, pp. 352–359, 1997.

[JIN 07] JINGUO L., YUECHAO W., BIN L. *et al.*, "Current research, key performances and future development of search and rescue robots", *Frontiers of Mechanical Engineering in China*, vol. 2, no. 4, pp. 404–416, 2007.

[KAN 07] KANCHANA K., TUNPAN A., AWAL M.A. *et al.*, A multimedia communication system for collaborative emergency response operation in disaster-affected areas, Technical Report, Internet Education and Research Lab, Asian Institute of Technology, Thailand, April 2007.

[KO 09] KO A., LAU H.Y.K., "Robot assisted emergency search and rescue system with a wireless sensor network", *International Journal of Advanced Science and Technology*, vol. 3, pp. 69–78, February 2009.

[LAK 11] LAKSHMI-NARAYANAN R.G., An architecture for disaster recovery and search and rescue wireless networks, PhD Thesis, Department of Electrical and Computer Engineering, University of Massachusetts Lowell, June 2011.

[LUO 07] LUO H., MENG X., RAMJEE R. *et al.*, "The design and evaluation of unified cellular and ad hoc networks", *IEEE Transactions on Mobile Computing*, vol. 6, pp. 1060–1074, September 2007.

[MAC 08] MAC R., KEANE M.T., COLEMAN G., "A wireless sensor network application requirements taxonomy", *Proceedings of the Second International Conference on Sensor Technologies and Applications, Sensor Communication*, pp. 209–216, 2008.

[MAJ 09] MAJID S., KAZI A., "Cluster-based communications system for immediate post-disaster scenario", *Journal of Communications*, vol. 4, no. 5, pp. 307–319, 2009.

[MAN 95] MANTEGNA R.N., STANLEY H.E., "Scaling behavior of an economic index", *Nature*, vol. 376, pp. 46–49, 1995.

[NAR 12] NARAYANAN R.G.L., IBE O.C., "A joint network for disaster recovery and search and rescue operations," *Computer Networks*, vol. 56, pp. 3347–3373, 2012.

[PRA 06] PRATT K., MURPHY R.R., STOVER S. *et al.*, "Requirements for semi-autonomous flight in miniature UAV's for structure inspection", *AUVSI Unmanned Systems North America Symposium*, pp. 1–13, 2006.

[RAM 04] RAMOS-FERNANDEZ G., MATEOS J.L., MIRAMONTES O. *et al.*, "Levy walk patterns in the foraging movements of spider monkeys", *Behavioral Ecology and Sociobiology*, vol. 55, pp. 223–230, 2004.

[RHE 08] RHEE I., SHIN M., LEE K., *et al.*, "On the Levy-walk nature of human mobility", *Proceedings of the IEEE INFOCOM*, pp. 924–932, 2008.

[RUB 08] RUBNER O., HEUER A., "From elementary steps to structural relaxation: A continuous-time random-walk analysis of a super-cooled liquid", *Physical Review E*, vol. 78, p. 011504, 2008.

[SAH 07] SAHA S., MATSUMOTO M., "A framework for disaster management system and WSN protocol for rescue operation", *Proceedings of the IEEE*, pp. 1–4, October 2007.

[SIL 04] SILVER W., *Ham Radio for Dummies*, John Wiley & Sons, 2004.

[SUZ 06] SUZUKI H., KANEKO Y., MASE K. *et al.*, "An ad hoc network in the sky, SKYMESH, for large-scale disaster recovery", *Proceedings of the Vehicular Technology Conference*, pp. 1–5, September 2006.

[VIS 96] VISWANATHAN G.M., AFANASYEV V., BULDYREV S.V. *et al.*, "Levy flight search patterns of wandering albatrosses", *Nature*, vol. 381, pp. 413–415, 1996.

[WAT 12] WATTS A.C., AMBROSIA V.G, HINKLEY E.A., "Unmanned aircraft systems in remote sensing and scientific research: classification and considerations of use", *Remote Sensing*, pp. 1671–1692, 2012.

[WU 01] WU H., QIAO C., DE S. *et al.*, "iCAR: integrated cellular and ad hoc relaying systems", *IEEE Journal on Selected areas in Communications*, vol. 19, pp. 2105–2115, October 2001.

[YAN 02] YANCO H.A., DRURY J.L., "A taxonomy for human-robot interaction", AAAI Technical Report FS-02-03, 2002.

[YIC 08] YICK J., MUKHERJEE B., GHOSAL D., "Wireless sensor network survey", *Computer Networks*, vol. 52, pp. 2292–2330, 2008.

7

The Evolution of Intelligent Transport System (ITS) Applications and Technologies for Law Enforcement and Public Safety

7.1. Introduction

Intelligent transport systems (ITS) refers to the application of information and communication technologies (ICT) for road transportation to support and enhance various applications. The main concept is to integrate computers, electronics, wireless communications, sensors and navigation systems like global navigation

---

Chapter written by Gary STERI and Gianmarco BALDINI.

satellite systems (GNSSs) to enable the collection and distribution of information to and from the vehicles. This information can be analyzed and used for various purposes including the capability to improve the efficiency of road transportation. In some cases, the term "ITS" has been associated with vehicle telematics. Even though we can say that these two terms are different, they are related to similar concepts and technologies.

There are already various applications based on the ITS concept, like intelligent fleet management system, smart parking, restricted access and so on. A survey on ITS applications has already been presented in various references, including [SHE 11]. Another important survey on vehicular *ad hoc* networks is [HAR 08], which was one of the first papers to identify the main opportunities (e.g. new safety applications) and related challenges (e.g. security and privacy) for the deployment of VANETs. In addition, government agencies are actively working on this area through initiatives and actions plan to support the development and deployment of ITS applications and projects. In Europe, the ITS action plan [CEC 10] was instrumental in defining the framework for the development of ITS applications in Europe. In the USA, a strategy plan for ITS has been defined in [UST 14].

In this chapter, we focus on the ITS applications and technologies which can support law enforcers' capabilities. More specifically, we investigate how wireless technologies for vehicle-to-vehicle (V2V) and vehicle-to-infrastructure (V2I) communications, like dedicated short range communications (DSRC), can be used to support law enforcers. Another important technology that has gained widespread use in recent time in commercial road transportation is localization based on GNSS, which can be used to track the position of vehicles. Location information can be an important asset in many ITS applications, like fleet management and transport of dangerous goods, even if it is less feasible for collision avoidance because of the limited accuracy of GNSS (e.g. 5–10 m) [BOU 08].

A set of ITS applications where law enforcers have an important role are the so-called "regulated applications" (created to address a specific regulation), like the regulated application of the digital tachograph (DT), which is one of the case studies described in this chapter. While many regulated applications in different parts of the world have been in existence for some time (e.g. the tachograph concept goes back to 1950), only the recent development of ITS technologies has created the possibility to enhance the design of current and future regulated applications to take advantage of the new technologies.

In general, the role of ITS and related technologies (e.g. VANET) is to augment the capabilities of public safety to address their specific tasks (e.g. emergency crisis, law enforcement), which are described in detail in section 7.2. In addition to law enforcers' tasks (including the regulated applications mentioned above), road transportation is an important element of public safety operations to ensure fast moving of officers and delivery of materials in case of emergency or to provide relief to victims. Here we have to make a distinction between the design and deployment of ITS technologies for commercial purposes and for public safety as the operational requirements of public safety organizations could be quite different. Public safety organizations require performance, resilience and security for all the services and equipments they use for different reasons. Firefighters need to operate in difficult environments (e.g. forest fires), emergency medical services require priority in traffic to ensure the fast transport of a patient to the medical facilities. Law enforcers require security and integrity in the transmitted data, which may contain sensitive information. More details in the ITS requirements of public safety organizations are provided in section 7.2.

In many cases, the potential of ITS for public safety organizations is still not fulfilled for various reasons, including the specific requirements in terms of performance, resilience and security of public safety organizations. The initial step is to identify these requirements as described in section 7.2. An overview of the current regulatory and standardization activities in ITS in relation to public safety activities is presented in section 7.3, which also includes a section on wireless communication technologies. An essential element for the successful deployment of any new technology is a clear understanding of the impact on existing organization and processes. In comparison to the consumer (e.g. cars) or commercial sectors (e.g. trucks), the public safety sector has specific organizational structures and processes to ensure an effective response to emergency crisis, infringements and so on. How the new ITS technologies impact the existing processes is an aspect which is often overlooked. In addition, the specific phases of the lifecycle of ITS telematics equipment (e.g. bootstrap, enrolment, periodic checks and end-of-life) must be taken into consideration. All these aspects are taken up in section 7.4.

Finally, we show how all the elements identified and described in the previous sections can be applied to real-world case studies (sections 7.5 and 7.6). In one of them, the authors are directly involved: the revision of the DT regulated application in Europe. Here we discuss how the introduction of wireless communication, support for GNSS and the evolution of the security framework are related to ITS and organizational constraints.

7.2. Public safety organizations and requirements

The term “public safety” includes a wide range of organizations dealing with various activities including law enforcement, medical emergency, fire-fighting, and others.

From [SAF 06] and [BAL 14], we can identify the following main functions for public safety organizations:

1) *Law enforcement.* Prevent, investigate, apprehend or detain any individual, which is suspected or convicted of offenses against the criminal law.

2) *Emergency medical and health services (EMHS).* Supportive care of sick and injured citizens in the field and the ability to transfer people in a safe and controlled environment, when possible.

3) *Border security.* The police or a specialized guard usually performs border security, in order to control transported goods, check people crossing the border, and so on.

4) *Environment protection.* Protect the natural environment.

5) *Fire-fighting.* Extinguish hazardous fires.

6) *Search and rescue.* Locate lost or missing persons and transport them to a place of safety.

7) *Emergency crisis.* Deal with emergency crisis both human made (e.g. a terrorist attack) or natural (e.g. flooding).

While the number of potential scenarios, where public safety organizations can operate can be quite wide, in this chapter we will focus only on the scenarios related to road transportation.

Historically, law enforcement in road transportation has been mainly focused on the first two functions (*law enforcement* and *emergency medical and health services*) with the objective to support road safety. Public safety organizations involved in other functions use road transportation as a means to support their activities. For example, fire-fighters use the road to reach the place where they must extinguish a fire. There are also many regulated applications, where public safety organizations are directly involved to ensure the conformance to regulation. One of these regulations is the DT in Europe, which records the driving time, breaks and rest periods undertaken by professional drivers for all vehicles above 3.5 tons, and involves millions of trucks in the European roads.

The main technological requirements for public safety in road transportation can be classified in the following categories:

– *Support for security.* Public safety officers usually have to access highly sensitive data (e.g. criminal records) to support their operational activities. In some cases, these data can be accessed by law enforcers in any situation. In other cases, access to data is allowed only under special circumstances. For example, access to criminal records should be possible in any situation, while access to medical records by emergency medical officers (e.g. doctors) may only be possible in the case of an accident (with previous consent of the victim based on specific regulations). Security includes at least the following functions: authentication, confidentiality, non-repudiation, authorization and integrity. The deployment of some or all these security functions is depending on the specific ITS use case. For example, messages from a vehicle may only need to be signed for integrity to be sure that they have not been changed, while in other cases the messages should be encrypted for confidentiality as well. The provision of the security functions depends on the specific type of application, the type of data and the role of the public safety organization. For instance, law enforcers (e.g. highway police) "may request integrity and confidentiality", while emergency healthcare may only request integrity on specific messages. In general, we should envisage that public safety organizations have a higher level of access to data and services in comparison to a generic individual (e.g. driver). This means that systems deployed on the road should support (at least) the authentication and authorization of members of the specific public safety organization.

– *Support for wireless connectivity.* Public safety organizations use wireless communication services for a variety of services. A survey on the use of wireless communications in public safety organizations is available in [BAL 14]. The reason for wireless rather than fixed is just because public safety officers operating on the road must have mobile connectivity. Currently, the main uses of wireless communications are to:

- provide voice communication among law enforcers. For example, it may include the coordination to address a specific crisis or action (e.g. blocking a fugitive) or for generic coordination between the headquarters and the law enforcers during patrolling;

- provide limited data connectivity for the exchange of messages. For example, data related to criminal records;

- notify a car accident or crisis through the eCall initiative (see [ECI 14]).

The relevance of wireless communication for emergency vehicles (EVs) has also been highlighted in [RAT 13]. While wireless communications have limited data connectivity capabilities at this moment, new wireless communication standards have been developed to support wideband and broadband data connectivity, like Long Term Evolution [FER 13]. We define wideband connectivity within a range of 100 Kbits and 1 Mbits, while broadband connectivity is above 1 Mbits. Different wireless communication systems could support mobility in different ways. For example some short range communication systems could not support mobility beyond certain speeds. The advantage of long term evolution (LTE) is that high performances can be maintained at speeds even up to 120 km/h, envisaging the employment for different kinds of vehicles and mobility rates.

With wideband and broadband data connectivity, new applications for public safety organizations can be envisaged like:

– remote healthcare support, where doctors in the field (e.g. during a road accident) can transmit images or medical data of a patient to a medical centre or a hospital for an improved diagnosis and medical support;

– collection and transmission of images from the field (i.e. on the road) to the headquarters for an easier identification of a vehicle non-compliant with road regulations.

Other wireless communication standards specifically used for wireless communications, like the ones defined by the Connected Vehicles program (see [HAR 14]) or the Car2Car consortium (see [C2C 14]), can also be employed to support public safety organizations. These specific standards and technologies will be discussed in section 7.3.1.

– *Support for applications.* Public safety officers are using and will use various software applications to support their operational tasks. These applications include analytical tools for the collection and analysis of data, image processing tools to analyze images and extract useful information (e.g. plate number) and so on. An overview of the current and future ITS applications for law enforcement in the road transportation sector is provided in section 7.3.3.

7.3. The evolution of intelligent transport system (ITS) applications and the role of public safety organizations

7.3.1. *Regulatory and standardization activities*

The following regulatory and standardization activities are identified and described in this section:

– Car2Car consortium;

– European Commission ITS applications action plan;

– US Connected Vehicles program;

– Australian gatekeeper.

Figure 7.1 shows how the different activities relate to the main requirement areas of public safety organization in ITS. Details on these relationships are described in the following sections.

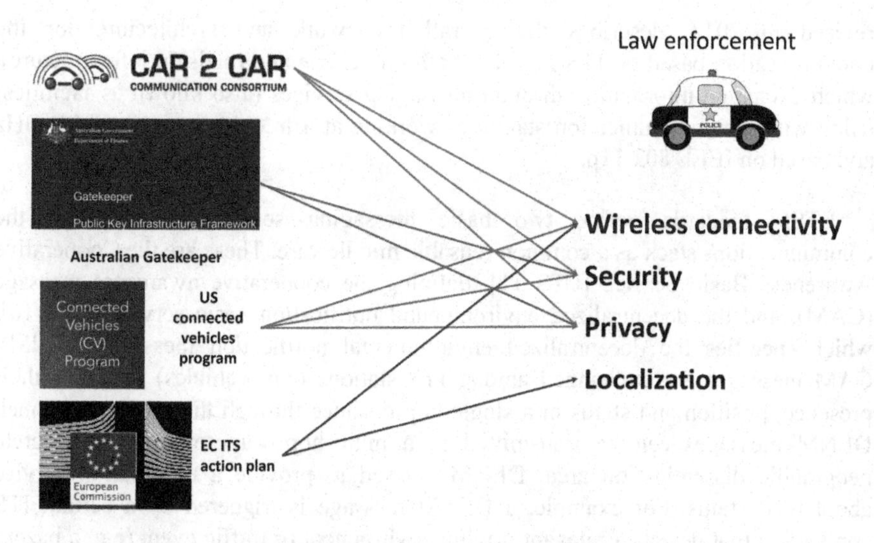

Figure 7.1. *Different regulatory and standardization activities in the world and how they relate to the main needs of public safety organization in road transportation*

7.3.1.1. *Car2Car consortium*

The Car2Car consortium [C2C 14] is a non-profit, industry driven organization initiated by European vehicle manufacturers and supported by equipment suppliers, research organizations and other partners. Car2Car is specifically focused on cooperative intelligent transport systems (C-ITS), where the word cooperative is related to the exchange of information through wireless communication technology among telematic devices, which are both installed in vehicles and in a roadside infrastructure. The term vehicle-to-vehicle (V2V) communication is used for communication among telematic devices in vehicles, and the term vehicle-to-infrastructure (V2I) communication is used for communication among telematic devices and the infrastructure. Therefore, the Car2Car consortium is focused on the specific set of ITS applications, which can exploit a direct exchange of information in the field: collision avoidance, notification of hazards and so on.

The Car2Car consortium is working closely with standardization bodies like ETSI for the definition of the technical specification and related standards. Various standards describing the architecture of Car2Car have been drafted by ETSI TC ITS. The full list and description of the technical specifications is out of scope of this technical report, so we identify only the main ones, which are related to use cases where law enforcers are participating. [ETI 10], originally drafted in 2010 and revised in 2014, describes the overall framework and architecture for the communication based on ITS stations (both on vehicle and roadside infrastructure), which exchange information through messaging services (also known as facilities) using wireless communication standards working at a frequency range of 5.9 GHz and based on IEEE 802.11p.

ETSI standards define two basic messaging services included in the communications stack as a common reusable middleware. These are the Cooperative Awareness Basic Service [ETS 11], defining the cooperative awareness message (CAM), and the decentralized environmental notification basic service [ETS 10], which specifies the decentralized environmental notification message (DENM). CAM messages are exchanged among ITS stations (e.g. vehicles) to notify their presence, position and status in a single hop distance through the wireless channel. DENM messages can be transmitted in a multi-hop way to cover a concrete geographic dissemination area. DENM is used to provide a notification service about road status. For example, a DENM message is triggered by a certain ITS application that detects a relevant driving environment or traffic event (e.g. a hazard on the road).

While the Car2Car architecture has been mostly defined for the commercial market, the role of law enforcers has been identified clearly in the various technical deliverables and associated standards. One of the main methods to support law enforcement is the use of the *emergency vehicle to car communication* function, which gives priority to police or EVs in situations affecting the safety of people, where every minute is crucial. At the moment, road users are obliged to make way for ambulances or police cars when they hear the alarm of the siren. However, the alarm does not give precise information on the direction of the emergency vehicle (EV) or its position unless it is completed with a visual detection by the driver. In the Car2Car framework, EVs notify their presence and location through CAM messages where the profile *emergency vehicle* is used. In more general terms, this profile can be used to support the operational needs of law enforcers in Car2Car enabled applications.

To summarize, the Car2Car consortium has put in place mechanisms to support law enforcers through a specific profile (i.e. emergency vehicle). Even if the set of applications supported by Car2Car is smaller than the overall set of ITS applications, the deployment of Car2Car technologies could improve the operational capabilities of law enforcers in the road in the future.

7.3.1.2. *European Commission ITS applications action plan*

The directive 2010/40/EU [CEC 10] was adopted on 7 July 2010 to accelerate the deployment of ITS applications across Europe. It aims to establish interoperable and seamless ITS services while leaving Member States the freedom to decide which systems to invest in. The directive drives the European Commission ITS action plan, which focuses on:

1) the provision of EU-wide multimodal travel information services;

2) the provision of EU-wide real-time traffic information services;

3) data and procedures for the provision, where possible, of road safety related minimum universal traffic information free of charge to users;

4) the harmonized provision for an interoperable EU-wide eCall;

5) the provision of information services for safe and secure parking places for trucks and commercial vehicles.

In 2014, the European Commission published a report [CEC 14] to describe the progress of the ITS action plan. The report confirmed the successful completion of the technical work for the eCall and the timely progress of the application for information services for safe and secure parking places.

The support for law enforcers is an important objective of the ITS action plan. Both the eCall application and the secure and safe parking places are directly related to the work of law enforcers. The eCall application is an emergency call generated either manually by vehicle occupants or automatically via activation of in-vehicle sensors. When activated, the in-vehicle eCall system will establish a voice connection directly with the relevant public safety answering point (PSAP), this being either a public or a private eCall centre operating under the regulation and/or authorization of a public body. At the same time, a minimum set of incident data (MDS) will be sent to the eCall operator receiving the voice call.

The secure and safe parking places application has the objective to support European truck drivers on their need to have access to information on the availability of secure parking places for their trucks as well as to be able to make use of a timely pre- and on-trip reservation service. In this case, law enforcers can benefit from the improved level of safety of a secure parking of commercial trucks.

These two applications are part of the Action Area 3: road safety and security in the ITS action plan and further future activities are foreseen to support improved levels of security in road transportation in Europe.

7.3.1.3. *US connected vehicles program*

The U.S. department of transportation (DOT) connected vehicles research program is a multimodal initiative, which has the objective to provide a framework for safe and interoperable wireless communications for V2V and V2I like the Car2Car initiative, but within a different regulatory and standardization context.

One of the most recent documents describing the US connected vehicles program is [HAR 14], which include an analysis of the US DOT's research findings in several key areas including technical feasibility, privacy and security, and preliminary estimates on costs and safety benefits. The connected vehicles program uses the DSRC standard based on SAE J2735 [HED 08].

The standard specifies three sets of messages [SAE 10]:

– Basic safety message (BSM) is the primary message set proposed to send data among vehicles and between vehicles and the infrastructure. While the BSM is mainly developed for safety applications, the data in the message may also be used by other connected vehicle applications.

– Roadside alert message (RSA) is the message used in the various traveler information applications, specifically in the emergency vehicle alert message to inform mobile users of nearby emergency operations. In this use, a message from either the public safety on-board units (OBUs) or the infrastructure is broadcasted to the mobile user to provide information on traffic conditions.

– Probe vehicle message (PVM) is used by multiple applications. Vehicles gather data on road and traffic conditions at intervals. This data are combined into messages and transmitted to the roadside units (RSUs), and from there to information service providers. This probe data provide information on traffic, weather, and road surface conditions.

The connected vehicles program supports law enforcers and their applications through specific values in the BSM and the RSA messages. As described in [SAE 10], the combination of BSM and RSA messages can be used to get priority in traffic for EVs, provide information to travelers about hazards conditions or current emergency crisis (e.g. car accident), provide information about weather conditions and so on. From this point of view, SAE J2735 offers a wide set of message types, which could support a variety of law enforcement applications.

7.3.1.4. *Australian gatekeeper*

A description of the support of ITS to law enforcers is presented in [YOU 07]. The objective of the report is to investigate the potential of ITS technologies to enhance the effectiveness and efficiency of police enforcement activities in Australia. The report identifies a number of technologies and makes a comparison between the approaches based on existing technologies and the new potential approaches based on new technologies. For example, wireless communication technology like DSRC could be used to check the speed of vehicles on the road. The report covers various technologies used in ITS including camera, analysis of images, analytics tools and so on.

Australia has not developed a specific program for ITS like the US connected vehicles program and it is probably going to use the technologies and standards defined in other geo-political areas. An area where the Australian Government has adopted a specific strategy is related to security, where the Australian Gatekeeper framework is an essential pillar. The Gatekeeper is a whole-of-government suite of policies, standards and procedures that govern the use of public key infrastructure (PKI) for the authentication of agencies and their clients.

The Australian Gatekeeper PKI Framework governs the use of PKI in government, where adopted, for the authentication of internal and external clients (organizations, individuals and other entities).

More specifically, the national telematics PKI framework, administered by the Transportation Certification Australia (TCA), incorporates physical and electronic security requirements. The security requirements relating to encryption and non-repudiation use the accredited service provider symantec (eSign), accredited under the Gatekeeper Framework. The National Telematics Framework has a number of applications deployed that are all used for enforcement, like the intelligent access program (IAP) and intelligent speed compliance (ISC). The implementation of the electronic work diary is also under the National Telematics Framework, and it will use the same security requirements.

The security requirements are important because they are needed to present evidence in court that can demonstrate the necessary chain of evidence from data generation and collection to its processing and storage.

Currently a number of law enforcers' organizations in Australia use the Gatekeeper. The list is available on [TCA 15] and it includes the Department of Defence, Medicare Australia and the Australian Tax office among others.

7.3.2. *Wireless communication technologies used by law enforcers*

This section provides an overview of the wireless communication technologies used by law enforcers in the current and the future ITS context.

7.3.2.1. *CEN-dedicated short range communication*

The DSRC developed by CEN (European Committee for Standardization) is a standard for one-way or two-way wireless communication over limited distances (around 30 m), specifically designed for ITS applications. It operates in the ISM band (5.8 GHz), with a free lower band at 5.795–5.805 GHz and an upper band at 5.805–5.815 GHz, providing two channels each.

Typical CEN DSRC components are road side units (RSUs) and OBUs with transceivers and transponders, characterized by a low cost of production (especially for on-board equipment that does not contain an active radio transmitter) and a very good pointing accuracy. Main applications of DSRC are for toll payment, but the possibilities can be extended to collision avoidance, intersection warning, signaling and so on.

The three-layered architecture of CEN DSRC (physical, data link and application) is defined by the following standards:

– EN 12795:2003 – Road transport and traffic telematics – DSRC – DSRC data link layer: medium access and logical link control;

– EN 12834:2003 – Road transport and traffic telematics – DSRC – DSRC application layer;

– EN 12253:2004 – Road transport and traffic telematics – DSRC – physical layer using microwave at 5.8 GHz.

In addition to them, we can mention the standard EN 13372:2004 (DSRC profiles for RTTT applications), which defines interactive DSRC systems based on two-way and broadcast communication services.

Actually, the term "DSRC" identifies a series of different standards for short-range communication. Besides the European CEN definitions, in North America we find an implementation that combines the standard American Society for Testing and Materials (ASTM) E2213-03 for physical and medium access layers with the standard IEEE 1455-1999 for the application layer. In Japan, DSRC is defined by the Association of Radio Industries and Businesses (ARIB) STD-T75 standard for physical, data link and application layers (the latter is specifically defined in ARIB STD-T88).

7.3.2.2. *Vehicle-to-vehicle communications at 5.9 GHz*

Communications access for land mobiles (CALM) M5, standard ISO 21215, is a mid-range V2V communication technology, operating at 5.9 GHz. It extends the

IEEE 802.11 WLAN protocols suite to meet the requirements of ITS by incorporating the wireless access in vehicular environments (WAVE) protocol stack.

WAVE addresses all the layers of the ISO/OSI model except session and presentation (see Figure 7.2). Physical layer and medium access control (MAC) sublayer are defined by the IEEE 802.11p standard, which is actually based on 802.11a. Moreover, we find IEEE 1609.4, which defines an extension of the MAC sublayer management entity (MLME) both for data and management planes, in order to provide channel coordination and multi-channel synchronization. The result is the implementation of *ad hoc* communication over 10 MHz control and service channels (CCHs and SCHs), using OFDM up to 27 Mbps.

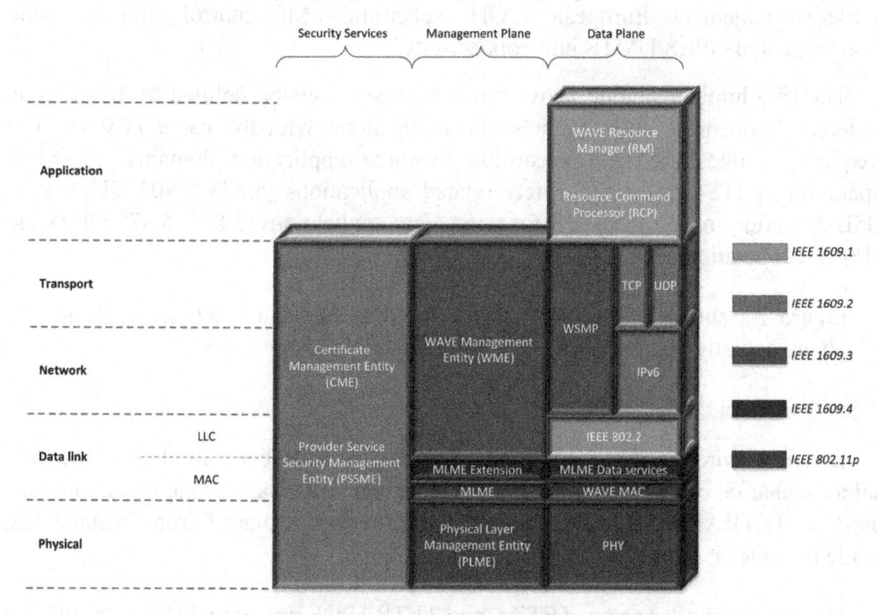

Figure 7.2. *WAVE protocol stack*

Logical link control (LLC) sublayer (based on IEEE 802.2), network and transport layers are defined by IEEE 1609.3, which supports IPv6, TCP and UDP for what concerns data services besides the WAVE short message protocol (WMSP). For management services, the standard defines the WAVE management entity (WME), responsible, for example, for service requests and channel access assignment, data delivery and IPv6 configuration. Finally, at the application layer, IEEE 1609.1 allows remote applications, known as resource manager applications (RMAs), to communicate with a resource command processor (RCP) residing in an

OBU through a resource manager (RM) residing in an RSU. Next to the previous standards, the IEEE 1609.2 implements security services for processes and applications running in the different layers. These services are divided into *security processing services* (to secure communications) and *security management services*, with certificates and keys management provided by the certificate management entity (CME) and the provider service security management entity (PSSME).

WAVE alone is optimized for single-radio technology on US channel plan and does not include interconnectivity with external networks like GSM or UMTS. The goal of CALM M5 is indeed the extension of WAVE in order to include cooperation with regional DSRC systems (integration with CEN DSRC OBUs [ISO 10]), domain border management, European 5 GHz spectrum, EMC control, multiple radio interfaces, and GPRS/UMTS interconnectivity.

The ISO implementation above has a European version defined by ETSI using different frequencies and channels allocation along with the name ITS-G5. The frequency ranges are split according to three application domains: ITS-G5A operation of ITS-G5 for ITS safety related applications (5.875–5.905 GHz), ITS-G5B operation in European ITS for non-safety applications (5.855–5.875 GHz) and ITS-G5C operation of ITS (5.470–5.725 GHz).

Figure 7.2 shows the collocation of IEEE 802.11p and 1609 family standards, which are actually at the basis of ITS communication technologies.

7.3.2.3. *Narrowband dedicated cellular networks*

Different wireless communication technologies have been employed in public safety scenarios by organizations from Europe and America. Among them, the most used are TETRA, TETRAPOL and APCO25: the first two are European standards, while the latter is from USA.

Despite the similar name, TETRA and TETRAPOL are quite different from each other. TETRA (acronym for terrestrial trunked radio) is a standard develop by ETSI from 1995 and dedicated to professional mobile radio (PMR) users like public safety organizations, military and civil services, government, emergency and so on. Since users and application scenarios can be very different, TETRA is designed for high coverage scalability, ranging from local to wide (national) areas, offering the possibility for individual, group and broadcast calls. The maximum data rate in release 1 is 28.8 Kbits, while release 2 features up to 473 Kbits/s and the TETRA enhanced data service (TEDS). Figure 7.3 shows the stack of TETRA protocols for the mobile station (MS), which supports circuit mode calls and short data. Data services, provided also with support to the Internet Protocol (IP) packet data, can be activated by internal (running in the MS) or external (terminals connected to the MS) data applications. The architecture is split into control and user planes (C-plane and U-plane respectively), where the first one is responsible for signaling information for control messages and data, while the second one implements circuit mode calls and data. The protocol stack for the base station (BS) is similar for physical, data link and network layers.

Originally developed for the needs of the French National Gendarmerie, TETRAPOL is a proprietary solution from EADS (now Airbus Defense & Space) based on FDMA technology. It offers calls and messaging services and, like TETRA, the possibility to interface with PSTN/PABX networks. However, TETRAPOL has not been accepted by ETSI as a standard and its interoperability with TETRA is now addressed by the FP7 Inter System Interoperability for Tetra-TetraPol Networks project (ISITEP) [ISI 15].

APCO-25 (Project 25 or P25), is a suite of standards from the US Telecommunications Industry Association (TIA), employed mainly in North America for public safety, security and commercial applications. It supports both digital voice and data transmission at a low bitrate (9.6 and 12.5 kbps for Phase I and II respectively), but it is also compatible with analog FM radios. Phase II of APCO-25 uses both FDMA and TDMA with 12.5 KHz bandwidth (the same as TETRAPOL, while TETRA operates with 25 KHz), but there is no compatibility with the European standards described before.

Besides these three standards that cover the most of the PMR scenarios, we can find other solutions usually employed in more specific contexts, like satellite networks in out of coverage areas, or avionic and marine communications in VHF band. The digital mobile radio (DMR) ETSI standard represents the digital evolution of analog PMR networks and, like P25 Phase II, uses TDMA mode with 12.5 KHz bandwidth.

In addition to these technologies, there are the commercial cellular systems like GSM/GPRS and UMTS, which perhaps offers broader band and more digital services. If these differences are not very strong in the first three generations of cellular network, with the 4G LTE system we can really see an important step forward. The next section will describe this technology and its potentialities exploitable also for public safety and law enforcement applications.

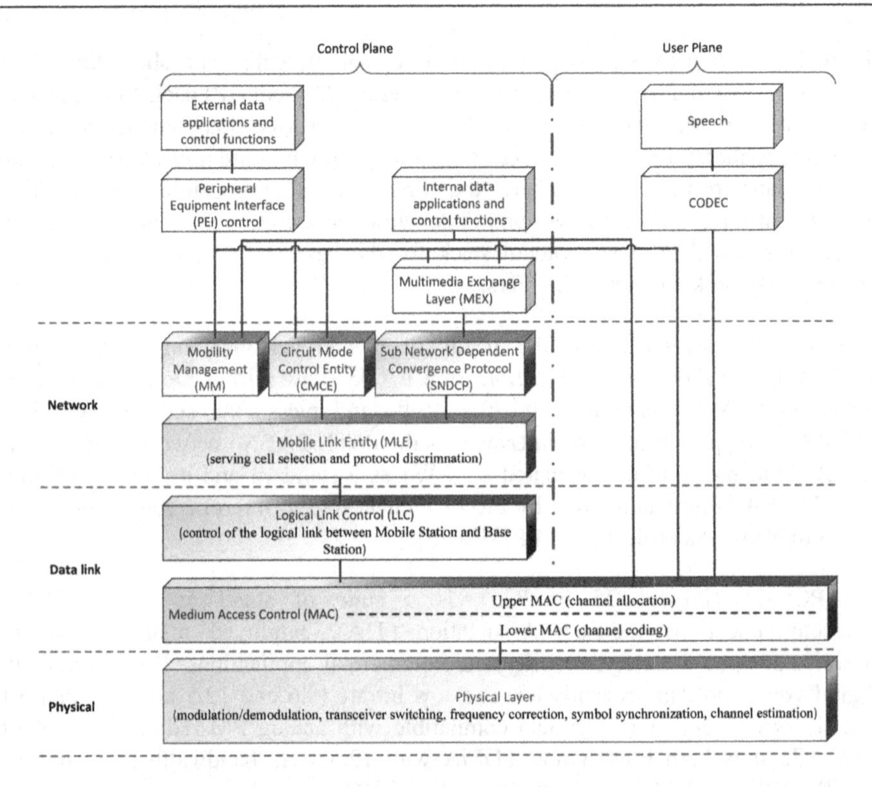

Figure 7.3. *TETRA mobile station protocol stack*

7.3.2.4. *New generation broadband cellular networks: LTE*

The fourth generation of cellular networks, represented by the UMTS LTE technology, provides mobile broadband connectivity at an order of magnitude of tens or hundreds megabits per second, according to the last LTE-advanced releases. The high bit rate offers the possibility of exchanging a huge amount of digital contents and of implementing many dedicated services for law enforcers [FER 13]. The protocol stack for an LTE user equipment (UE) is shown in Figure 7.4. For the evolved NodeB (eNB), the architecture is quite similar from the physical layer up to the first half of the network layer (RRC protocol), while the mobility management entity (MME) presents a different stack.

One of the most interesting features for law enforcement applications is probably the device-to-device (D2D) communication system, introduced by the third generation partnership project (3GPP) release 12 of LTE. With D2D communication, UEs are able to establish direct links even without the presence of a base station such as an eNB, filling the gap between standard cellular approach and *ad hoc* networks.

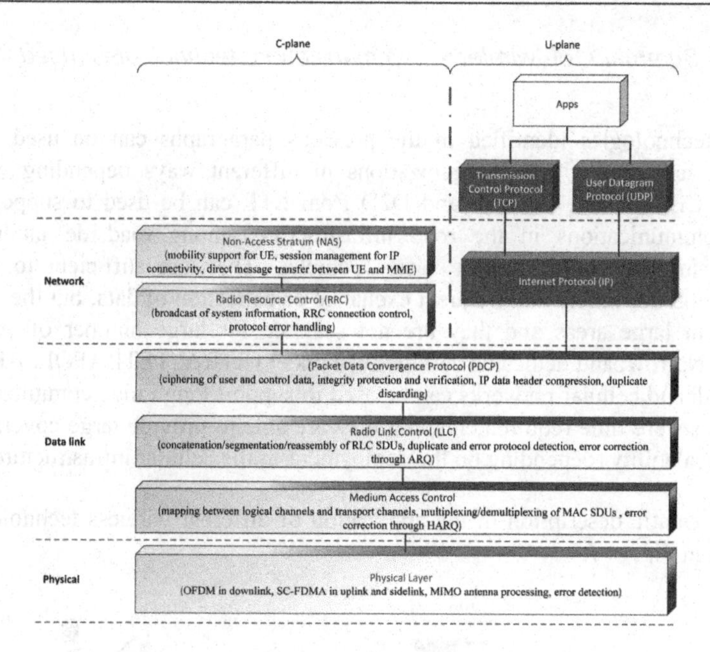

Figure 7.4. *LTE user equipment protocol stack*

The use of LTE D2D for law enforcement and ITS applications in general, requires a high level of security both for protection and for resilience of direct communications. In this sense, standardization activities are currently not very evolved and some possible solutions have been proposed in literature. For example, the authors in [GOR 14] propose a secure D2D protocol in which references to encryption keys, to be used during the communication, are already exchanged with the beacon frame. The reason behind this is the minimization of packet exchanges during the establishment of a connection: indeed, as the authors show in the paper, the interference in the channel generated by UEs acting as beaconing devices is a key component in the probability of establishing a direct link. At the same time, since the security is embedded in the beacon, this probability is also affected by the possibility of sharing encryption keys, which, for resiliency reasons, are deployed according to a random distribution scheme, exactly like for *ad hoc* networks.

It is clear how the introduction of D2D in cellular networks really represents a bridge and, probably in the near feature, a strong bond with *ad hoc* networking. The fusion of these two approaches can also be present in the technologies employed: in [ASA 13] for example, LTE and Wi-Fi Direct are used for D2D communication in a cellular infrastructure. Clearly, this kind of integration can introduce problems related to packet delays and different transmission rates which, however, can be resolved thanks to the performance of the LTE system.

7.3.2.5. *Summary of wireless communication technologies used by law enforcers*

The technologies identified in the previous paragraphs can be used by law enforcers and public safety organizations in different ways depending on their features. CEN-DSRC, Car2Car and D2D from LTE can be used to support short range communications in the road infrastructure among roadside facilities or facilities installed in the public safety vehicles. They are sufficient to support requirements for timely and frequent exchange or collection of data, but they cannot be used in large areas and they are not scalable for large number of vehicular systems. Narrowband dedicated cellular networks (TETRA, TETRAPOL, APCO25) and broadband cellular networks can be used to support long range communications with less severe time requirements, but they are able to provide large coverage and support scalability (depending on the deployment of the cellular infrastructure).

A schematic description of the application of different wireless technologies is provided in Figure 7.5.

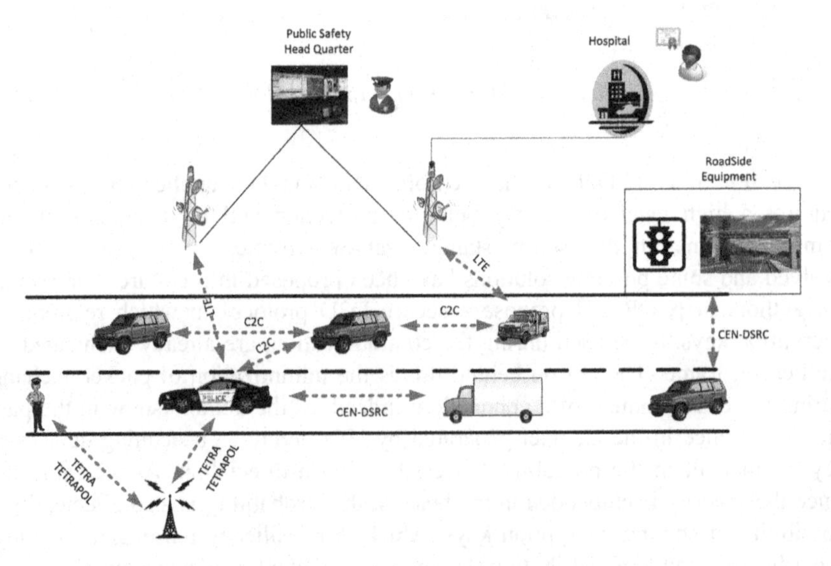

Figure 7.5. *Application of different wireless technologies in the road transportation domain*

7.3.3. *List of current and future ITS applications*

In this section, we identify the current and future ITS applications where the role of law enforcers is significant, where law enforcers will be directly involved or where law enforcers can benefit.

The main sources used in this section to identify such list of applications are the United States connected vehicle reference implementation architecture (CVRIA) [CVR 14], the European Telecommunication Standards Institute [ETS 09], the ITS action plan from the European Commission and various others from literature references.

The connected vehicle reference implementation architecture (CVRIA) is a large set of research activities centered around a vehicle or a mobile device that is equipped with communications and processing allowing those equipped platforms to be aware of their location, their status, to communicate with each other and with the surrounding infrastructure (from [CVR 14]). The CVRIA has defined an exhaustive list of connected vehicles applications, which are divided into four main areas:

1) *Environmental*. This area includes all the applications related to sustainable travel for energy efficiency and decrease of pollution and the applications related to weather conditions. Most of these applications are not related to law enforcement.

2) *Mobility*. This area includes all the applications, which are used to improve the mobility of the road transportation infrastructure including the operational capabilities of the law enforcers. In this area, CVRIA has defined a specific subcategory for public safety.

3) *Safety*. This area includes the safety applications focused on the safety of the drivers and the pedestrians. This area includes the specific vehicle to vehicle safety applications, the vehicle to infrastructure safety and the transit safety. Some of the applications in this area are related to law enforcement.

4) *Support*. This area includes the support functions and management of the infrastructure including the system monitoring and credentials management.

ETSI has defined a list of ITS applications in [ETS 09]. The list of applications or uses cases defined in the technical report is:

1) *Cooperative road-safety*, which includes all the use cases/applications related to collision avoidance, hazards warning, vehicle warning and so on.

2) *Traffic efficiency*, which includes all the use cases/applications related to regulatory speed limits, electronic tolling and so on. This area also includes many regulated applications.

3) *Others*, which include various use cases related to infotainment, vehicle maintenance, fleet management and so on.

The ITS action plan identifies various applications both regulated and commercial, which are already defined or they will be defined in the future. In the literature, various surveys have also identified similar applications, but we will refer mostly to [CVR 14] and [ETS 09] unless specified.

From the previous sources, we can identify the following list of applications, which are related to law enforcement activities identified in section 7.2. The list of applications is presented in Table 7.1, which identifies the wireless communications technologies (presented in section 7.3.2) that can be used in each application.

Application Area	Relation to law enforcement activities	Source	Wireless Communications
Border Security (including border management systems or container checking). This application area includes the law enforcers activities related to trade, and the enforcement of customs and immigration laws at the borders. For example, the identification and checking of trucks.	Border security	[CVR 14]	Narrowband dedicated cellular networks (TETRA, TETRAPOL, APCO25)
Commercial Vehicle Roadside operations to improve the efficiency and safety of roads and conformance to laws. This area includes the applications related to the monitoring of the work hours of commercial drivers (e.g. the case study of the Digital Tachograph) and other tax, and safety regulations.	Law Enforcement	[CVR 14]	CEN-DSRC, C2C, LTE

Automatic Crash Notification. This includes the capabilities to support a vehicle, which has been involved in a crash or other distress situation. For example, he European eCall.	Emergency Medical and Health Services (EMHS)	[CVR 14] eCall	LTE
Emergency Communications and Evacuation in case of an emergency crisis. This application area includes all the applications related to response to emergency crisis including traffic prioritization for emergency vehicles, broadcast messages, hazards notification and so on.	Emergency crisis, Environment protection, Fire-fighting	[CVR 14]	C2C
Alerts for Emergency crisis zone. After a crisis happened, zone alerts are used to control or divert traffic. This application area includes all the activities needed after the response phase of an emergency crisis. This includes the notification of risk areas, diverting traffic, hazards notification and traffic management.	Emergency crisis	[CVR 14]	C2C, LTE
Emergency Vehicle Preemption to give priority to public safety	Law Enforcement, Emergency Medical and Health Services (EMHS),	[CVR 14] and	C2C

vehicles. This is the specific application of granting priority to the emergency vehicles for any reason. The case for response for emergency crisis has been addressed in the previous application area.	Emergency crisis	[ETS 09]	
Emergency Vehicle Alert to notify that a public safety vehicle is approaching the scene. This application is the notification to the other vehicles and their drivers that an emergency vehicle is approaching. This application can be paired to the Emergency Vehicle Preemption.	Law Enforcement, Emergency Medical and Health Services (EMHS), Emergency crisis	[CVR 14] and [ETS 09]	C2C
Collision Avoidance. This application includes all the use cases to avoid collisions among vehicles, including forward collision warning, situational awareness and lane change warning. This is one of the most important application from a market point of view, because it can prevent car accidents on a large scale.	Law Enforcement	[CVR 14] and [ETS 09]	C2C
Warning about hazards in an area. This application area is related to the	Law Enforcement	[CVR 14]	C2C, LTE

notification to the drivers of hazards in the road infrastructure or environmental conditions.			
Geo-fencing and zone warning. This application area is related to the capability to forbid vehicle access or provide warnings when entering a specific area. This application area includes the exclusion of transport of dangerous goods (e.g. chemicals) from specific areas (e.g. urban areas).	Environment protection, Law Enforcement	[CVR 14] and [ETS 09]	C2C, LTE

Table 7.1. *Applications and law enforcement activities*

7.4. Operational and deployment aspects

In this section, we discuss the organizational and deployment aspects for ITS and law enforcements. We envisage that the new technologies will have not only a technological impact but they will also change the existing processes and organizational structures. We can identify the following areas where ITS applications will have an impact:

1) Law enforcers must be equipped with new telematics devices and learn how to use them. An obvious impact is the definition of new training processes to acquire the knowledge to use the new technologies.

2) Specific capabilities like traffic prioritization and profiles like EV described above will require specific authentication procedures to ensure that only law enforcers can have access to these capabilities. In addition, law enforcers will have access to a new set of data and services, which also require authentication. Therefore, new procedures for the authentication of personnel must be put in place for the use of telematic equipment. An example of authentication for law enforcers using smartcards is described in the case study of the DT in section 7.5.

3) Law enforcers will use many different wireless communication systems in the future ITS evolution. Communication systems deployed for law enforcers in mobile

vehicles, should be able to support the different wireless communication standards. This will require multi-standard and multi-frequency RF equipment.

4) Security is an important requirement for law enforcers and public safety organizations. In addition to the authentication requirement described before, sensitive data should be protected by confidentiality. Even more important is the integrity of the data transmitted over wireless channels. This requires the design and deployment of security solutions and the deployment of cryptographic material (like certificates), which must be installed and maintained in the OBUs of the vehicles and the roadside equipment. As described in the case of the DT (section 7.5), similar security mechanisms for the generic citizen and the law enforcers could be designed and deployed to support an economy of scale, but the access profiles for law enforcers should be clearly defined.

7.5. Case study of the digital tachograph application

7.5.1. *Background*

Currently, the digital tachograph (DT) is a device that records the driving time, breaks and rest periods undertaken by professional drivers. The purpose of the DT is to store relevant information to ensure that the data recorded are retrievable, intelligible when printed out, reliable, and that they provide an indisputable record of the work done by both the driver over the last few days and by the vehicle over a period of several months. Since May 2006 all new vehicles over 3.5 tons, except for those exempt, have had to be supplied with a DT in Europe.

The original tachographs were mechanical and they progressed to the early electronic units around 1985. With the advent of digital electronics, following a proposal from the European Commission and opinions expressed by the Economic and Social Committee, a decision was taken to replace the current system with a digital version. The DT was clearly less vulnerable to illegal acts by users to distort the data and allowed for easier and better control of driver's hours by operators and the enforcement authorities.

The current version of the DT is based on the European Commission regulation (EC) no. 1360/2002 of June 2002 (which has been amended various times and more recently in (EC) Regulation no. 3821/85 in 2009). It is composed of a recorder unit in the truck, called vehicle unit (VU), connected to a motion sensor (MS), which is used to calculate the travelled distance in relation to time, as described in Figure 7.6. Since the DT is based on a regulation, the application is implemented in the same way and on the same technical specifications across the member states of Europe.

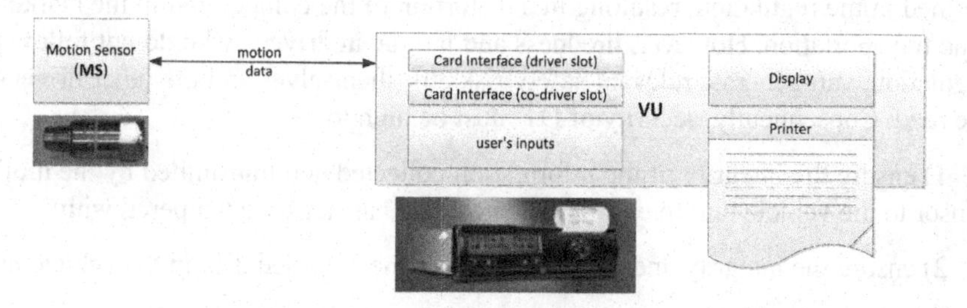

Figure 7.6. *Structure of the digital tachograph*

The MS is usually screwed into the vehicle's gear box. The detection of the movement is provided by a Hall-effect position sensor located in the motion sensor itself. The Hall-effect position sensor is a non-contact device that detects the small electromagnetic variations of the movement of the gear and converts them into electrical signals, which are processed by an electronic board in the MS to calculate the driving time. The electronic board sends real-time pulses through a cable to the VU, which processes the pulses to calculate the driving time. The latter is provided through various means (printout, remote download, access to smartcards that are inserted into the VU during an inspection by law enforcers) to law enforcers to ensure that the driver is conformant to the regulation. A smartcard, similar in size to a credit card and provided with a microchip, can be of the four different types each with different levels of access to the services and data of the DT, as shown in the table below.

Driver card	Used by drivers to activate the digital tachograph application by inserting it in the VU during working times
Company card	Used by companies to protect company-related data in the digital tachographs. It allows a company to download the information from the DT in order to carry out checks on driver's hours as required by the legislation and to maintain the required records for examination by Transport Officers
Workshop card	Used by workshops personnel to perform the periodic calibration of the DT to ensure that the motion sensor correctly records the driving time through the detection of the movement in the gear
Law enforcer or *control card*	Used by law enforcers to display or download the data recorded by the VU

Table 7.2. *List of smartcards for the digital tachograph*

The infringement of the regulation can result in economic advantages for the infringers. For example, the driving time can be extended in comparison to what defined in the regulation, resulting in a distortion of the competition in the market of road transportation. However, tiredness and fatigue in drivers, who do not follow the regulation, can generate relevant safety risks for themselves and the other drivers in the road. Consequently, security of DT must be high to:

1) ensure the integrity of the information collected and transmitted by the motion sensor to the vehicle unit to ensure that such data has not been tampered with;

2) ensure the integrity and confidentiality of the recorded data in the vehicle unit;

3) support access control mechanism to ensure that only an entity authenticated through the smartcard can have authorized access to data and services on the basis of his/her role;

4) ensure the confidentiality of the data related to the privacy of the driver.

Security solutions in the DT are described in detail in its technical specifications [DIT 09]. The main elements are depicted in Figure 7.6.

The DT system is based on a PKI (detailed in [DIT 09]), where the Joint Research Centre of the European Commission is responsible for the main root CA (called ERCA) that provides support for two cryptographic mechanisms:

1) symmetric key cryptography for pairing of the VU with the MS;

2) public key cryptography for VU and Cards interaction and Data Download function.

7.5.2. *New version of the digital tachograph*

New regulations published in February 2014 (Reg. 165/2014) [REG 14] introduce a completely renewed DT for the 2016–2018 horizon. This new recording equipment will offer functionalities that are more advanced:

1) connection of tachographs to GNSS services to record the location data for the starting and ending points of the daily working period and every three hours of cumulative time;

2) wireless communication based on CEN-DSRC standard to support law enforcers in targeted roadside checks. This function, also called *remote communication* or *remote control*, it is used to collect early detection data necessary to detect commercial vehicles infringing the regulation;

3) harmonized interface allowing the data recorded or produced by the tachograph to be used for ITS applications.

In this section, we will focus on the design and implementation of the second functionality.

With the current version of the DT, in order to access the data stored in the recording unit and analyze them, law enforcers are obliged to stop every vehicle, limiting the number of checks and detection of possible infringements to the regulation. The introduction of targeted road side checks allows for a substantial increase in the percentage of commercial vehicles that can be checked.

Operational requirements for remote controls have driven to the adoption of the CEN-DSRC communication systems at 5.8 GHz as the preferred option in comparison to the other communication systems identified in section 7.3.2. This does not exclude the use of other wireless communication technologies to complement the remote communication. For example, cellular wireless communication can be used by the law enforcers located at the roadside equipment to tell other law enforcers to stop a vehicle.

The CEN-DSRC standard was defined for the Electronic Tolling application, which has an operational scenario similar to the remote communication of the DT, where in a limited time, the reader must capture the essential data from a moving vehicle. In the DT, the amount of data to be transmitted is less than 300 bytes in less than 100 milliseconds. This means that all the information essential for the law enforcer must be contained in this limited set of data. The regulation also requires that the data shall be accessed only by law enforcers and workshops for testing purposes, requiring for encryption and signature of the data and thus reducing the amount of information that can be transmitted.

The two use cases shown in Figure 7.7 are representatives of future scenarios for ITS, where law enforcers can check the conformance to regulation through similar wireless readers hosted in their vehicles or along the road. In alternative to the DT, the reader could also be installed on gantries over the road and then connected to a remote center. In this case, the aspects of targeted control are less evident and the reaction time by law enforcers could be less effective because it will be managed from the remote control center. In addition, data will be transmitted over networks and services provided by private companies, increasing the risk of confidentiality vulnerabilities for early detection information.

Use case 1

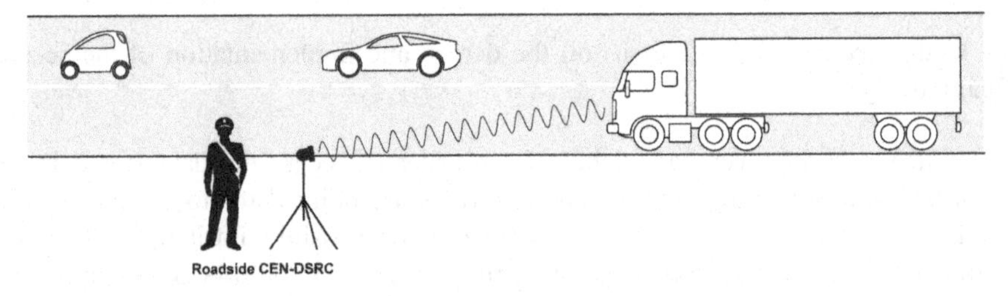

Use case 2

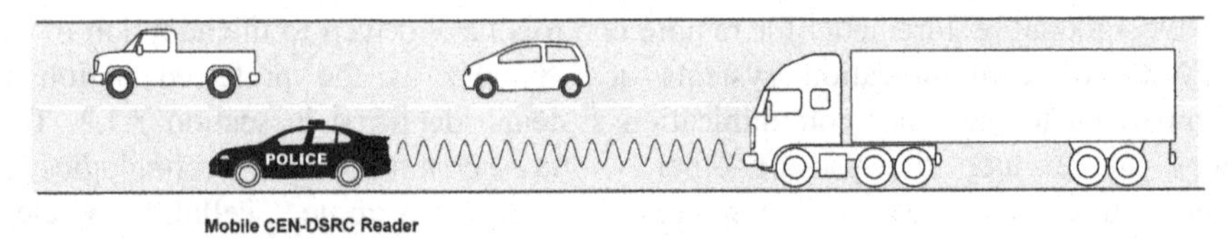

Figure 7.7. *Use cases for remote communication*

The data to be transmitted are defined in the regulation 165/2014 [REG 14] for the following fields:

1) the latest security breach attempt;

2) the longest power supply interruption;

3) sensor fault;

4) motion data error;

5) vehicle motion conflict;

6) driving without a valid licensce;

7) card insertion while driving;

8) time adjustment data;

9) calibration data including the dates of the two latest calibrations;

10) vehicle registration number;

11) speed recorded by the tachograph.

Most of the data are related to potential infringements of the regulation due to tampering with the sensors used to record the position, tampering with the recording unit, speed infringements and so on. A specific field may not be enough to give a clear indication to the law enforcer about an infringement of the regulation, but the combination of the fields can increase the evidence and support the decision process of the law enforcers to stop the vehicle.

The law enforcers which detect a non-compliance can notify a colleague using wireless communications like the TETRA system described in section 7.3.2.3 to request the stop of the non-compliant vehicle.

Once a vehicle is stopped, the law enforcers can have access to the wider set of data stored in the recording unit. The data for early detection is correlated with the wider set of data stored in the recording unit. In fact, the early detection data is calculated periodically from the wider set of data. Both the early detection data and the wider set of data are expressed in ASN1.0 format and they will be defined in the new set of technical specification, which are currently expressed as an annex and a set of technical appendices.

Access to data is performed through smartcards. Law enforcers use their smartcards to access the data received in the reader from the CEN-DSRC module in the commercial vehicle. The smartcard is used to authenticate the law enforcer and grant him/her access to the early detection data. The same smartcard is used by the law enforcer in the subsequent phase when the truck is stopped to access the wider set of data stored in the recording device.

From the case study of the new version of the DT presented in this chapter provides indications on how law enforcers may use ITS systems and technologies in the future:

1) existing applications, which are now not "connected", could be extended to support wireless communication as described the DT. The result is to considerably improve the effectiveness of law enforcers, who can check a much larger number of commercial vehicles;

2) the deployment of wireless communication technologies in ITS must be coherent with the operational processes and security constraints which are defined for law enforcers organizations. For example, transmitted or accessed data could be sensitive and should be kept confidential. Existing authentication systems (e.g. smartcards in the DT) must be integrated with the new wireless communication systems.

The amount of transmitted data does not need to be extended in size. A limited set of data is often what is needed to support law enforcement activities. This is also due to the fact that the time to complete a transaction is short, because of the speed of the vehicles involved in the transaction. Cellular networks with their capability to support broadband communications will have a role in ITS applications, but in many cases, without completely replacing narrowband communications.

7.6. Case study on traffic prioritization for emergency vehicles

Traffic prioritization for EV is an application where EVs can request a specific priority to other vehicles for traffic management. This application is enabled by both Car2Car and Connected Vehicles program through specific fields in the exchanged messages (e.g. CAM).

In [CVR 14], among the other traffic signals applications in the Mobility area (see section 7.3.3.), we find the emergency vehicle preemption (EVP), whose purpose is to give traffic priority to emergency first responder vehicles in order to facilitate their movement through intersections by clearing the queues. This is basically done with the modification of traffic signal timing which in traditional systems relies on line-of-sight preemption using infrared emitters or strobe lights. In the EVP system from CVRIA, we have the interaction of different objects that exchange messages using DSRC. The main objects involved and their roles are described below:

– the emergency vehicle *On-Board Equipment* (OBE) is the processing and communication unit that resides in an emergency vehicle and supports the connected vehicle applications related to public safety operations. It receives inputs from the emergency personnel and sends a *local signal preemption* request to the *Roadside Equipment* (RSE) placed in proximity of the intersection;

– the RSE implements the *RSE Intersection Management* communicating, via DSRC, with approaching vehicles and ITS infrastructure in order to coordinate the traffic signal operations in the intersection. It forwards the preemption request to the *ITS Roadway Equipment*, which will provide the intersection status;

– the ITS Roadway Equipment, which also contains the *Roadway Signal Control* (signal controllers, heads and detectors), consists of all the sensing devices distributed along the roadway, such as traffic detectors, environmental sensors, traffic signals, highway advisory radios, dynamic message signs, CCTV cameras and video image processing systems, grade crossing warning systems, etc. With the data coming from this equipment, it is able to give safe supporting information to the driver of the EV, and it is responsible for notifying the *Traffic Management Center* (TMC) about requests for signal prioritization, preemption or pedestrian calls;

– the Traffic Management Center is the monitoring entity for traffic and road network including urban and suburban areas, highways, freeways, etc. It receives and sends information to the ITS Roadway Equipment and RSE in order to perform monitoring and traffic management operations according to local and environmental conditions. It implements the *TMC Signal Control* capability, which allows traffic managers to control signalized intersections sending *signal control commands* to the ITS Roadway Equipment. In case of emergency, it receives from the *Emergency Management Center* an *emergency traffic control request*, a special request to obtain the preemption of the current traffic control strategy for one or more intersections so as to facilitate EV route and guarantee safe emergency operations;

– the Emergency Management Center is the entity responsible for all the security and public safety actions related to an emergency operation, including the management of sensor and surveillance equipment for roadway infrastructure (bridges, tunnels, etc.) and public transportation system. After having asked the TMC for a traffic control request, it receives an *emergency traffic control information* containing the status of the strategy implemented upon the request. It provides the *Emergency Routing* functionality to the EV OBE, suggesting the best route according to the traffic information received by the TMC and the *emergency vehicle tracking data* (current location from the EV).

The EVP system therefore is not only a simple modification of traffic signal timing but it is based on the coordination of multiple entities offering full control of the traffic on wide areas. Figure 7.8 shows an explanatory scenario with some of the entities involved.

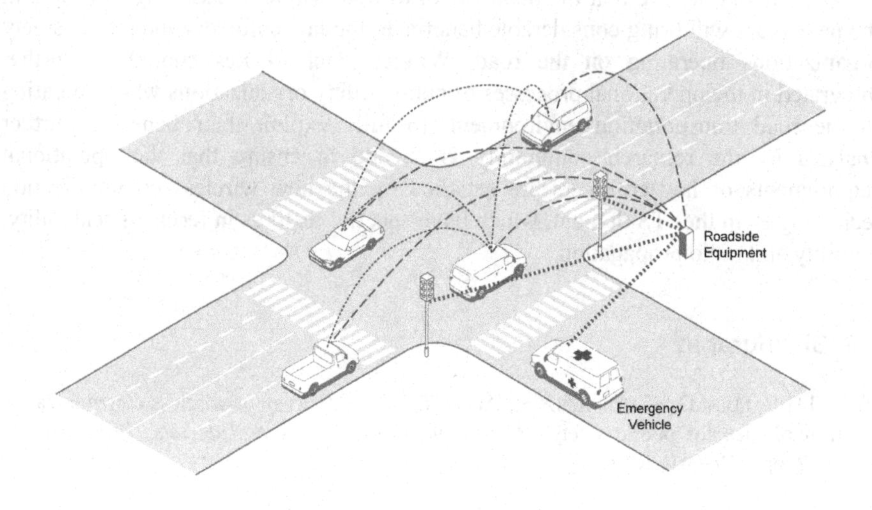

Figure 7.8. *Traffic signal preemption by an emergency vehicle*

All the interactions are aimed to enhance the emergency operations making the route for the EV as fast as possible but, at the same time, there is a considerable effort to guarantee the safety of all the vehicles and pedestrians on the roads.

7.7. Conclusions

In this chapter, we provided an overview on how the evolution of wireless communications used in the ITS domain can support the activities of law enforcers and public safety organizations and their operational needs. In addition, we described how wireless communication technologies used or planned for future adoption by public safety organizations (e.g. TETRA and LTE) can also be applied to the ITS domain. We identified the set of ITS applications, where wireless communication plays an important role for public safety organizations. In this context, security requirements are crucial. Then we presented the revision of the current DT application with wireless communication technology, and we described what would be the operational impact and the advantages for law enforcers. In particular, we showed how the introduction of ITS applications like the DT could make the control on regulation infringements much more efficient, helping to increase the safety of the roads and of people in general. Finally, in the EVP case study, we showed how the interaction of multiple ITS objects deployed on roadways could achieve a full control of traffic and vehicles coordination even in emergency scenarios, making the link between ITS infrastructures and public safety more evident.

The authors believe that the deployment of new wireless technologies in ITS in the next years will bring considerable benefits to the law enforcers and public safety organizations operating on the road. Wireless technologies should be further integrated in the operational processes of public safety organizations when operating in the road transportation environment, to fully exploit their benefits. Further analysis by the research community is needed to ensure that the operational requirements of law enforcers are satisfied by the new wireless communication technologies in the ITS domain, which have specific features in terms of scalability, mobility and radio propagation.

7.8. Bibliography

[BAL 14] BALDINI G., KARANASIOS S., ALLEN D. *et al.*, "Survey of wireless communication technologies for public safety", *Communications Surveys & Tutorials IEEE*, vol. 16, no. 2, pp. 619–641, 2014.

[BOU 08] BOUKERCHE A., OLIVEIRA H.A., NAKAMURA E.F. *et al.*, "Vehicular ad hoc networks: a new challenge for localization-based systems", *Computer Communications*, vol. 31, no. 12, pp. 2838–2849, 2008.

[CEC 10] COMMISSION OF THE EUROPEAN COMMUNITIES, Directive 2010/40/EU of the European Parliament and of the Council of 7 July 2010 on the framework for the deployment of Intelligent Transport Systems in the field of road transport and for interfaces with other modes of transport, White paper, 2010.

[CEC 14] COMMISSION OF THE EUROPEAN COMMUNITIES, Report from the commission to the European parliament and the Council Implementation of Directive 2010/40/EU of the European Parliament and of the Council of 7 July 2010 on the framework for the deployment of Intelligent Transport Systems in the field of road transport and for interfaces with other modes of transport, White paper, 2014

[CVR 14] CONNECTED VEHICLE REFERENCE IMPLEMENTATION ARCHITECTURE, available at http://www.iteris.com/cvria/html/about/connectedvehicle.html, 2014.

[C2C 14] CAR2CAR COMMUNICATION CONSORTIUM, available at https://www.car-2-car.org/index.php?id=5, 2014.

[DIT 09] DIGITAL TACHOGRAPH EUROPEAN ROOT POLICY, available at http://dtc.jrc.it/erca_of_doc/JRC53429_ERCA_CP_v2_1.pdf, 2009.

[ECI 14] eCALL INITIATIVE, European Commission, available at http://ec.europa.eu/digital-agenda/en/ecall-time-saved-lives-saved, 2014.

[ETI 10] EUROPEAN TELECOMMUNICATIONS STANDARDS INSTITUTE, "Intelligent Transport Systems (ITS), Communications Architecture", *EN 302 665 V1.1.1*, 2010.

[ETS 09] EUROPEAN TELECOMMUNICATIONS STANDARDS INSTITUTE, "Intelligent Transport Systems (ITS), Vehicular Communications (VC), Basic Set of Applications, Definitions", *TR 102 638 V1.1.1*, 2009.

[ETS 10] EUROPEAN TELECOMMUNICATIONS STANDARDS INSTITUTE, "Intelligent Transport Systems (ITS); Vehicular Communications; Basic Set of Applications; Part 3: Specifications of Decentralized Environmental Notification Basic Service", *ETSI TS 102 637-3V1.1.1*, 2010.

[ETS 11] EUROPEAN TELECOMMUNICATIONS STANDARDS INSTITUTE, "Intelligent Transport Systems (ITS); Vehicular Communications; Basic Set of Applications; Part 2: Specification of Cooperative Awareness Basic Service", *ETSI TS 102 637-2 V1.2.1*, 2011.

[FER 13] FERRUS R., SALLENT O., BALDINI G. *et al.*, "LTE: the technology driver for future public safety communications", *IEEE Communications Magazine*, vol. 51, no. 10, pp. 154–161, October 2013.

[GOR 14] GORATTI L., STERI G., BALDINI G. *et al.*, "Connectivity and security in a D2D communication protocol for public safety applications", *International Symposium on Wireless Communications Systems (ISWCS)*, pp. 548–552, 26–29 August 2014.

[HAR 14] HARDING J., POWELL G.R., YOON R. *et al.*, "Vehicle-to-vehicle communications: readiness of V2V technology for application", *Report No. DOT HS 812 014*, Washington DC, 2014.

[HAR 08] HARTENSTEIN H., LABERTEAUX K.P., "A tutorial survey on vehicular ad hoc networks", *IEEE Communications Magazine*, vol. 46, no. 6, pp. 164–171, June 2008.

[HED 08] HEDGES C., PERRY, F., "Overview and Use of SAE J2735 Message Sets for Commercial Vehicles", *SAE Technical Paper 2008-01-2650*, 2008.

[ISI 15] INTER SYSTEM INTEROPERABILITY FOR TETRA-TETRAPOL NETWORKS PROJECT (ISITEP), available at: http://isitep.eu/, 2015.

[ISO 10] INTERNATIONAL ORGANIZATION FOR STANDARDIZATION, "Intelligent transport systems – communications access for land mobiles (CALM) – M5", *ISO 21215:2010*, 2010.

[RAT 13] RATHOD P., KAMPPI P., "User requirements of Emergency Response Vehicles: Preliminary findings from a current research study", *2013 International Conference on Connected Vehicles and Expo (ICCVE)*, pp. 418–424, 2–6 December 2013.

[REG 14] EU REGULATION, Regulation no. 165/2014 of the European Parliament and of the Council of 4 February 2014 on tachographs in road transport, available at http://eur-lex.europa.eu/legal-content/EN/ALL/?uri=CELEX:32014R0165, 2014.

[SAE 10] SAE INTERNATIONAL, "DSRC Implementation Guide – a guide to users of SAE J2735 message sets over DSRC", 2010.

[SAF 06] SAFECOM (US communications program of the Department of Homeland Security), Public safety Statements of Requirements for communications and interoperability v I and II, internal report, 2006.

[SHE 11] SHENG-HAI A., BYUNG-HYUG L., DONG-RYEOL S., "A survey of intelligent transportation systems", *Third International Conference on Computational Intelligence, Communication Systems and Networks (CICSyN)*, pp. 332–337, 26–28 July 2011.

[TCA 15] TRANSPORTATION CERTIFICATION AUSTRALIA, "Directory of Gatekeeper Accredited Service Providers", available at http://www.finance.gov.au/policy-guides-procurement/gatekeeper-public-key-infrastructure/directory-of-gatekeeper-accredited-service-providers, 2015.

[UST 14] ITS STRATEGIC RESEARCH PLAN 2010-2014, available at http://www.its.dot.gov/strat_plan/, 2014.

[YOU 07] YOUNG K.L., REGAN M.A., "Intelligent transport systems to support police enforcement of road safety laws", *Research and Analysis Report*, 2007.

8

Communication Technologies for Public Warning

8.1. Introduction

The United Nations International Strategy for Disaster Reduction (UNISDR) defines early warning systems (EWS) as "the set of capacities needed to generate and disseminate timely and meaningful warning information to enable individuals, communities and organizations threatened by a hazard to prepare and to act appropriately and in sufficient time to reduce the possibility of harm or loss" [UNI 09, INT 09]. This definition comprises the four interlocking elements indicated in Figure 8.1, namely (1) risk knowledge, (2) monitoring and warning, (3) dissemination and communication, and (4) response capability. Each element

---

Chapter written by Cristina PÁRRAGA NIEBLA.

must function efficiently, as the failure of any of them could result in a failure of the whole system [INT 09, INT 15, FOU 15, UN 06, UNI 06, INT 13, EFF 00]. The dissemination and communication component of the EWS model proposed by UNISDR is commonly known as a public warning and its effectiveness is unavoidably linked to communication technologies.

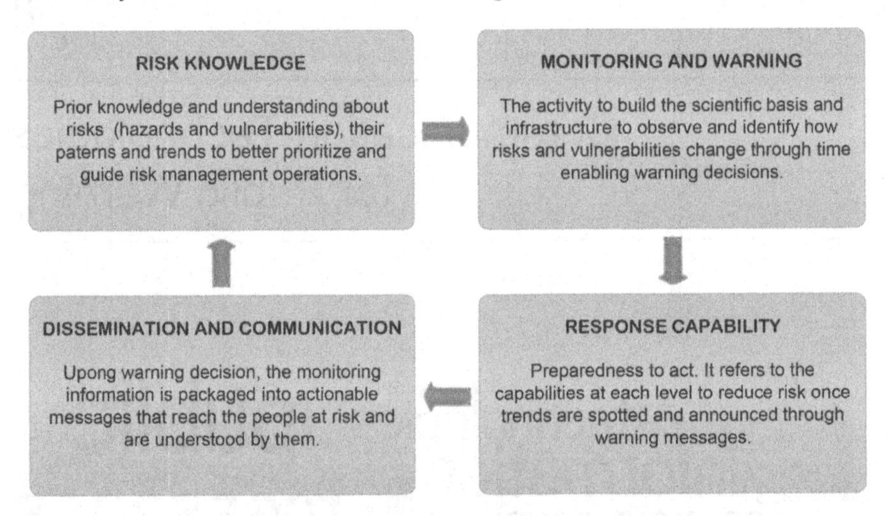

Figure 8.1. *People-centred early warning system model*

The continuous growth of mobile communication networks and smart portable devices along with TV broadcast, streaming markets and social media has definitely affected every day life. The public demands continuous and ubiquitous broadband connectivity, staying connected and accessing content anywhere at any time through a variety of devices with high processing, storage and multimedia presentation capabilities. Although traditional means for public warning are still useful, the information era we are living imposes high demands in the information flow during emergency and disaster situations in a society that is becoming more and more "information-hungry". Indeed, this large availability of heterogeneous network infrastructures and media has proven to be a key element in public warning. The relief of recent natural disasters (e.g. Katrina and Sandy hurricanes, Fukushima earthquake and subsequent tsunami) [INT 09, INT 13] showed that populations can only be alerted efficiently if various communication media are used simultaneously.

8.2. Requirements on public warning technologies

The UNISDR and the International Federation of Red Cross and Red Crescent Societies often use the term "people-centred early warning systems" to emphasise that effective warning systems are those that recognize human needs and human behaviour and are developed with the participation of communities [INT 09, INT 15, FOU 15, UN 06, UNI 06, INT 13, EFF 00]. In other words, public warning is effective if the following conditions are met:

1) the citizens at risk receive, notice, understand and trust warning messages on time;

2) the citizens at risk are prepared to act, i.e. are sufficiently familiar with warning procedures and trained to act upon.

The use of contemporary technologies is one important step to adapt public warning concepts to human behaviour and maximize penetration of warning messages, as people's communication habits are largely influenced by the technologies they can access, especially with regard to personal mobile and portable devices. However, there are a variety of factors to be considered to maximize the effectiveness of public warning systems; the majority of these factors will turn into requirements to the communication technologies applied for public warning.

In addition, the checklist proposed by the Third International Conference on Early Warning for developing EWS [UNI 06] points out a number of cross-cutting issues formulated as recommendations. Among these recommendations, there is one that directly influences the public warning component of EWS: it is recommended to link all hazard-based systems where possible to enhance economies of scale, sustainability and efficiency, i.e. to adopt a "multi-hazard approach". As stated in [UN 06], "dissemination and telecommunication mechanisms must be operational, robust and available every minute of every day and tailored to the needs of a wide range of different threats and different user communities [...]". The warning lead times range from seconds to weeks (depending on the type of hazard): "[...] where common needs exist, it may be possible to make use of the same communications systems for more than one type of warning information".

8.3. Influencing factors in the effectiveness of public warning

8.3.1. *Warning decision process*

From the perspective of the warning message issuer, i.e. the person responsible for deciding to issue warning messages, there are a number of decision gates that precede the dissemination of a warning message as discussed below.

8.3.1.1. *Whether/when to warn*

The decision "whether to warn" is commonly supported by available emergency plans derived from past experiences or risk analysis of expected incidents and is typically driven by factors such as the certainty of the available information, expected warning impact, etc. Possibly, myths related to (1) warning impact (e.g. creating panic thus worsening the situation) and (2) the long term consequences of false alarms may create reluctance in the decision whether to warn. There is sufficient scientific evidence to abandon both myths [INT 15, INT 13]. Research shows that public panic occurs seldom and only in very specific conditions such as in close physical space, limited or inadequate escape routes and imminent and clear threat [INT 15, INT 13]. With regard to false alarms, most authors recommend warning in case of doubt; the public is more tolerant to false alarms if they are explained in a valid and rational manner [MIL 90]. Furthermore, given the exposure to additional information sources that can spread rumours, it is preferable to warn and state the certainty of the information than remaining silent and letting rumours spread [CEN 09].

Eventually, the decision "whether to warn" develops into a different but related decision "when to warn". Once the risk is quantified and the warning need is identified, the alert message should be issued as soon as possible given the tight lead times of some hazards. However, low certainty in the available information may cause that warning officials wait for more data to increase the certainty. If the warning is issued too early, the available information may not be sufficient to provide accurate recommendations for protective actions; however different types of warning messages can be issued depending on the criticality of the situation given the knowledge stage at every moment in the course of the emergency: information, alert and warning. Hence, several messages should be issued including more details as they become available (see section 8.3.1.3) [MIL 90, CEN 09].

Figure 8.2 shows a generic representation of the time at which a warning message can be issued, related to the false alarm probability (or the certainty with regard to the hazard onset). The generic terms "seasonal risk", "forecast" and "fact" have been applied to represent the type of messages that can be issued at the represented periods (information, alert, warning), related to the certainty about the hazard onset. As time gets closer to the actual hazard onset, the certainty associated with a forecast (or hazard detection) increases, but the lead time decreases. This trade-off has been pointed out in the literature regarding the effectiveness of alert systems, highlighting also the impacts on the risk perception (and consequential reaction to the alert) of the population [EFF 00, GLA 04, GRA 07]. The generic representation in Figure 8.2 will be tailored to each hazard: for example, it is fully applicable to the severe rainfall case, but for the tsunami case the "seasonal risk" and "forecast" slices will be removed as a tsunami cannot be forecast, only the time it will hit land. Also, the time frame between the "hazard onset" and the "hit population" events will differ from hazard to hazard.

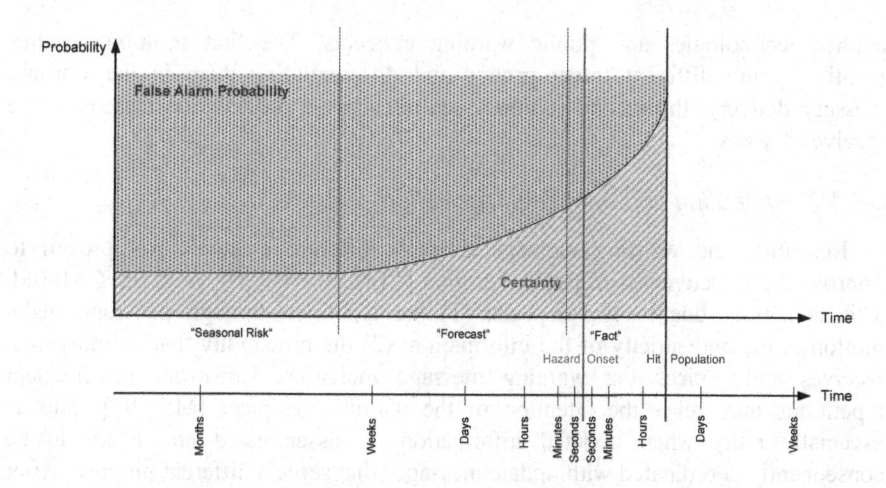

Figure 8.2. *Warning lead time vs. false alarm and certainty*

8.3.1.2. *Where and who to warn*

Warning messages should be addressed to all people that are at risk due to an occurred or expected hazard, i.e. all people located at a geographical area that is or may be affected by the hazard. This, however, is somewhat further influenced by the governance hierarchies in the respective country with regard to public warning. In other words, civil protection agencies may warn areas under their own jurisdiction, but may need clear escalation procedures or cooperation frameworks to undertake public warning in cross-regional and cross-border disasters.

Assuming that the governance aspect is solved, the definition of the risk area boundaries may depend on the type of incident, existing emergency plans and additional information (e.g. weather forecast). Nevertheless, the public in the risk area should not be understood as a whole group, but as a set of groups possibly having different needs. The warning messages should address all groups at risk under consideration of these needs. The differentiation between "where" and "who" to warn now becomes more concrete. From a general perspective, the level of risk in a community is related to its geographical location with regard to the hazard and its area of spread. However, the vulnerability of groups within the community at risk is an additional aspect that is more related to social and personal aspects (such as sensory impairments, age, belonging to specific social crowds, etc.) and need to be taken into account too.

The warning system should provide solutions to address all groups at risk, either by disseminating group-specific messages or by adapting the message presentation to the different groups' needs. Each solution imposes different requirements in the applied technologies and public warning concepts. The first solution requires identifying the different target groups and differentiating them in the warning message delivery; the second solution requires adapted presentation features at the receiver devices.

8.3.1.3. *Repeating and updating information*

Repeating the warning message through different channels has proven to improve the effectiveness of public warning [CEN 09, PPW 04, AUS 05, CAL 08]. This is mainly due to two aspects: (1) the repetition through different media reinforces the authenticity of the information; (2) the probability that an individual receives and notices the warning message increases. However, too frequent repetitions may relax the attention of the warning recipient [MIL 95]. This is especially risky when updated information is disseminated and needs to be consequently coordinated with update messages that serve a different purpose. After having sent the first warning message, it is important to maintain the communication with the public, updating information when it becomes available until there is sufficient evidence to consider the situation "all clear" [INT 15, INT 13]. As soon as the "emergency situation" is released, a message should be issued stating the end of the risk situation and return to normality [PPW 04, COO 11]. The frequency to provide update messages should be adapted to the dynamics of the concrete incident [MIL 90].

8.3.2. *The warning message*

The content, style and presentation of the warning message can significantly contribute to the understanding and trust of the individuals at risk receiving the message [MIL 90]. Furthermore, the presentation of warning messages plays an important role in assuring that warning messages are noticed.

8.3.2.1. *Warning message style*

Best practice literature recommends that warning messages be formulated according to the following principles:

1) *Specificity*: describing the situation in a precise, non-ambiguous manner and avoiding information omission [EFF 00].

2) *Consistency*: the warning message should be consistent on its own content and with other warning messages relating to the same incident [MIL 90, FIT 91].

3) *Accuracy*: the warning message should be accurate in the spelling and description of information [FIT 91].

4) *Certainty*: when the warning message is presented in the form of speech and/or video, the spokesperson should be certain of the message and apply appropriate body language.

5) *Clarity*: the warning message should be understandable by the targeted audience [MIL 90]. This implies that clear and simple words as well as standard terminology should be applied, whereas technical language, codes, acronyms or jargon should be avoided [EFF 00]. With the purpose of fostering clarity, additional guidelines related to the message syntax are also recommended in the literature, e.g. most important information should be provided first, so that the use of headlines and concise sentences is recommended [EFF 00].

8.3.2.2. *Warning message content*

The warning audience is more likely to trust the warning message and implement the recommended protective actions when the alert message includes sufficient information, including at least the following aspects [EFF 00, PPW 04, PER 07]:

1) *Hazard information*: this includes a brief description of the hazard event, its intensity level and likelihood (or certainty). The terms likelihood and certainty is related to the accuracy of the information that is available.

2) *Target location*: this refers to the specification of the location that is or could be at risk. This is essential for the warning recipients to understand if they are in or out of the risk area. The target location should be described in terms of geographic areas or recognisable landmarks (e.g. transportation routes, jurisdictional boundaries or recognisable geographical features).

3) *Time information*: the expected incident onset time and the expected time extent of the risk should be included in the warning message. Furthermore, a statement should be included regarding to the relevant time to take protective actions.

4) *Guidance*: clear statements and guidance information on the recommended protective actions must be included in the warning message.

5) *Source*: mentioning the institutional identity of the warning source is essential for the credibility of the message and trust of the warning audience.

6) *Further information sources*: disseminating warning messages often yields an overload of the Public Safety Answering Points (PSAPs) for the purpose of validating the message. To mitigate this undesired effect, it is recommendable to include sources of further information in the warning message (e.g. web page, radio channel).

7) *Sequence number*: a crisis situation after a hazard onset evolves with time, so that update warning messages will be issued with regard to the same incident. It is therefore important that the warning messages corresponding to the same incident

are correctly numbered to allow the warning recipients sorting the information appropriately.

8.3.2.3. *Warning message presentation*

The target of public warning systems is to alert and inform the potentially affected population about a situation of risk. Warning messages target all affected (or potentially affected) citizens, including groups with special needs. People with cognitive, sensory and physical impairments face significant limitations to access the warning information. In addition, people with no cognitive, sensory or physical impairments may be impaired anyway by situational factors, such as being in a noisy environment. However, one of the people-centred EWS principles is that all affected citizens must be addressed. This requirement clearly involves the appropriate presentation of warning messages.

The Federal Communications Commission (FCC) released a Notice of Proposed Rulemaking (NPRM) in 2004 to review the Emergency Alert System (EAS) in the USA, as the technologies applied to date for public warning seldom addressed the needs of disabled people. As a result, the FCC rules were amended to ensure that persons with disabilities have access to public warnings. It is worth highlighting that this amendment drove, among others, the recommendation that "all wireless device users would benefit from a multi-modal approach to providing accessible, wireless emergency alerts emergency alerts communication" [JOH 10]. One of the immediate consequences is the support of accessible alert delivery to individuals with hearing and sight impairments by the Common Alerting Protocol (CAP), facilitating the achievement of "*functional equivalency*" [JOH 10].

Multimodal approaches to deliver warning messages (i.e. exploiting various modalities and their combination) can increase the likelihood that the warning information will be noticed and understood by the target audience, including but not limited to groups with special needs [LAN 10, SUL 10]. Multimodal delivery involves two major aspects: the interface design at the user device and the related content transport in the underlying network. Multimodal interface design enables compensating impairment in one modality (namely hearing, vision, touch and cognition) by using other modalities (for example a deaf person receiving a text alert instead of speech). By extrapolation, multiple impairments can be mitigated to some extent by the multimodal interfaces performance gain [LAN 10]. However, addressing the full spectrum of disabilities is not as simple as delivering a message in the form of speech or text. Sensory disabilities can manifest in different intensities and cognitive disabilities are more complex to address; the use of visual symbols to depict threat and appropriate action will complement the warning message content, given their potential to address cognitive impairments (and linguistic impairments in general) [SUL 10, BRY 10]. In particular, the need of a basic vocabulary for children and adults with cognitive impairments was identified in [SUL 10] and the need of speech-generating devices (SGD) was stressed in wireless devices to cope with cognitive impairments.

Hence, multi-content delivery of warning messages is desirable, which can be challenging for underlying networks due to capacity limitations, especially if warning messages need to be delivered in short time. Alternatively, a client-based approach that locally transforms notifications into accessible formats is proposed in [SUL 10]. This approach has two main advantages: it allows efficient transport of the contents over networks and it avoids official targeting of messages to specific disabilities by subscription; the latter is in fact an issue due to personal privacy issues [SUL 10].

A large number of new multimodal options are currently available in advanced mass market devices that could be exploited to deliver warnings to people with different impairments (and impairment combinations) [LAN 10]. A distinction is made between programmable and proprietary devices. Programmable devices, with configurable interfaces, are more flexible for the demands of accessibility, e.g. fixed and mobile phones, smart devices, computers, laptops and PDAs. Proprietary devices refer to those for which interfaces can be configured and upgraded by the manufacturer, such as TVs, set top boxes (STBs) and radio receivers. Programmable devices have a great potential to fulfill multimodal warning requirements. Non-programmable devices can be also applied for warning purposes, within their intrinsic accessibility limitations, e.g. small, low-contrast and fixed font sizes, complex menus and small inaccessible controls.

Empirical studies in the literature provide interesting insights in the accessibility capabilities of mobile (wireless) devices. For example, a study focused on deaf people [MUE 10] concluded that warning delivery in the form of text alone is not sufficient for addressing all deaf people, as many people who are deaf have little understanding of English. Instead, combining text messages with American Sign Language (ASL) videos provided excellent results. Also, the study carried out in [CRA 10] showed that auditory information is preferred above tactile information by sight impaired people. Accessibility limitations for SMS and web browsing for disabled participants were identified in [JOH 10]. Concerns were raised on the quality of the synthesised speech and the effectiveness of the warning attention signal and vibration cadence; the two latter aspects are also relevant issues for situational impaired people. The study presented in [BRI 10] also stressed other factors that deal more with accessibility to the devices as such: the type and severity of the disability, together with cost of adaptive technology, will influence the use of specific devices by disabled people.

8.4. Requirements on communication technologies

8.4.1. *Translating public warning needs into requirements for communication technologies*

As previously discussed, effective public warning systems are those that achieve that the citizens at risk (1) receive, notice, understand and trust warning messages on time and (2) are prepared to act. First, the reception of warning messages is exclusively related to the use of communication means and technologies and the accessibility of the targeted audience to such communication means and technologies. Secondly, the terms "notice", "understand" and "trust" warning messages can be influenced by the factors discussed in the previous section (the decision process and the warning message style, content and presentation). Most of the presented recommendations in terms of decision process and warning message will be translated into requirements for the (communication) technologies used for public warning. Finally, the preparedness of communities to act upon warning reception implies educational and training programmes that are part of the public warning system as a whole.

8.4.1.1. *Receive the warning message on time*

In order to ensure that warning messages are received on time to allow the people at risk implement protective actions, several influencing factors are relevant. A multi-hazard approach is considered optimal in [UNI 06]. In this respect, the public warning communication system will be prepared to transmit warning messages with regard to different hazards having different lead times, including those with very low lead times. The latency in the warning message transmission should not be the limiting factor. This translates into requirement 1 in Table 8.1.

The decision process "where/who to warn" is also relevant. All individuals at risk will be addressed. On the one hand, this means that the individuals in the area at risk will be warned. On the other hand, this means that all groups at risk will be warned, including those with special requirements and regardless of their habits in the use of information, communication and entertainment technologies. For the communication system, this implies requirements 2 to 10 in Table 8.1. Finally, the decision process "repeating and updating information" involves that warning messages will be repeated with certain periodicity to increase awareness and trust. Also, updates to the situation will be transmitted in a consistent manner as soon as new information is available. Hence, requirements 5, 6 and 11 in Table 8.1 apply.

8.4.1.2. *Notice the warning message on time*

Reception of a warning message does not necessarily imply that the people at risk notice it immediately. The warning message presentation is here the essential factor. Attention calling mechanisms ("wake-up" features) must be incorporated to foster awareness. Such wake-up features must encompass sufficient resources to address all groups of individuals at risk, including cognitive, sensory, physical and situationally impaired groups. This implies requirements 9 and 14 in Table 8.1.

8.4.1.3. *Understand the warning message*

Best practice studies prove that individuals are more likely to understand warning messages if the style and content guidelines as developed in the previous section are met. The style guidelines do not impose requirements in the communication system, but the recommended content completeness does imply capacity requirements. Furthermore, people with special needs must also understand the warning message, implying requirements in the presentation capabilities of the used technologies. Given that several channels may be used for warning dissemination, the coherence of information coming from different channels is also of high importance. Hence, requirements 9 to 13 in Table 8.1 apply.

8.4.1.4. *Trust the warning message*

Similarly, the trust in received warning messages deals with the style and completeness of information. Regarding the latter, the identity of the issuing authority as well as the indication on where to find further information for verification are essential elements. The coherence aspect between messages received through different channels is also critical for trust. Furthermore, given that several channels may be used for warning dissemination, the coherence of information coming from different channels is also of high importance. Hence, requirements 12 and 13 in Table 8.1 apply.

8.4.1.5. *Be prepared to act*

People may not respond to first warnings immediately. The trend to be expected is that people will first cross-check and try to validate the information until it makes sense to them [INT 15]. Ways to minimize this latency are related to long term educational and training programmes, i.e. build the action capacity upon warning reception of people already in early stages (at school) and undertake training exercises with the communities. In the course of an emergency, message repetition will help. Here, requirements 5 and 6 in Table 8.1 are relevant.

Req. No	Description
1	The communication technologies used for public warning will have a low transmission time (50% of the citizens will be warned within 3 min; 97% will be warned within 5 min; in very rapid emergencies, warning should happing within a few seconds [ETS 15a]);
2	The communication technologies used for public warning will have uquitous coverage;
3	Public warning systems will make use of several communication technologies (multi-channel approach) to reach the people at risk in any context;
4	The technologies used to disseminate warning messages will be accessible to the population (high penetration of technologies used to warn);
5	The technologies (channels) used to disseminate warning messages will have high availability;
6	The technologies (channels) used to disseminate warning messages will be highly reliable;
7	The technologies (channels) used to disseminate warning messages will support flexible addressing and multi-casting options;
8	The technologies (channels) used to disseminate warning messages will have the ability to apply geographical limits to the reception or delivery of warning messages (geocasting);
9	The technologies used to disseminate warning messages will support several presentation modes (multimodal presentation);
10	The technologies used to disseminate warning messages will support the presentation of warning messages in several languages;
11	The technologies used to disseminate warning messages will support the delivery of multi-media content;
12	The technologies used to disseminate warning messages will have the capability to carry sufficient information;
13	The delivered warning messages over different channels will be consistent with each other (identical, if possible);
14	The technologies used to disseminate warning messages will support attention-calling features (wake-up effect);
15	Attention-calling features will be distinguishable from other notifications;
16	Public warning systems may be intrusive;
17	The technologies used to disseminate warning messages will provide security features (authentication, authorization and integrity).

Table 8.1. *Requirements on communication technologies*

8.4.1.6. *Further requirement sources*

The European Telecommunications Standards Institute (ETSI) issued a technical specification on the requirements for communications from authorities/organizations to individuals, groups or the general public during emergencies [ETS 15a]. Furthermore, the International Standardization Organization (ISO) recently issued a requirements document for public warning [ISO 15]. The list of requirements in Table 8.1 is completed in accordance with these two additional sources.

8.4.2. *Validation model and criteria*

A method to evaluate the effectiveness of public warning systems is suited to validate the effectiveness of existing systems and to identify gaps for further improvements. The evaluation model proposed in [MUL 14] provides a qualitative/quantitative method to evaluate the capability of communication technologies to meet the high level requirements discussed in the previous sections. A metric is assigned to evaluate the capability of a warning channel to fulfil each high level requirement as indicated in Table 8.2. These metrics are evaluated separately to enable identifying concrete gaps, avoiding masking effects by multiplicative approaches. The effectiveness of warning channel *m* is defined by the set (R_m, CT_m, N_m, A_m, T_m).

Metric	Description
R_m	Capability of channel *m* to deliver the warning message to the population;
CT_m	Capability of channel *m* to deliver the warning message to targeted citizens;
N_m	"Wake-up" capability of channel *m*;
A_m	Capability of channel *m* to provide sufficient information to the citizen to understand and trust it;
T_m	Capability of channel *m* to deliver the warning message in a timely manner.

Table 8.2. *Warning channel evaluation metrics*

The R_m metric is defined by the product of the factors defined in Table 8.3 and equation [8.1].

$$R_m = C_m \cdot MP_m \cdot TD_m \cdot NS_m \cdot RL_m \cdot RD_m \quad [8.1]$$

Metric	Description
C_m	Coverage – the portion of the territory where the technology is available;
MP_m	Market penetration of the related receivers;
TD_m	Portion of time during the day that the specific device is used;
NS_m	Impact of requiring subscription liable to pay costs, service activation or cost-free subscription to receive the alert message, according to the criteria in Table 8.7;
RL_m	Portion of time that the channel is available to the citizen due to the percentage of the time that people are at home, on weekend house, at work, etc.
RD_m	Availability of the communication channel during the disaster (availability, robustness and reliability metric).

Table 8.3. *R_m metric factors*

RL_m is a averaged over different locations *j* according to equation [8.2], where α_j is the expected time share that a person will spend in different locations, considering empirical statistics [EUR 04]. RD_m is averaged over different types of disasters, where the weight assigned to each disaster *n*, β_n considers empirical statistics [PRE 08].

$$RL_m = \sum_{j=1}^{J} \alpha_j RL_{m,j} \quad [8.2]$$

$$RD_m = \sum_{n=1}^{N} \beta_n RD_{m,n} \quad [8.3]$$

The CT_m metric captures the capability of channel *m* to provide different addressing (multi-casting) options. CT_m is an average of different addressing options, see equation [8.4]. A value $CT_{m,i}$ is assigned to each type of addressing option (unicast, multicast, broadcast and geocast) according to the rule specified in Table 8.7.

$$CT_m = \frac{1}{N} \sum_{i=1}^{N} CT_{m,i} \quad [8.4]$$

The N_m metric evaluates the capability of the channel m to attract the attention of the warning message recipients. N_m is calculated as the product of two parameters: the “wake-up” capability WU_m and the exclusive alert capability, EA_m, i.e. the capability to make warning notifications that can be easily distinguished from other notifications. Both parameters are evaluated according to the rule in Table 8.7.

$$N_m = WU_m \cdot EA_m \tag{8.5}$$

The A_m metric captures the capability of the warning channel to provide sufficient information in the warning message by averaging the contribution of three key components as in equation [8.6]: ML_m (multi-lingual capability of channel m), SA_m (capability of channel m to provide source identity) and SN_m (capability of channel m to transport complete information and to address citizens with special needs). These parameters are evaluated according to Table 8.7.

$$A_m = \frac{ML_m + SA_m + SN_m}{3} \tag{8.6}$$

The T_m metric captures the warning message delivery latency in channel *m*. Exact transmission delay values are not considered here; instead, the order of magnitude of the transmission delay that can be expected is considered, as in Table 8.7.

8.5. Communication technologies for public warning

Public warning can be implemented with any means of communication. This includes very advanced technologies, such as internet-based solutions like social media, but also the good old "mouth to mouth". Indeed, depending on the level of development, the most suitable solutions may be human-based if the penetration of technologies is not sufficient for technologies to be effective warning channels. This is also reflected in the effectiveness evaluation model discussed in the previous section by the multiplicative impact of the technology penetration parameter MP_m.

The rapid development of communication technologies in the past 2 decades is a great opportunity for public warning purposes, but should not mean a complete abandoning of simple traditional public warning means, as technology may also fail during emergency situations or may not be accessible to all groups at risk in the same way (due to cost or limitations in the operation of advanced technologies by vulnerable groups, such as elderly or disabled people) [INT 15, INT 13]. The key of effective public warning resides in the complementarity of communication means, so that communication habits and accessibility to technology of different groups at risk do not mean a barrier for the objective of warning those in danger. An overview of communication technologies that can be used as warning channels is provided in Table 8.4. Two categories are differentiated: (1) direct (referring to technologies typically owned or operated by the warning authority) and (2) intermediary (referring to technologies typically used for other more generic purposes, but can be used for public warning by interfacing the public authority, i.e. the technology or service provider acts as intermediary between the public warning authority and the targeted audience) [BER 14].

Warning channel	Description	Cat.*	Audience
Sirens	Single or multi-tone local systems.	D	Broadcast
Audible/ visual signs	Maroons, klaxon, foghorns, flashing lights, etc.	D	Broadcast (local area)
Tannoys	Loudspeakers transmitting voice messages in fixed location or installed on mobile platform.	D	Broadcast (local area)
Electronic billboards	Electronic displays in public places, e.g. gantry signs on motorways/railway and flight information boards.	I	Context-dependent
Security systems	Smoke detectors, indoor audible evacuation systems, etc. can be also used for public warning.	I	Context-dependent
Pager	Hand held text message to receiving units.	I	Broadcast
TV broadcast (wired, wireless or satellite)	TV broadcast can be used in many ways (in less or more invasive manner): embedding warning information in the news content of TV programmes and/or videotext, interrupting programme to disseminate the warning message as “breaking news” or making use of data carrousel to display overlaid banners or pop-up windows to the TV programme.	I	Broadcast
Radio broadcast	The warning message is read on-air (embedding warning in radio news); the Radio Data System (RDS) can also be used.	I	Broadcast
Terrestrial fixed telephone	Fixed telephones can be used to reach Public-safety answering points (PSAPs); at the same time, PSAPs can take the role of calling registered users in case of emergency, for example making use of auto-dialers.	D	Unicast
Mobile networks (land and satellite)	Mobile networks can be used as fixed networks. However, they provide further features, such as SMS and cell broadcast (CB). Satellite networks provide larger broadcast reach.	I	Unicast, Multicast, Broadcast
Satellite navigation	The satellite-based augmentation system (SBAS) can be used to deliver public warning messages.	I	Broadcast, Geocast
Internet	Through data networks, the public can access warning information. Warning providers make use of internet connectivity to push notifications (email, mobile applications) or the public checks information in web sites or social media.	I	Unicast, Multicast, Broadcast

* Cat.: stands for “Category”: (D) Direct, (I) Intermediary

Table 8.4. *Overview of communication technologies for public warning*

8.6. Evaluation of communication technologies for public warning

The evaluation methodology described in section 8.4.2 will serve the effectiveness comparison of different communication means that can be used for public warning purposes. Indeed, public warning systems will be in most cases based on a multi-channel approach. Hence, complementarities can be spotted by using this evaluation method to compose a complete and effective public warning system. The evaluation of the metrics presented in section 8.4.2 follow a quantitative/qualitative approach in [MUL 14]. Values are applied to the different parameters that define the metrics based on qualitative criteria. The parameters C_m, MP_m and TD_m are an exception, as they are subject to country-specific statistics. A few European countries and technologies are taken as model for an example evaluation of the effectiveness parameters described in the previous sections in Tables 8.4 and 8.5. It will be noted that the evaluation is an example and does not relate to the actual technologies deployed in the selected model countries to avoid any conflicts. This section is only intended to understand how the evaluation model can be used.

Country	Parameter	Sirens	Loud-hailers	Radio	TV	Sat TV	Fixed Phones	Mobile Phones		PND	Internet
								Mobile	Smart Mobile		
DE	MP_m	1	1	0.98	0.95	0.38	0.66	1	0.16	0.15	0.8
	C_m	0.1	0.1	1	1	1	1	0.85	0.999	1	0.97
UK	MP_m	1	1	0.98	0.97	0.25	0.52	1	0.23	0.15	0.82
	C_m	0.1	0.1	1	1	1	1	0.91	0.996	1	1
FI	MP_m	1	1	0.92	0.96	0.25	0.49	1		0.15	0.85
	C_m	0.8	0.8	1	1	1	1	0.75	0.996	1	0.96
CZ	MP_m	1	1	0.94	1	0.26	0.36	1		0.15	0.66
	C_m	0.85	0.85	1	1	1	1	0.6	0.999	1	0.92

Table 8.5. *Evaluation of MP_m and C_m parameters for several channels in four example countries*

A number of parameters corresponding to the R_m metric relate to empirical statistics that are country-specific, e.g. market penetration of technologies. Other metrics are also based on empirical statistics, but inherent to the technology itself, not related to geographical spread. For the sake of clarity, region-dependent parameters are compiled in Table 8.5 for a selection of European countries based on the data in [EUR 04, PRE 08, IWS 13, ITE 09, ATW 11, ITU 10, ALL 10, TOM 11] with regard to device market penetration. Coverage data on siren networks and loudhailers are extracted from [EC 11, SCP 11, HAL 06]; furthermore, in the case of Germany and UK, the respective coverage has been set to 10%, given that only local networks are deployed in risk areas (nearby of nuclear power plants or chemical industries). It is also assumed that all households having a radio, TV or fixed phone are inside the respective coverage area, not to penalize twice. In terms of 3G and DLS, the statistics in [IDA 09] are applied.

Portable navigation devices (PND) are also considered as potential public warning channels [DE 15]. The transmission of warning information would however not be implemented through the GPS satellites, but over the satellite-based augmentation system (SBAS), e.g. the European Geostationary Navigation Overlay Service (EGNOS) over Europe, which is based on three of geostationary satellites. EGNOS offers good performances in all European member states except for the parts of that are located in the northern, southern and eastern extremes of the EU territory [HTT 15a]. Although this is planned to improve in future, the current situation does not provide sufficient coverage in high latitudes, affecting UK and Finland in the example cases in Table 8.5. In general, central Europe has EGNOS coverage with 99.9% availability, while Finland and UK achieve 99.6% [ESA 15a, ESA 15b].

The TD_m, NS_m, RL_m and RD_m parameters are compiled in Table 8.6. While Table 8.5 was focused on the receiver devices, Table 8.6 makes further distinction in terms of services that can be used to deliver the warning message (and may run in the same device). In this context, CB refers to cell broadcast and "Smart+PND" refers to smart mobile devices incorporating a satellite navigation receiver.

The statistics in [EUR 05, EUR 04] and [EUR 03] have been applied to determine the TD_m parameter for TV, radio and fixed phones. With regard to mobile phone-based services, it has been assumed that 90% of the population has an activated mobile phone in the nearby during 16 h/day and that 50% of the population switches it off or puts it aside during night. Regarding internet services, the statistics on the habits of people in [EUR 04] have been applied to determine the RL_m parameter. The RD_m parameter has been set taking into account the statistics regarding the amount of people affected by different disasters from 1980 to 2008 in [PRE 08].

For the rest, the criteria to evaluate different parameters are detailed in Table 8.7.

Parameters		Sirens / Louad-hailer	Radio	TV	Sat TV	Fixed Phones	Mobile Phones					Internet
							SMS	CB	Smart	Smart + PND	PND	
TD_m		1	0.14	0.14	0.14	0.84	0.77	0.77	0.77	0.77	0.06	0.08
NS_m		1	1	1	1	0.01	0.01	0.5	0.01	1	1	1
RL_m	(1)	1	1	1		1	1	1	1	1	0.5	1
	(2)	1	1	1	1	1	1	1	1	1	0.5	1
	(3)	1	1	1	1	1	1	1	1	1	0.5	1
	(4)	1	1	1	1	1	1	1	1	1	0.5	1
	(5)	1	0.5	0.5	0.5	1	1	1	1	1	0.5	0.5
	(6)	0.5	0.5	0.5	0	0	0.5	0.5	0.5	1	1	0
	(7)	0.5	0.5	0.5	0.5	0.5	0.5	0.5	0.5	0.5	0.5	0.5
	Av.	0.92	0.91	0.91	0.25	0.89	0.92	0.92	0.92	0.95	0.53	0.88
RD_m	(8)	1	1	1	1	1	1	1	1	1	1	1
	(9)	1	0.5	0.5	0.5	0.5	0.5	0.5	0.5	1	1	0.5
	(10)	1	1	1	1	1	1	1	1	1	1	1
	(11)	1	1	0.5	0.5	0.5	0.5	0.5	0.5	1	1	0.5
	(12)	1	1	1	1	1	1	1	1	1	1	1
	(13)	0.5	0.5	0.5	0.5	0.5	0.5	0.5	0.5	1	1	0.5
	Av.	0.95	0.82	0.69	0.69	0.69	0.69	0.69	0.69	0.99	0.99	0.69

(1) Home (2) Weekend home (3) Work / study place (4) Other's home	(5) Restaurant (6) Travelling (7) Other (8) Wildfire	(9) Storm (10) Drought (11) Flood (12) Extreme temperature	(13) Earthquake Av: weighted average

Table 8.6. *Evaluation of TD_m, NS_m, RL_m and RD_m parameters for several warning channels*

Par.	Value	Criteria
NS_m	0.01	Subscription (liable to pay cost) is required to receive warning messages
	0.5	Service activation or cost-free subscription is required to receive warning messages
	1	No subscription and no service activation is required
$RL_{m,j}$	0	Channel *m* does not allow delivery of the message at location *j*
	0.5	Channel *m* may allow delivery of the message at location *j*
	1	Channel *m* allows delivery of the message at location *j*
$RD_{m,n}$	0	Channel *m* is not resilient to OR is affected by overload during event *n*
	0.5	Channel *m* is partly resilient to event *n*
	1	Channel *m* is resilient to event *n*
$CT_{m,i}$	0	Channel *m* does not support casting option *I*
	1	Channel *m* supports casting option *i*
WU_m	0	Channel *m* has no 'wake-up' effect
	0.5	Channel *m* has 'wake-up' effect that can be de-activated by the user
	1	Channel *m* has 'wake-up' effect that cannot be de-activated by the user
EA_m	0.5	Channel *m* is normally used for other purposes than warning
	0.75	Channel *m* is provides a special notification means for warning
	1	Channel *m* is used exclusively for warning
ML_m	0	Channel *m* can provide the warning in one language
	0.5	Channel *m* can provide the warning in a limited set of languages
	1	Channel *m* can provide the warning in several languages
SA_m	0	Channel *m* cannot provide source identity
	0.5	Channel *m* can technically provide source identity but it is not implemented
	1	Source identity is provided or not required for channel *m*
SN_m	0	The technology supports warning in a single mode with a set of tones
	0.25	The technology supports warning in a single mode with text or speech
	0.5	The technology supports 2-mode warnings (text, speech)
	0.75	The technology supports 3-mode warnings (text, speech, graphics)
	1	The technology supports multi-mode warnings (text, speech, graphics, video)

Table 8.7. *Qualitative/quantitative parameter evaluation criteria*

It is worth highlighting a number of aspects that yield the proposed evaluation values in Table 8.7. The values for the NS_m parameter reflect that requiring subscription or service activation has a negative impact in the warning penetration, especially if this implies a cost for the citizen. This disadvantage could be mitigated by informative campaigns or in an *ad hoc* manner during emergency situations, where the citizens become more aware of their own risks. Concretely, the value of 0.01 for the subscription case and 0.5 for the service activation case are justified in [KLA 11] and [LAN 10], respectively. The large gap between values 0.5 and 1 in the evaluation of the AM_m parameter is meant to stress that technology allowing multi-language delivery but having to cope with capacity ties can only transport a limited amount of languages to ensure timely delivery. However, broadband systems or client-based approaches that solve the multi-language issue with different approaches than the pure increase of data to be transmitted can be scored with 1. The criteria used to evaluate the SA_m parameter reflect the importance of providing the identity of the issuing entity in the willingness to act of citizens upon warning reception. Finally, the evaluation criteria for the parameter SN_m stress that accessible public warning systems are by far more effective when the population at risk as a whole is targeted, according to the discussion in [JOH 10, LAN 10, BRY 10, MUE 10]. The highest score is given to systems with high level of inclusion, i.e. addressing sensorial but also cognitive impairments [SUL 10, BRY 10].

8.7. An overview of public warning systems

8.7.1. *MoWaS – the public warning system in Germany*

The public warning system in Germany, MoWaS (*Modulares WarnSystem*, Modular Warning System) is highly influenced by the Federal characteristic of the civil protection responsibility. Indeed, civil protection is handled in Germany in a shared manner between the Federal and the State level, where the Federal level is responsible for those risks that can become a case of defence; the rest are the responsibility of the State level and may scale down into regional level depending on the range of the risk [FED 15b, HTT 15b].

The preceding system to MoWaS, SatWaS (*Satelliten-basierte WarnSystem*, Satellite-based Warning System) was deployed under the responsibility at Federal level. SatWaS connects the warning terminals located in situation centres to a warning server over satellite, which is at the same time connected over satellite to other situation centres and the so-called "*Multiplikatoren*" (which could be translated into English as "repeaters") in a star topology. These repeaters are dissemination nodes connected to specific media broadcasters, internet providers, and paging services (e*Message), press agencies and the Deutsche Bahn AG (the German national railway operator) [FED 15a, FED 15b, HTT 15c, HAR 11]. Given that SatWaS was a warning system at Federal level, it could only be applied in case of threats under the responsibility at the Federal level.

For a complete public warning coverage in all types of threats, also those that are under the responsibility of State, even regional levels, SatWaS evolved towards MoWaS, the Modular Warning System. The main objective behind MoWaS is to allow any official in charge of civil protection to activate all warning channels connected to MoWaS in his area of responsibility in a decentralised manner and without discontinuity of media use [FED 15b]. MoWaS allows addressing the warning messages at scalable areas (municipalities, regions, States and the Federal Republic), enabled through a user interface based on a geographical information system (GIS). Through its compatibility with the Common Alerting Protocol [OAS 04], the de-facto standard for description of warning messages and other protocols (such as POCSAG, typically used for paging services), the MoWaS system can deliver warning messages to a variety of dissemination channels. Furthermore, its modular architecture enables low-complex integration (from the technical perspective) of further channels.

Currently, MoWaS disseminates warning messages over media broadcasters to radio and TV, pagers and internet connected devices (e.g. PCs, tablets or mobile phones). In June 2015, the application for mobile devices called NINA was launched by the Federal Office of Civil Protection and Disaster Relief in Germany to deliver warning messages to mobile devices with internet connection. MoWaS is actually operational since 2013 in the whole German Republic, after a pilot phase that started at State level in a few States. However, MoWaS is continuously under development, exploring potential further channels that could be integrated to the warning system. Some examples are [MOW 15]:

1) cell Broadcast;

2) regional warning systems (such as KatWarn [HTT 15d] offering warning messages to mobile devices at regional level via SMS, email or App);

3) wireless connected sirens over the radio signal dedicated to authorities and organizations with security and safety duties (BOS);

4) smoke detectors connected via POCSAG.

8.7.2. *NL-Alert – the public warning system in the Netherlands*

The Netherlands was pioneer in Europe in the deployment of a public warning service based on Cell Broadcast, in compliance with the ETSI standard EU-Alert [ETS 15b], to overcome the weaknesses of the previous warning system based on a network of sirens. Indeed, the use of Cell Broadcast for public warning, even if technically simple to solve, requires agreements with network providers to activate the Cell Broadcast channel in the deployed base stations. This is normally not done by default due to resource allocation (and hence economic) reasons. Furthermore, it is required to activate the Cell Broadcast reception option in the mobile devices to receive the warning messages, also referred to as "*opt-in*" requirement.

The Netherlands has succeeded in setting agreements with three commercial mobile networks operators to activate the Cell Broadcast channel and went operational in 2012. The support of NL-Alert in mobile networks has been actually mandated under the Dutch Telecom law since December 2014. Furthermore, the *opt-in* requirement has been relaxed with time, as several smart phone manufacturers set the reception of NL-Alert warning messages as default. Still, mobile phone users can deactivate the function [KUB 13]. The Ministry of Security and Justice in the Netherlands (responsible for public warning) deployed a nation-wide public awareness campaign in 2012 and 2013 advertising the NL-Alert warning service in national TV and social media to minimize the risk of citizens deactivating the service [KUB 13].

An added difficulty to Cell Broadcast is that it is supported over 2G and 3G networks in the Netherlands, but not yet over 4G networks due to backward compatibility issues of the relevant standards. Nevertheless, 4G solutions to deliver Cell Broadcast-like warning messages in the Netherlands will be implemented before the end of 2015. Furthermore, additional improvements to include rich media in warning messages over 4G networks are planned [SAN 15].

8.7.3. *BE-Alert – the public warning system in Belgium*

Belgium has deployed a multi-channel public warning system, where the warning channel or channels to be used from the available ones may be selected on the basis of the crisis situation characteristics by the warning officials. The warning service requires user registration and offers the following warning delivery means [HTT 15e]:

1) call automatons to fixed and/or mobile telephones;

2) SMS to mobile phones;

3) text warning messages per email;

4) text warning messages per fax;

5) *Twitter* and *Facebook* messages in the municipality, province and national crisis centres' profiles in such social media.

In all cases, the channels applied require the active registration to the service or following the social media profiles of the relevant authorities. Nevertheless, it is under consideration to further extend the system to include Cell Broadcast [HTT 15e].

8.7.4. *SAIP – the future public warning system in France*

With a first deployment in 1950s, the public warning system in France, the RNA (*Réseau National d'Alerte*, National Alerting Network), was composed of a network of fixed and mobile sirens and speakers mounted on vehicles. A project to consolidate, complete and improve the public warning systems in France has been started in 2009: SAIP (*Système d'Alerte et d'Information des Populations*, Public Warning and Information System.

In addition to the RNA system, SAIP will deploy additional means including national and local media, call automatons (voice calls, SMS, fax and email – all of these services being subject to subscription), broadcasting service, i.e. warning operators of broadcasting services and relying that information will be further disseminated by them (e.g. for variable message signs of radio stations). Furthermore, SAIP will be activated by a software application integrating a map-oriented interface that enables the authorities at various levels to securely trigger alert channels. Additionally, the alerting network will be reinforced by the use of public media (France Television and Radio France) for alerting dissemination as well as for providing additional information [HTT 15f].

The deployment of the warning service over RDS (Radio Data System) and Cell Broadcast are also being considered in SAIP, given its efficiency in the broadcast of warning messages; in the case of RDS, the warning message is distributed to all compatible radios in the coverage area of radio repeaters in the risk zone; in the case of Cell Broadcast, only one message needs to be sent over each base station in the affected area to reach all citizens holding an activated mobile phone. However, the related legal framework to deploy Cell Broadcast as public warning channel is still a barrier that needs to be solved [HTT 15g].

SAIP has an alerting and an information function. The alerting function is meant to warn the population in case of imminent or immediate danger; the information function is meant to provide the population with some security advices to be followed. In order to obtain further information, citizens are requested to listen to the public channels of radio or TV [HTT 15f]. SAIP is currently in the deployment phase. A first pilot was carried out in the department of Rhône and a second one included Bouches-du-Rhône. Full deployment is planned by 2020 [HTT 15f].

8.7.5. *Integrated Public Alert and Warning System (IPAWS) in USA*

IPAWS is an alerting system that is conceived in order to implement "an effective, reliable, integrated, flexible and comprehensive system to alert and warn the American people" [EXE 06]. IPAWS is developed in a public-private partnership (PPP) between the Federal Emergency Management Agency (FEMA) together with the Department of Homeland Security and several public and private industry partners, so as to use as many communication channels as possible to disseminate warning messages in a rapid and authenticated way. Furthermore, the IPAWS solution is applicable at local, territorial, state and federal levels.

In fact, IPAWS is intended to be an improved version of the Emergency Alert System (EAS) (see below) diversifying its use and modernising the architecture and communication channels to be used for warning dissemination. Additionally, the system is intended to integrate and assure seamless transmission of the message through all national alert and warning networks, like the National Oceanic and Atmospheric Administration (NOAA) weather radio [HTT 15h]. IPAWS also provides the open platform for emergency networks (OPEN). This is a web service that enables the creation and routing of warning messages by third-party applications, systems, networks and devices, assuming compatibility with the Common Alerting Protocol (CAP) [ETS 15b]. Warning messages fed into OPEN are then further disseminated via radio or television broadcast, NOAA Weather radio, Internet-based systems, mobile phones and other dissemination services.

The Emergency Alert System (EAS) is a national multi-channel warning system in the U.S. operating since 1994 and delivering warning messages at nation or area specific levels over: TV and Radio Broadcasts, cable TV, satellite broadcast systems (such as Sirius Satellite Radio, World space, XM Satellite Radio, DAB) and wire line video service [HTT 15i].

The use of Cell Broadcast in IPAWS is integrated by the Commercial Mobile Alert Service (CMAS), specified in [ATI 15] CMAS-compatible handsets have the Cell Broadcast function activated, so that users do not have to opt-in. They can be identified by a CMAS compatibility label. Furthermore, a unique tone signal and vibration cadence for warning messages must be provided in such devices for inclusive warning [HTT 15j].

8.7.6. *Earthquake Early Warning (EEW) and J-Alert in Japan*

The design of public warning systems in Japan are highly influenced by the major risks experienced in the area: earthquakes causing tsunamis, where the lead times of the hazard are extremely short. Depending on the distance of the earthquake epicenter to the coast, the lead time of a tsunami can be in the range of seconds. An advanced infrastructure of seismic sensors (over 4,000) deployed over and around Japan (in the ocean) feed an automatized early warning system at the Japan Meteorological Agency that calculates the epicentre and the intensity of the earthquake to warn the population in case the intensity exceeds an established threshold [JMA 15].

The dissemination of warning messages can be done through different channels: the "Area Mail" service, similar to Cell Broadcast, combined with a paging channel. The paging channel is devoted to provide very rapidly (within 4 s) basic information (earthquake or tsunami), and the mobile device will only display a pre-defined message. A second notification is then disseminated with further information using the "Area Mail" service [SAN 15, RMD 15]. Moreover, a satellite-based public warning system (using the Superbird-B2 satellite operating in Ku- and Ka-bands [HTT 15l]) is available since 2007 that allows authorities to broadcast warning messages to local media and also directly to citizens by means of a connected loudspeakers system [HTT 15k].

An interesting feature of the warning messages issued by the EEW is that Emergency Warning Broadcast system automatically turns on and tunes the TV and radio in the areas of risk and tunes them into relevant channels that provide the warning message. Furthermore, the warning message is delivered in several languages and TVs show a flashing window with information about the epicentre of the earthquake and the areas of risk. Moreover, the mobile phones that are compatible with this warning system will be notified with a pop-up window on the handset screen (no need to open a message) and a special ring tone [RMD 15]. Hence, the system is intrusive and inclusive.

8.8. Conclusions

Public warning systems have gained a great momentum in the last decade, taking advantage of the great advances in personal communication and entertainment devices and their proliferation in society. The accessibility to such devices has boosted a culture of always staying connected and informed that needs to be responded also in crisis situations by public warning systems, giving people at risk the needed information to act and protect themselves.

Indeed, many countries have upgraded their public warning systems in the last years with somewhat similar approaches: multi-channel warning is understood as a must to reach the maximum of people at risk. The complementarity of different communication means to fulfil public warning requirements and address different segments of society drives the adoption of multi-channel approaches. Still, significant differences are observed in already deployed or planned solutions, e.g. the combination of used channels, level of intrusiveness and suitability of applied solutions to make public warning accessible to all segments of population, including the most vulnerable ones. Furthermore, each country must face its own civil protection administrative and governance challenges, which may significantly influence the adopted solutions. In addition, the use of personal communication and entertainment systems for public warning purposes requires private-public partnerships (PPPs), agreements between public administrations and operators or the establishment of regulations. It is worth noting that interoperability of public warning systems is still a challenge to be solved, especially in the European Union, where the challenges related to interoperability are higher due to decentralised organization of civil protection (that has also large advantages with regard to other aspects) and solutions are highly fragmented.

These pages have provided an overview of requirements for public warning systems, communication and entertainment systems that can be applied for public warning purposes, an evaluation method that can be applied to assess the effectiveness of public warning systems and a few examples of existing or planned systems. The evaluation model presented in these pages can be applied by administrations to identify gaps in the fulfilment of public warning requirements and design public warning solutions consequently. Furthermore, the qualitative nature of the proposed approach allows adapting the evaluation scores to the perceived importance of each parameter by the designer for those parameters that are not subject to statistics.

8.9. Bibliography

[ALL 10] ALL ABOUT SYMBIAN, "Smartphone situation in Europe", available at http://www.allaboutsymbian.com/news/item/11342_comScore_data_shows_smartphone.php, 2010.

[ATI 15] ATIS 0700006 Commercial Mobile Alert Services (CMAS) via GSM/UMTS Cell Broadcast Service Specification, 2015.

[ATW 11] ADVANCED TELEVISION WEBSITE, Research: satellite TV to overtake cable by 2015 in Western Europe, available at http://www.advanced-television.com/index.php/2011/05/20/research-satellite-tv-to-overtake-cable-by-2015-in-western-europe/, 2011.

[AUS 05] AUSTRALIAN GOVERNMENT, Emergency management Australia Evacuation planning, 2005.

[BEA 15] BE ALERT, Alert in emergeng situations, available at: http://be-alert.be/de/, 2015.

[BER 14] BERIOLI M., DE COLA T., RONGA L.S. *et al.*, "Satellite assisted delivery of alerts: a standardisation activity within ETSI", *Proc. of 7th Advanced Satellite Multimedia Systems Conference and the 13th Signal Processing for Space Communications Workshop (ASMS/SPSC)*, Livorno, Italy, September 8–10, 2014.

[BRI 10] BRICOUT J., BAKER P., "Leveraging online social networks for people with disabilities in emergency communications and recovery", *International Journal of Emergency Management*, vol. 7, no. 1, 2010.

[BRY 10] BRYEN D., "Communication in times of natural or man-made emergencies: the potential of speech-generating devices", *International Journal of Emergency Management*, vol. 7, no. 1, 2010.

[CAL 08] CALIFORNIA EMERGENCY MANAGEMENT AGENCY, Alert and Warning, Report to the California State Legislature, 2008.

[CEN 09] CENTERS FOR DISEASE CONTROL AND PREVENTION, Crisis and Emergency Risk Communications: Best Practices, 2009.

[COO 11] COOMBS W.T., *Ongoing Crisis Communication: Planning Managing and Responding*, 3rd edition, Thousand Oaks: SAGE, 2011.

[CRA 10] CRANDALL B., BENTZEN B., MYERS L., "Evaluation of emergency egress information for persons who are blind", *International Journal of Emergency Management*, vol. 7, no. 1, 2010.

[DE 15] DE COLA T., PÁRRAGA NIEBLA C., "Alerting over satellite navigation systems: lessons learned and future challenges", *IEEE Communications Magazine*, pp. 178–185, May 2015.

[EC 11] EUROPEAN COMMISSION, *Humanitarian Aid & Civil Protection Website*, http://ec.europa.eu/echo/civil_protection/civil/vademecum/menu/2.html, 2011.

[EC 15] EUROPEAN COMISSION, EGNOS, http://ec.europa.eu/enterprise/policies/satnav/egnos/programme/index_en.htm, 2015.

[EFF 00] EFFECTIVE DISASTER WARNINGS, Report by the Working Group on Natural Disaster Information Systems Subcommittee on Natural Disaster Reduction; National Science and Technology Council Committee on Environment and Natural Resources, November 2000.

[EME 15] E*Message, http://www.emessage.eu/, 2015.

[ESA 15a] EUROPEAN SPACE AGENCY (ESA), GALILEO System Simulation Facility, available at http://www.gssf.info/GSSF%20subpage%20GSSF%20analyses%20-%20Visibility.htm, 2015.

[ESA 15b] EUROPEAN SPACE AGENCY, EGNOS Fact Sheet, Safety of Life for Aviation, available at http://download.esa.int/docs/Navigation/factsheet_SoL.pdf, 2015.

[ETS 15a] ETSI, TS 102 900: Emergency Communications (EMTEL) – European Public Warning System (EU-ALERT) using the Cell Broadcast Service, 2015.

[ETS 15b] ETSI, TS 102 182: Emergency Communications (EMTEL) – Requirements for communication from authorities/organizations to individuals, groups or the general public during emergencies, 2015.

[EUR 03] EUROSTAT, "Cinema, TV and radio in the EU. Statistics on audiovisual services. Data 1980-2002", European Commission. Theme 4: Industry, trade and services, 2003.

[EUR 04] EUROSTAT, "How Europeans spend their time. Everyday life of women and men. Data 1998-2002", European Commission. Theme 3: Population and Social Conditions, 2004.

[EUR 05] EUROSTAT, "Consumers in Europe. Facts and Figures. Data 1999-2004", European Commission. Theme 3: Population and Social Conditions, 2005.

[EXE 06] EXECUTIVE ORDER 13407, Public Alert and Warning System, Federal Register, vol. 71, no. 124, June 2006.

[FCC 15a] FCC, available at: https://www.fcc.gov/guides/emergency-alert-system-eas, 2015.

[FCC 15b] FCC, available at: https://transition.fcc.gov/pshs/services/cmas.html, 2015.

[FED 15a] FEDERAL OFFICE OF CIVIL PROTECTION AND DISASTER ASSISTANCE, Alert of the Public via the Satellite-Based Warning System (SatWaS), available at http://www.bbk.bund.de/SharedDocs/Downloads/BBK/EN/booklets_leaflets/Flyer_Satellite-based-Warning-System.pdf?__blob=publicationFile, 2015.

[FED 15b] FEDERAL OFFICE OF CIVIL PROTECTION AND DISASTER ASSISTANCE, Protection and Aid for the Population, available at http://www.schutzkommission.de/SharedDocs/Downloads/BBK/EN/booklets_leaflets/Protection_a_Aid_%20f_the_Population.pdf?__blob=publicationFile, 2015.

[FED 15c] FEDERAL MINISTRY OF THE INTERIOR, Civil protection and disaster management, available at: http://www.bmi.bund.de/EN/Topics/Civil-Protection/Civil-Protection-isaster-Management/civil-protection-disaster-management_node.html, 2015.

[FEM 15] FEMA, available at: https://www.fema.gov/integrated-public-alert-warning-system, 2015.

[FIT 91] FITZPATRICK C., MILETI D.S., "Motivating public evacuation", *International Journal of Mass Emergencies and Disasters*, August 1991.

[FOU 15] FOUR ELEMENTS OF PEOPLE CENTRED EARLY WARNING SYSTEMS, Early Warning Systems – A Public Entity Risk Institute Symposium available at: www.unisdr.org, 2015.

[GLA 04] GLANTZ M.H., "Early warning systems: do's and don'ts", February 6, 2004; Report of Early Warning Systems Workshop, Shanghai, China, October 20–23, 2003.

[GRA 07] GRASSO V.F., "Early warning systems: state-of-the-art analysis and future directions", United Nations Environment Programme (UNEP), 2007.

[GUN 15] GUNTER'S SPACE PAGE, Superbird 4, available at : http://space.skyrocket.de/doc_sdat/superbird-4.htm, 2015.

[HAL 06] HALONEN A., VERBOEKET R., HEDIN S., "Study report on alarm systems and early warning in Baltic Sea Region", Eurobaltic Civil Protection Project, 2006.

[HAR 11] HARITZ M., *The Public Alert System in Germany*, 8th GESA Conference, June 2011.

[IDA 09] IDATE CONSULTING & RESEARCH, Broadband Coverage in Europe, Final Report, Survey 2009, 2009.

[INT 09] INTERNATIONAL FEDERATION OF RED CROSS AND RED CRESCENT SOCIETIES, World Disasters Report 2009; Focus on Early Warning, Early Action, available at www.ifrc.org, 2009.

[INT 13] INTERNATIONAL FEDERATION OF RED CROSS AND RED CRESCENT SOCIETIES, World Disasters Report 2013; Focus on technology and the future of humanitarian action, available at: www.ifrc.org, 2013.

[INT 15] INTERNATIONAL FEDERATION OF RED CROSS AND RED CRESCENT SOCIETIES, Community early warning systems: guiding principles, available at: www.ifrc.org, 2015.

[ISO 15] ISO 22322:2015 SOCIETAL SECURITY, Emergency management – Guidelines for public warning, 2015.

[ITE 09] INTERNATIONAL TELEVISION EXPERT GROUP, German TV market Report 2009, available at: http://www.international-television.org/tv_market_data/tv-market-germany.html, 2009.

[ITU 10] INTERNATIONAL TELECOMMUNICATIONS UNION ITU ICT Data and Statistics, Mobile Cellular Subscriptions, 2010.

[IWS 13] INTERNET WORLD STATS, European Union Internet Directory, available at: http://www.internetworldstats.com/eu/, 2013.

[JMA 15] JAPAN METEOROLOGICAL AGENCY, available at: http://www.jma.go.jp/jma/indexe.html, 2015.

[JOH 10] JOHNSON J., MITCHELL H., LAFORCE S. *et al.*, "Mobile emergency alerting made accessible", *International Journal of Emergency Management*, vol. 7, no. 1, 2010.

[KAT 15] KATWARN, Warning and information system for the public, available at: http://www.katwarn.de/, 2015.

[KLA 11] KLAFFT M., MEISSEN U., "Assessing the economic value of early warning systems", *Proceedings of the 8th International Conference on Information Systems for Crisis Response and Management (ISCRAM 2011)*, Lisbon, Portugal, 8–11 May 2011.

[KUB 13] KUBBEN P., "NL-Alert", presentation in EENA annual conference, Riga, Latvia, 17-19 April 2013.

[LAN 10] L P., HOSKING I., "Inclusive wireless technology for emergency communications in the UK", *International Journal of Emergency Management*, vol. 7, no. 1, 2010.

[MIL 09] MILETI D.S., Warning messages and public response, Social science research findings & applications for practice, Internal report, August 2009.

[MIL 90] MILETI D.S., SORENSEN J.H., Communication of emergency public warning: a social science perspective and state-of-the-art assessment, Internal report, August 1990.

[MIL 95] MILETI D.S., Factors related to flood warning response, U.S. Italy Research Workshop on the Hydrometeorology, Impacts, and Management of Extreme Floods Internal report, Italy, 1995.

[MIN 15] MINISTÈRE DE L'INTÉRIEUR,, available at: http://www.interieur.gouv.fr/, 2015.

[MOW 15] MOWAS, available at: http://www.bbk.bund.de/SharedDocs/Downloads/BBK/DE/Publikationen/Broschueren_Flyer/Flyer_MoWaS.pdf?__blob=publicationFile, 2015.

[MUE 10] MUELLER J., MORRIS J., JONES M., "Accessibility of emergency communications to deaf citizens", *International Journal of Emergency Management*, vol. 7, no. 1, 2010.

[MUL 14] MULERO CHAVES J., PÁRRAGA NIEBLA C., "Effectiveness evaluation model for public alert systems," *Disaster Advances*, vol. 7, no. 8, August 2014.

[OAS 04] OASIS COMMITTEE, CAP: Common Alerting Protocol v. 1.0 Specification, Internal report, March 2004.

[PER 07] PERRY RONALD W, LINDELL MICHAEL K., *Emergency Planning*, John Wiley & Sons, 2007.

[PPW 04] PARTNERSHIP FOR PUBLIC WARNING (PPW), An introduction to public alert & warning, available at: www.partnershipforpublicwarning.org, 2004

[PRE 08] PREVENTIONWEB, Europe Disaster Statistics. Region Profile for Natural Disasters from 1980-2008, available at: reventinweb.net, 2008.

[PUY 15] PUY-DE-DÔME STATE SERVICES, warning system and population information, available at: http://www.puy-de-dôme.gouv.fr/le-systeme-d-alerte-et-d-information-des-a3500.html, 2015.

[RMD 15] RISK MANAGEMENT AND DISASTER PREVENTION SOLUTIONS, available at: https://www.ntt-review.jp/archive/ntttechnical.php?contents=ntr200812sf2.html, 2015.

[SAN 15] SANDERS P., WILLIAMS D. "EENA Operations Document – Public Warning", *European Emergency Number Association Event*, available at http://www.eena.org/uploads/gallery/files/pdf/2015_07_15_PWS_Final.pdf, 2015.

[SCP 11] SWISS CIVIL PROTECTION AUTHORITY, Alerting and informing the population, available at: http://www.bevoelkerungsschutz.admin.ch/internet/bs/en/home/themen/alarmierung.html, 2011.

[SUL 10] SULLIVAN H., HÄKKINEN M., DEBLOIS K., "Communicating critical information using mobile phones to populations with special needs", *International Journal of Emergency Management*, vol. 7, no. 1, 2010.

[TER 15] TERRADAILY, available at http://www.terradaily.com/reports/Japan_Launches_Alert_System_For_Tsunamis_And_Missiles_999.html, 2015.

[TOM 11] TOMTOM WEBSITE, *Three Brands Rule PND Market*, available at: http://www.yourtomtom.com/347/three-brands-rule-pnd-market.html, 2011.

[UN 06] UNITED NATIONS, Global Survey of Early Warning Systems: An assessment of capacities, gaps and opportunities toward building a comprehensive global early warning system for all natural hazards, Internal report, September 2006.

[UNI 06] UNISDR, "Developing Early Warning Systems: a Checklist", *3rd International Conference on Early Warning (EWC III)*, March 27-29, 2006, Bonn, Germany.

[UNI 09] UNISDR, Terminology on Disaster Risk Reduction, available at: www.unisdr.org, 2009.

9

Enhancing Disaster Management by Taking Advantage of General Public Mobile Devices: Trends and Possible Scenarios

9.1. Introduction

When a crisis occurs, particular infrastructures that are specific to the emergency situation are deployed in order to support rescue operations. Indeed, those normally

Chapter written by Olivier SEBASTIEN and Fanilo HARIVELO.

used may have suffered from the disaster resulting in an altered service or may not be available at all. This process concerns physical (logistics) as well as immaterial (information) items delivery. Nowadays, special emphasis has been put on the latter as it has been proven [CAM 12, TAN 09, LEF 14] that it is a major requirement to ensure that the former is realized properly and more generally to address the crisis. Generally, required information aims at answering the following basic questions: what (happened)? When (did it happened)? Where (is the disaster located)? And possibly, how (can it be reached)?

This is why research and development have led to creating technologies that can manage information acquisition and delivery in harsh environments: specific communication networks, rugged equipment, solar-powered devices, etc. Those means are dedicated to risk and emergency experts who are specially trained to develop under difficult circumstances and the tools they rely on are designed to cope with this context.

On the other hand, a new generation of mobile consumer electronic devices has appeared during the past decade: smartphones, tablets, connected devices, etc. They have been adopted all over the world at an astonishing rate so that Gartner claims that more than one billion units of smartphones were sold during 2014 [GOA 15]. This phenomenon was made possible by the combination of several factors: decrease of prices, improvement of many multimedia features and of course availability of an appropriate communication infrastructure. All of this has been the support for new usages which are very popular among the population, such as social networking, the permanent possibility of reaching somebody and the ability to produce information in addition to consulting what was produced by others (as it was in the past). Current development work aims at making digital personal devices more "aware" of user environment [KHA 13], in order to have them automatically assist him/her in each encountered situation.

Therefore, we may legitimately wonder whether it would be possible to take advantage of this new trend to gather information and communicate with victims when a crisis occurs. The idea is not to replace the professional equipment and networks but to add new ways of interacting with the terrain of a disaster at every stage of the phenomenon. In this chapter, we thus aim to present how this can be achieved, what the parameters that efficiency rely on are and discuss what issues can be encountered and why.

To see a realistic picture of the situation, we propose the following path: we will start by depicting an overview of crisis management as it is in mid-2010s, that is to say the actors and the way information technologies play a major role in their duty. Then, we will move onto mobile equipment and proximity communication aspects in order to show what can be expected from non-dedicated devices and network protocols in information gathering and *in situ* transmission. Finally, we will focus on how mobile devices and proximity network-assisted crisis management operations can be led.

9.2. Crisis management overview

The goal of this section is to shed light on the main processes that are implemented by the various actors of risk management, quite independently from how they achieve them. We will also evoke the difficulties that can be encountered. This will allow us in the forthcoming section do draw a link with the new trend we would like to describe using general public devices and infrastructures. We will start with listing the different actors involved.

9.2.1. *Actors*

As far as actors are concerned, a separation can be made between risk professionals and the population.

9.2.1.1. *Risk professionals*

This category of actors involves people that have acquired dedicated skills to manage emergency situations (including civilian volunteers that reside in the area) as well as stakeholders, who have to take decisions in the context of a disaster.

Professionals operating physically on the terrain are experts in various domains: rescue teams, firemen, emergency service unit (ESU) members, etc. Their profile of course depends on the type of disaster that must be addressed. They have two points in common. Firstly, the fact that they have been specifically trained to act accordingly under harsh circumstances and to make use of professional-level procedures and devices. Secondly, they evolve in the area of the crisis, that is to say they do not have an overview of the situation, they work at a very localized level.

Stakeholders on the other hand have a globalized view in order to allow them to take decisions in addressing the crisis. What is critical to them, more than the former, is to have reliable information in the shortest time possible. For example, distribution of the population density at the time a disaster strikes is precious information to set priorities in emergency resource allocation because it will increase chances of rescuing more people within a single terrain operation.

Both terrain experts and stakeholders share the fact that they are very few in comparison to the population. This is a drawback as well as an advantage as they are clearly identified and a clear communication channel between them exists most of the time.

9.2.1.2. *Population*

People living in the area where a disaster occurs constitute most of the actors. They are the potential victims as an unknown amount of them at the time a phenomenon strikes are to be impacted: they may be injured, buried, flooded or even if they are safe, they may need supplies (food, water, covers, etc.). They await a response from the professionals mentioned before.

One of the main aspects here is that people are permanently in the area where the crisis happens: before, during and after. Moreover, most of them have a good knowledge of this area.

We therefore aim at taking into account this fact as an advantage in the following section that will describe how general public mobile devices can be used to enhance disaster management for each stage of the process.

9.2.2. *Different stages*

In this section, we will list the different stages of the crisis according to [TAN 09] but adding the period before its occurrence too. For each of them, we will focus on how consumer mobile equipment can assist victims and experts and what the limitations raised from a conceptual point of view are.

9.2.2.1. *Before a crisis: early warning system*

A disaster may not necessarily be expected or even if it can be predicted some time before the strike, location of the areas that are to be the most damaged along with the precise nature of the threat cannot be expected. This is why, as described by [CHA 11], stakeholders have to anticipate by proceeding to:

1) risk assessment;

2) safe areas identification;

3) creation of disaster-proof building codes;

4) early warning system implementation.

Information of course plays a major role to achieve these tasks and professional high-end techniques and technologies, as mentioned in [LEF 14], are required.

But general public devices can also play a role as far as task 4) is concerned. Indeed, they first have the advantage to be permanently located on the terrain, which means that there is no costly deployment to be done. Second, there are many of them, generally from hundreds to thousands for large cities.

This is why they can be used to detect the early signs of a disaster in a way that will be described hereafter. Of course, this does not replace the scientific tools but it is an auxiliary means of gathering early information about a trend (from a statistical point of view) rather than granular data. Indeed, such an expert network already exists: the disaster charter is based on a list of authorized users from all over the world. The idea is to build a complementary network involving non-specialized users to build an alert map similar to the one available at https://www.disasterscharter.org/web/guest/activations/activations-map

The next step of the process takes place once the disaster strikes.

9.2.2.2. *During crisis: victim localization and assistance system*

When the catastrophe occurs, the main priorities are to search and rescue victims while at the same time assist people that are not injured and guide them to a safe place.

From an information point of view, the questions for ESU teams are: who? From where? To where? Time is also a major constraint [TAN 09]. In this context, popular devices like smartphones or connected devices can help to acquire this information and communicate between professionals and the population. Indeed, nowadays devices embed lots of sensors which can be used to capture data on the immediate environment of their owner.

Actually, we could even think that it is already a widely spread process now due to the very common telephone service. But we have to take into account the fact that the normally available communication infrastructure may be totally or partially down at that time. Moreover, from a user point of view, a victim may not be able to make use of their mobile phone for many reasons (he/she may be unconscious, buried under rubble, unable to move arms or fingers, etc.).

An appropriate solution must therefore be designed to support population assistance in those harsh conditions. Propositions will be given in an upcoming section of this chapter.

9.2.2.3. *After crisis*

Once the crisis is over, that is to say emergency tasks have been accomplished at best and no other threat is expected (for example, the aftershock after an earthquake), time for resilience takes place. The goals are to rebuild what has been destroyed or damaged, access and analyze the disaster (human and material casualties), and take measures to improve the management if another catastrophe happens in the future.

The impact of general public mobile devices may be smaller here than in the two previous stages because resilience is not a matter of emergency and relies a lot on scientific data. However, they can be used to acquire auxiliary data to achieve tasks 1) and 2) of [CHA 11]. They can indeed be useful to gain a simple overview of areas that need a special attention by allowing users to report priorities (according to them) in their neighborhood.

The limitation here lies in the fact that too much non-sorted data may be acquired from such a system, as there is no validation from specialists.

9.2.3. *Resulting information chain*

Given this overview, we can propose a specific information chain, which is a variation of the information chain shown by [TAN 09]. The result is visible in Figure 9.1.

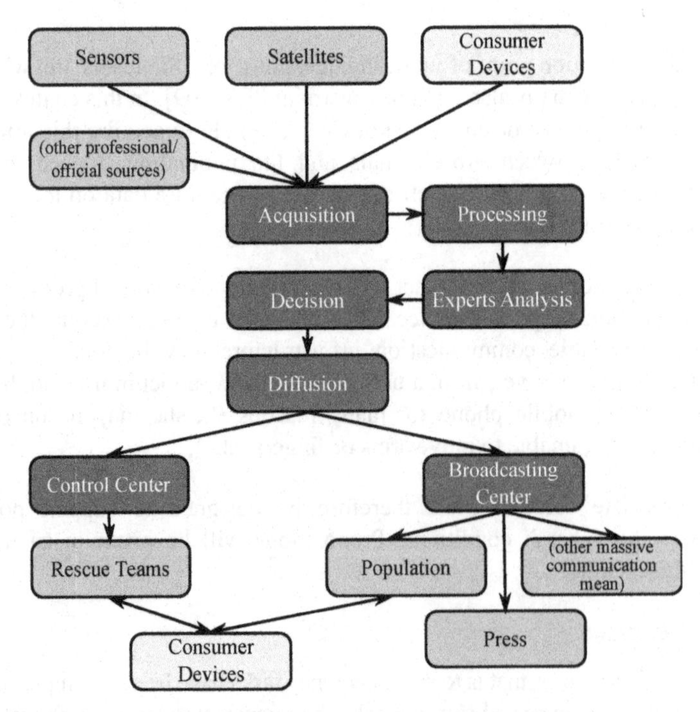

Figure 9.1. *Resulting information chain*

To build it, we include the inputs and outputs of the consumer mobile devices, as mentioned before, where they are useful.

The arrows show that communication can be mono-directional (for data acquisition purposes) as well as bi-directional (for victim localization and terrain communication), but the data processed is not the same according to the stage of the chain. Technical mechanisms which enable this in this particular context will be presented and discussed in a forthcoming section.

9.2.4. *Perimeter of action*

It is necessary to draw the limits of such a consumer-equipment-assisted process. First, we must realize that data obtained from the devices is not guaranteed to be exact, without mentioning that data directly given by users themselves (i.e. manually and voluntarily entered) can be erroneous too (on purpose or involuntarily). This is why our goal is less to provide a one-to-one communication system than to enable the creation of a map showing the spots where probability is high to have an event occurring concerning lots of persons, thus following a "best-effort" philosophy.

Second, this process raises the question of the interface between professional equipment, networks, procedures on one hand and consumer devices and infrastructures on the other. The expert environment has priority and must be secured from the consume's so that if the latter is faulty (or attacked), it does not bring down the former along. Consequently, implementation limits are also given by this rule: an information source must be dropped whether there is a probability that it can damage or alter the primary professional risk management environment.

Third, by nature, there is no obligation of result on the general public part of the system. It means that it may be partially or totally inoperative without creating a new threat in the context of the disaster.

That said, we will now focus on more technical aspects regarding the implementation of such a system.

9.3. Mobile equipment and proximity communication

In this section, we will present the technical features of average consumer devices nowadays, the communication networks that can be used to reach the goals described before and finally the architecture that links them.

9.3.1. *Mobile devices*

9.3.1.1. *Definition*

First, we have to define more precisely what we call mobile devices. Indeed, this phrase can match several types of devices as shown in [SEB 14]. We will focus on the following ones, as they are the more appropriate given the context of our chapter:

– smartphone: advanced mobile phone that is fitted with various input/output interfaces along with a good processing power and at least 2 network interfaces (mobile phone (3G/4G) and Wi-Fi);

– smart watch: a small computer housed in a watch-like format that is attached to the user's arm. It is generally equipped with biometric sensors and Bluetooth and Wi-Fi network interfaces. The most popular models work paired with a smartphone, hence a communication link between both;

– activity tracker: usually looks like a bracelet that measure biostatistics to keep logs of physical condition of the user. The user interface (UI) is very limited (a few buttons, no screen) but the device is generally configured and piloted from another paired device (smartphone or computer) that may not be always connected because the tracker is autonomous. The networking interface is generally Bluetooth.

Those mobile devices we rely on share common specifications in addition to being mobile: they embed sensors, an easy-to-use UI, decent processing power and one or more wireless networking interfaces. However, they are very different as far as autonomy, form factor and UI are concerned.

Finally, one important characteristic, which is a strong hypothesis for us, comes from the fact that they are attached to a person rather than group of persons or a place. We therefore made the assumption that one kind of device corresponds to one person. Categorization is indeed possible because each of them has a different signature. One person can wear a smartphone, a smart watch and an activity tracker at the same time. In order not to be counted as three different people, it is possible to rely on each device signature, such as the MAC Address. Moreover, smart watches and activity trackers must usually be paired with a smartphone to work, which gives it a central role in the personal area data and communication management.

That said, we can browse the technical features that are related.

9.3.1.2. *Features*

In this section, we will present what kind of useful data can be acquired and transmitted from embedded equipment. We will first deal with sensors.

9.3.1.2.1. Relevant sensors

The idea in this section is not to make an exhaustive list of the typical sensors provided in an average smartphone in 2015 but to point out the few ones that can be used for disaster management:

– GPS: allows the recording of the current position (given the fact that conditions are good enough);

– camera: allows the taking of still of images and video enabling transmitting a view of a devastated area. Nowadays, due to GPS, the location where the image has been taken is automatically recorded in the EXIF picture metadata;

– microphone: allows the user to record voice and communicate with others. Also allows capturing environment noise to try to define it if a victim is unconscious under rubble (in this case, GPS may be useless);

– speaker: allows us to communicate with others but also to stimulate an unconscious victim with specific sounds to wake him/her;

– light sensor: allows us to measure light at the surface of the device. Initially, this sensor was integrated to adapt display brightness to ambient condition, especially to dim it when the user holds it near his/her ear. In our context, it can be used to determine the kind of environment a person is in according to the time of the day. For instance, if there is very low amount of light on the sensor at noon, it may indicate that device owner is in a closed area;

– biometric sensors: usually measures heart beat and/or blood pressure. When sensors are attached to a watch or an activity tracker, measuring is permanent whereas when embedded in a smartphone, the user has to put a finger on the sensor to acquire a measure. The former is far more interesting for us in order to have an overview of the owner's health;

– accelerometer/gyroscope: allows us to determine the device's position. It is rather difficult to deduce a problematic situation for the owner from this information because there are lots of normal situations where the device could be either vertical or horizontal. However, the idea here is to compare this information between groups of users in a particular area to determine if there are similarities at a given time, possibly indicating people lying on the ground. Experiments are led to evaluate how it can be used to detect earthquakes [LAW 14].

These features can of course be used in combination. Indeed, in the pre-disaster stage, some recent studies [MIN 15] have been led to check whether accelerometers and GPS can be used to automatically detect seismic activity in a more reliable way than using only one sensor. Of course, the results will have to be updated according to technological evolution: sensor precision is expected to increase with newer devices [LAW 14].

We must also add the fact that other types of sensors can be embedded in devices, such as barometers, temperature or humidity sensors. But at the time this chapter was written, they were featured in a limited number of high-end devices, whereas the ones we listed above are very common.

Let us now focus on network interfaces.

9.3.1.2.2. Network interfaces

Nowadays, lots of consumer wireless communication standards are proposed with very different specifications. Actually, manufacturers tried to find the best compromise between range, speed, reliability, energy requirements, etc. This led them to build devices that embed several interfaces, such as smartphones, in order to use the most appropriate one for a given situation.

In the risk management context, this is a desirable feature as those devices can play a bridge role between a device having a short-range connectivity (a bracelet for instance) and a broader-range network.

Finally, common consumer network interfaces remain evolution of standards that exist since a dozen years [SEB 14]:

– UMTS and now long-term evolution (LTE) have the maximum range but an average speed;

– Wi-Fi is adapted to a domestic use, that is to say, it has an average range but at a rather high speed;

– Bluetooth was created to achieve personal-range communication. It has low range and speed but consumes less energy than the two other standards.

Sebastien *et al.* [SEB 14] depict the differences between these interfaces in terms of capacity, range, frequencies, etc. Of course, other wireless communication standards exist, however, they are quite rare or very specific, that's why we restricted ourselves to those above.

9.3.1.2.3. Software development concerns

Hardware features presented before give interesting perspectives regarding our goal but they have to be driven accordingly by the software. This raises many concerns.

The first one deals with software environment. Indeed, we must take into account popular platforms market shares to create a solution that works on most user devices. The idea there is to broaden the range of potential users that will involve in the system. Early 2015 figures [GOA 15] show that Google's environment Android is the leading mobile operating system (OS) with about 80% market share. Thus, its software development environment is to be selected. It has the reputation of being less restrictive than its main competitors (Apple iOS and Microsoft Windows Mobile) as far as installing programs and accessing on-board sensors and services are concerned. But competitors should not be put aside too. They can easily be involved in the system by using Android devices as gateways for them. Thus, despite being more closed environments, they can provide data without having to redevelop specific workaround solutions to overcome the limits imposed by their respective companies. Only the sensor-related features are to be conceived according to the device. Finally, agreements could be made with leading brands to have privileged access to the hardware.

The second concern is energy-related: as we deal with mobile devices powered by a battery, it is mandatory to ensure that the additional consumption provoked by the disaster-management solution working in background is not too high. This is a severe issue as polling continuously and permanently data from sensors causes a high energetic cost. This is why various energy-saving strategies have to be set in order to limit this cost. Moreover, restrictions may be applied directly by the OS for that purpose. Finally, new rules[1] are set as default with each major update released, which can prevent a risk-related application to work correctly once updated. We are therefore currently facing regular changes in applied policies regarding specific access to sensors and interfaces, making the creation of a ready-to-release software (i.e. not demanding specific technical operations from a user, like routing or jail-breaking the terminal) sometime difficult.

The last concern deals with the differences existing between devices. They are from different manufacturers, different generations and even at a component level, several references exist for a given specification. Indeed, OSs like Android demand dedicated specifications to ensure a given version can run on a given hardware. However, those are minimal specifications. Thus, an accelerometer may have a finer precision than another one, both running the same software. To address this fact, current tests have been (and still are) led (for example by [LAW 14]) to determine how this variability can impact the conclusion that can be made of the measurements.

1 An example of such a new rule is the "Doze Mode" that is expected to be released with the next version of Android, currently called Android M, prior to being given a commercial name. This particular mode automatically decides which applications should be deactivated or put on standby when user is not using the device (i.e. at night).

9.3.2. *Device-to-device communication technologies*

Device-to-device (D2D) communication is a solution to the ever-increasing number of users and the ever-growing traffic demand from applications and services. It permits users physically close to each other to communicate directly and efficiently without going through an access point (AP) or a cellular infrastructural node (e.g. Evolved Node B (eNB) of LTE that connects the UE to the network). Thus, communication is made opportunistically on local links. This approach has many advantages such as the emergence of new proximity services and applications for the user and enhanced networking (extended coverage, improved quality of service (QoS), traffic offload, special efficiency gain, etc.) for the operator. If not specifically addressed, public safety network is a privileged ground for D2D communication-based network. Indeed, it constitutes a fallback network when classical wireless and cellular networks are not available or fail.

Operations primarily involve communicating devices but can also call on infrastructural nodes or entities for coordination and control. D2D communication comprises two minimal phases: discovery and data communication. Devices in geographical proximity have to discover each other before being able to exchange data directly.

Two major and promising evolutions of existing technologies have emerged recently: Wi-Fi Direct [WIF 09] for wireless local area network (WLAN) and 3GPP LTE D2D for Telecommunication Network [GPP 13, GPP 14].

9.3.2.1. *Wi-Fi direct*

Classical Wi-Fi networks require a client device to connect to an AP to join the rest of the network. Wi-Fi Direct makes it possible for two or more devices to communicate directly [CAM 13]. The basic Wi-Fi specifies an *ad hoc* mode for an infrastructureless network. However, the support of this mode is limited in existing devices and its usage introduces difficulties such as a lack of efficient power saving mechanism and an unpredictable QoS. Wi-Fi Alliance that promotes Wi-Fi technology decides to rely on widely deployed infrastructure mode for Wi-Fi Direct. It consists of software update and does not require specific hardware. Former versions of the technology are supported, with their corresponding ranges and rates, except IEEE 802.11b.

Nearby communicating devices form a peer-to-peer (P2P) group that can be viewed as an equivalent to a traditional Wi-Fi infrastructure network. Multiple P2P groups can coexist in the same vicinity. The device assuming the role of AP, after discovery and negotiation phases, is named P2P Group Owner (GO). Other devices of the group are called P2P Clients. A role in a group is merely functional and dynamic. A device can be member of multiple groups and assumes different roles in every group. In this case, the Wi-Fi interface is time-shared between those distinct roles.

Along with the discovery phase, an upper layer protocol such as Bonjour and uPnP can handle service publication and discovery. If present, deduced available services are considered to decide whether the group formation continues or not. It is a cross-layer approach in the sense that higher layer information are transported, recognized and used in link layer operations.

Acting as an AP, P2P GO provides network configuration, gateway function, security support and power saving coordination. Assuming those burdens can dramatically draw P2P GO battery. To save energy, it can switch to sleep mode according to two mechanisms: opportunistic power save (OPS) and notice of absence (NoA). OPS leverages the sleep period of P2P clients deduced from transmission scheduling. NoA is at the P2P GO's own initiative: it informs other devices of periods of absence.

9.3.2.2. *LTE D2D*

Two main enhancements introduced in 3GPP LTE release 12 address D2D communication referred to as proximity services (ProSe) and group communication (one-to-many) in public safety network known as Group Call System Enablers for LTE (GCSE_LTE). They intend to reuse as much as possible technologies of LTE especially in MAC and physical features, easing through this way, the integration with existing LTE network. Operations can be performed under the supervision of the operator for in-coverage situations or in a complete autonomous manner in out-of-coverage situations [FEN 13, PAN 15]. Communication can be direct from one end-UE to another end-UE or multi-hop by the use of at least an intermediary relay.

Discovery phase can be open or restricted. In open scenario, UE, given an identity, can discover other UEs or publishes its presence and, can be discovered by other UE. We note that identity can comprise a group identity useful for group communication. The process is realized mainly in link layer using pre-defined parameters. For restricted scenarios, the core network of the LTE system evolved packet core (EPC) assists the discovery process by tracking involved UE locations and providing necessary information for them to discover each other. It means that UEs are identified, authenticated and authorized according to the operator's policy and user consent to be discovered.

ProSe communication takes place on an LTE-based path established between the UEs. It requires allocated resource that can be done in two different manners whether the UEs are under the supervision on the network or not. For in-coverage situations, resource allocation is centralized. Transmission schedules, and thus used resources, are maintained by the network. Transmission features, like power control, are also managed by the infrastructure. For out-of-coverage situations, pre-defined resources are allocated in a pool. Transmitting UEs contend for resources using random algorithms such as carrier sense multiple access with collision avoidance (CSMA/CA).

In addition to in-band ProSe communication in within licensed spectra, 3GPP LTE Release 12 provided a way to established outband communication through WLAN technology (e.g. Wi-Fi Direct). EPC handles WLAN discovery and configuration. Third-party developers are offered API to develop proximity-based and proximity-aware services and applications. Users can benefit from a rich variety of evolving services and applications. The current specification release restricts the use of group/one-to-many communication to public safety networks. A use case of such a service is, for example, the broadcast of information to all UEs in a disaster area.

9.3.3. *Mobile-based networks*

The two presented technologies are the most serious candidates for device-to-device communication in the market. Wi-Fi Direct specifications were released in December 2009, while the Release 12 of 3GPP LTE was functionally frozen in March 2015. Several handheld and mobile devices are already equipped with Wi-Fi Direct. Deployment was made easier because of the upward compatibility of the technology with the previous versions of Wi-Fi: it boils down to a software upgrade on existing hardware. LTE D2D deployment presents more difficulties. To mention two of them, we can cite the roaming situation and interference management when multiple operators operate simultaneously on outband frequencies.

Wi-Fi Direct inherits security concerns from Wi-Fi. LTE D2D provides robuster security. It covers a wider range (about 1 km) compared to Wi-Fi Direct (about 200 m). LTE D2D operates on licensed spectra, offering in this way a better and more predictable QoS and provides more reliability as resource usage and power control are managed by the network. Configuration is expected to be less cumbersome in LTE D2D as it is done upstream. Either the network controls D2D communication establishment or it provides pre-defined parameters to UEs for them to function autonomously. As operations are network-assisted, they will drain less energy from the UEs. LTE D2D was designed to be a more complete platform for the deployment and running of third party proximity-based applications and services.

9.3.4. *Architecture*

A specific architecture is required to allow consumer mobile devices to communicate with the rest of the infrastructure, more specifically the one that is used by disaster managers. Figure 9.2 shows this architecture.

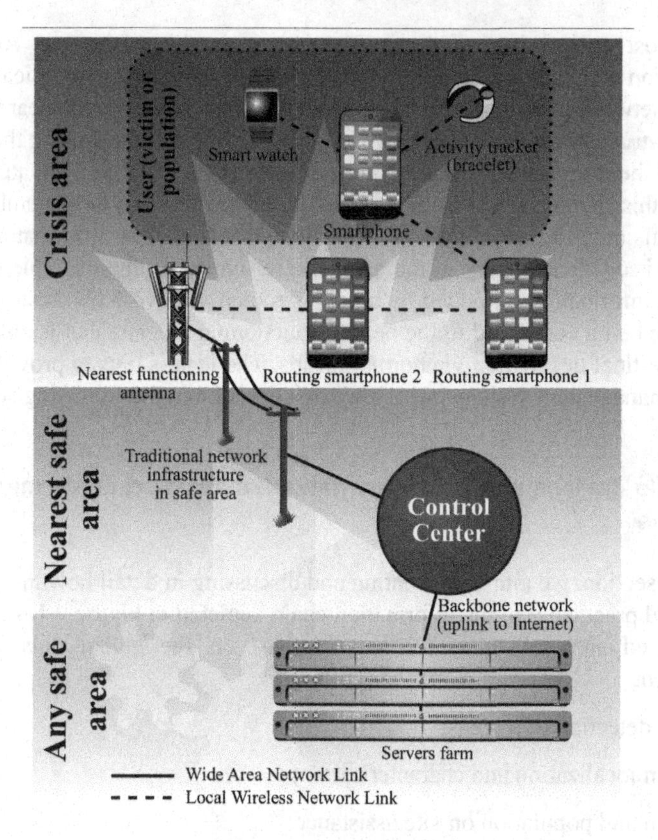

Figure 9.2. *Architecture*

Its goal is to allow information to be transferred bi-directionally between operators that are localized in three areas:

– crisis area: the place where population (including victims) is and where a disaster strikes. At this time, rescue units are also in this area but to communicate, they have specific professional tools that are built to work in those conditions, that is why they do not appear on the figure;

– nearest safe area: the place where the control center is in order to monitor the situation;

– any safe area: this is the place where the computing infrastructure is located. Actually, it could be anywhere in the world.

Stakeholders, for their part, are essentially localized in the nearest safe area, they may also be in capital cities, far from the crisis area.

The most critical part of the architecture is in the crisis area. Autonomous configuration and networking are therefore required to allow communication (Local Wireless Network Link in the figure) between mobile devices to the nearest working fixed infrastructure (Wide Area Network Link). This is achieved using the standards mentioned before. Implementation concerns will be discussed in an upcoming section of this chapter. From a global point of view, what must be remembered is the fact that this architecture allows us to create and maintain an information link that will adapt itself dynamically to the available resources. In the example depicted in the figure, information outputted by a user is routed via two other smartphones, the second one being connected to the nearest functioning antenna that is able to ensure relay to the final destination via normal means. The idea is thus to provide a service for crisis management systems, which we will describe in the following section.

9.4. Mobile devices and proximity networks-assisted crisis management operations

In this section, we aim at presenting and discussing in detail how mobile devices can be used practically in the information chain depicted in Figure 9.1. Four systems are proposed each of them being a response in the global duty of disaster management:

– early detection of a crisis;

– victim localization and characterization;

– victim and population on-site assistance;

– information gathering, dissemination and exploitation.

Each system shares a user profile completed during the installation process on the mobile devices. This profile allows us to specify user data and special needs: blood type, allergies, handicap, pregnancy, etc. A real name is not mandatory for people who care about privacy.

For each system, we will firstly present it, then discuss technical concerns and finally discuss the difficulties and/or limitations that can be raised and how to address them.

9.4.1. *Early detection system*

The first system is dedicated to early crisis detection by people living or passing by the concerned area.

9.4.1.1. *Description*

This warning system is based on capturing "visible" events that occur when a disaster strikes or is about to strike. For instance: fumes in a volcanic area, tremors or other warning signs. The idea is to take advantage of the fact that although not being specialists, people living in such an area are able to detect unusual events. Even passers by can notice particular signs that indicate something is about to happen.

Those signs can be detected in two ways, from the device owner's point of view:

– actively: one or more people send an alert signal because they have seen or heard something. In addition to the normal attitude for someone to trigger an alert when something dangerous happens, the development of social networks like Twitter, Facebook or Instagram has made people share with communities events that may not even be an immediate threat. That is to say, "odd" or unusual facts are to be reported on these online services. People are used to sharing information. Researchers and engineers working in the field of disaster management are thus willing to make use of this mechanism [TER 12].

– passively: the device autonomously captures data, which is processed to determine if something special is happening (for example, a specific pattern is detected in the values acquired by an accelerometer).

9.4.1.2. *Comparison with traditional approach*

At this stage, it is interesting to make a comparison between the proposed approach and the professional one in order to show how they are complementary.

As Table 9.1 shows, the proposed approach has many limitations but they are balanced by the facts that on one hand the number of data sources is much higher, and on the other hand by the cost issue. Indeed, in this situation, cost is largely shared by all the actors because consumer's devices are paid by the population, and so is their access to the network. Stakeholders have to fund the centralization system. Actually, they normally already have one, what remains to be done is to update it in order to support as an input what mobile devices can send.

	Consumer mobile devices-assisted data acquisition approach	Usual expert devices-driven data acquisition approach
Number of acquisition devices (=nodes)	~100 s possibly ~1000 s in urban area	~10 s
Data accuracy	Uncertain	Strong
Scientific level	Weak	Strong
Cost management	Shared with all actors	Paid by stakeholders
Obligation of result	Non	Yes
Robustness	Uncertain	High

Table 9.1. *Differences between the proposed approach and the usual one*

Finally, there is another reason that could justify such a warning system: it allows the gathering of information from an area not covered by expert/scientific devices. It is actually a current trend to create low-cost devices to detect disasters at an early stage [BHA 15]. This proposition thus also takes place in this particular context. But as shown in [HUA 12], the notion of "low-cost" is highly variable as in this article, the designed device costs $3,000 in regard to the typical cost of $25,000 for such a measuring station.

From these conditions, we can deduce that a mobile device-based disaster warning system can only be used to find an area that may be the subject of a crisis occurring from a statistical point of view only. The idea is to quickly draw the attention of stakeholders to a particular area that may not be equipped with professional devices, based on the fact that if hundreds of consumer devices report the same kind of data in a short time frame, there is chance that something is really happening. This allows experts to draw a map of hot spots. Those hot spots have a hierarchy according to the density of reports per area. Figure 9.3 shows an example of such a map.

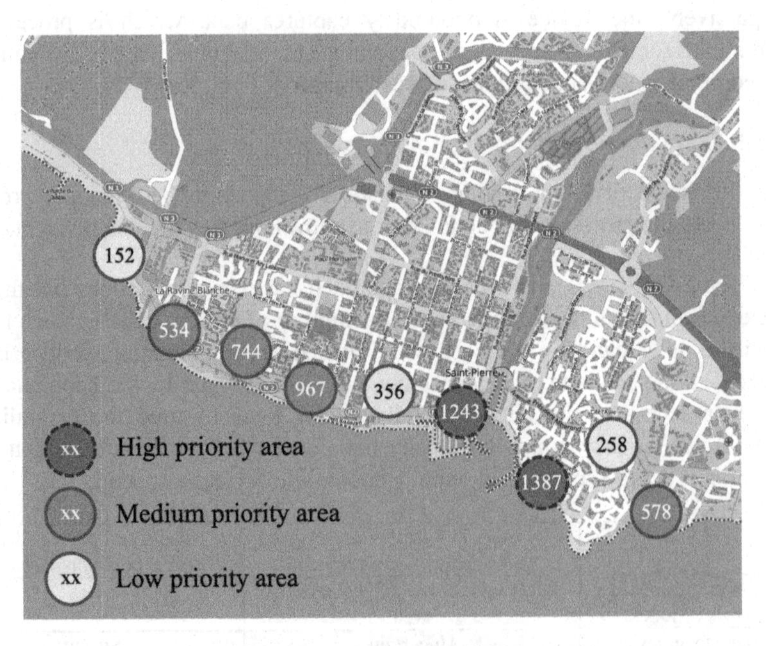

Figure 9.3. *Detection report map (background map provided by Open Street Map)*

This example depicts a fictive scenario: a tsunami striking Saint-Pierre, a town located in the south of Reunion Island, in the Indian Ocean, next to Madagascar.

Three levels of priority are defined, thus allowing stakeholders to find areas where lots of people are located and giving the most crowded area a higher probability of event occurring. Indeed, the more similar reports are emitted concerning a given area, the less chances are that information is erroneous (even if a massive malfunction remains possible).

However, this is all that can be expected as "highly probable": crisis characterization from this point of view (i.e. knowing what is happening in addition to knowing that something is occurring and where) is a "best effort" feature: reliability maybe poor for lots of reasons: relevancy of information provided by non-specialists, poor sensor precision, faulty devices, etc.

That said, we can focus on the technical concerns that are raised.

9.4.1.3. *Technical concerns*

To deal with this topic, we will split the discussion between the active and then the passive way of detecting an event. Then we will evoke encountered difficulties.

9.4.1.3.1. Active warning system: user triggered alert

This system is actually very simple: it consists of an application installed in the device that can be launched by a user to emit a warning message. We could note that it plays the same role as SMS (short message service). The difference here is that there is no need write a message or to select a recipient to report a problem to because conditions may not be good enough for the user to concentrate or even type-in all this information in case of emergency.

UI is, therefore, a major point in designing this application. Specific features can be built to facilitate reporting a problem in tough times: a simple visual interface with large buttons, voice-controlled interaction, gesture-based commands, etc. Doing so also allows the application to be operated by all profiles of users: young, old, disabled, illiterate, etc.

Finally, according to the situation, additional data (such as GPS position, time stamp, etc.) can also be sent with the warning message in a similar way to the passive warning system, which we will now present.

9.4.1.3.2. Passive warning system: sensors based detection

As seen in section 9.3.1.2.1, connected devices embed sensors that can be used in combination to detect and automatically report an unusual event: accelerometers, gyroscope and GPS as mentioned before. When a specific threshold is exceeded or a typical pattern appears, an alert message is emitted along with metadata consisting of:

– time stamp: date and time of the detected event to allow all the reports to be sorted;

– geographical position: device location based on multiple sensors: GPS, Wi-Fi and 3G cell-phone infrastructure. [ZOR 10] presents research work to achieve such a a positioning, even indoors;

– amount of brightness: this measure can give additional indication about conditions around the device to determine if they are consistent: for instance, if the device is located in a building in daytime, a low value of brightness may indicate that it is in a bag or a pocket as well as the fact that the owner is stuck in a lift.

Other data could possibly be captured such as a photo or a sound sample, but without active participation of the user, the resulting data may not have the required level of quality to be relevant.

This philosophy of passive detection of events by the device takes part in a more global one called "device sensing". It aims to allow a digital personal device to deduce the current situation its user is in from all the data it can obtain on its own: calendar, day and time, habits, sound, localization, sight, etc. From this, the device can automatically offer assistance for that particular situation. This is a long-term effort but new contributions are regularly being made in this direction. [LAN 10] presents the concept and proposes a framework dedicated applications. Three years later, [KHA 13] for developing a state-of-the-art of the situation by listing applications and the context in which they apply.

Risk management is a domain where there are still lots of developments to achieve before being able to release a ready-to-use solution. Indeed, there are specific parameters to take into account, even for a best-effort service. [RAD 14] presents an architecture that integrates many of them.

9.4.1.4. *Difficulties*

Implementing such an early detection system raises both technical and organizational questions.

Concerning the former, three issues can be listed. First, sensor-based detection requires the developer to know the specification of the embedded hardware: precision of a sensor may vary from one model to another. Moreover, newer systems have higher precision. This makes the process of aggregating and comparing the reports from users possessing very different devices performing the same task rather difficult. Experiments must be led to determine whether this could be a big issue or not. Secondly, a major energy issue is raised: the principle of early detection supposes that data is acquired from sensors continuously, which tend to lower battery life dramatically if a special policy is not applied. Users may not want to be involved in the early warning system otherwise. Finally, a smaller issue resides in the fact that researchers and developers have to access to specific programming modes to read raw sensors values for instance or to poll them in an usual way [MIN 15]. These modes are locked for security and commercial reasons for most users. In order to enable them for consumer use, it is necessary to work more closely with device and OS manufacturers, which is not always easy to do and also becomes an organizational issue.

From this point of view, another phenomenon can be noted: lots of dedicated applications aiming at managing disaster already exist as evoked by [TAN 14]. The most useful ones are restricted to a particular area (for instance, a country) and/or a particular type of disaster. Disaster Kit from the Japanese operator NTT Docomo is a good example of such a tool. It is dedicated to Japan and has been designed to manage earthquakes, which are likely to happen in this country.

These applications share the fact that they are (officially) disconnected from a public safety organization. For us, this is a requirement because data provided by user devices should be aggregated at the highest level of crisis management, that is to say national control centers. Otherwise, there is a risk that information does not reach stakeholders who are in charge of management. Therefore, a special effort is expected to connect the general-public-side operations to the professional-side ones at the highest level.

9.4.1.5. *Ethical concern*

The passive automatic detection system raises an ethical question: its efficiency is linked to the fact that it is permanently acquiring data about the environment the user develops in. This is why a privacy issue exists: the user may feel tracked by the system.

This issue is a lot more general than the way we present it as lots of famous applications are regularly pointed out by information technology specialists for accessing personal data that is not related to their role. There are lots of debates between users who care and those who do not.

In our context, the idea is to define a compromise: acquired data is not stored permanently, and a message is sent only when a special event occurs. Finally, the reports can be anonymous because from a stakeholder's point of view, only a statistical use matters.

9.4.2. *Victim localization and characterization system*

As for the early detection system, we will first describe the victim localization and characterization system and then discuss the various concerns that must be addressed to build such a system.

9.4.2.1. *Description*

This system aims at assisting search-and-rescue operations by reporting where victims are to the control center and what they are suffering from. To achieve this, an application, probably the same as the early detection system, is deployed on each personal device to send a message to authorities giving them the necessary information. The questions that must be answered here are: who, what and where?

In comparison to the previous system, the main difference consists of the fact that the transmitted report has a true signification on its own and not only from a statistical point of view. Each report stands for a person who needs care.

In this context, most concerns discussed for the previous system are still valid. We will thus focus on the items that are specific to this goal.

9.4.2.2. *Technical concerns*

Ideally, such an assistance request system should work automatically: if the connected devices detect that their owner is injured, unconscious or sick, they should take the decision to call for help.

But as of 2015, technology is not reliable enough to proceed in total autonomy as expected by [LAN 10]. Devices are not yet able to "sense" such situations, for plenty of reasons: biometric sensors are still rare, environmental sensors such as cameras or microphones cannot operate all around the user (because they move according to him/her) to target a useful point of interest and finally "collaborative work" between multiple personal devices is still poor. This is why a compromise must be made for the moment: user interaction is required but limited to the minimum.

Our proposition consists of a simple question/answer system with time-out and default answers if no answer is given. It can be finger or voice-operated and communicate with visual as well as aural information. Three steps are necessary to complete the operation of emitting an assistance request:

– Triggering a request: since device sensing is still a work in progress, chances are high that a victim has to be conscious to activate the process. He/she can operate the device directly or call the feature due to a keyword, in a similar way that digital personal assistants are invoked in a normal dialog (examples: "OK Google" for Android or "Hey Siri" for iOS). In rare, serious situations, paired devices could automatically detect that a request has to be sent. For instance, if a heartbeat monitor has no signal after accelerometers detected a brutal activity.

– Characterization of the request: the goal here is to answer the questions: who, what and where? Information is asked be a device that can handle sound (typically a smartphone). If there is no reaction from the user before a certain amount of time, particular sounds (high frequencies) are played to try to wake him/her, making the hypothesis that he/she may have fainted. As far as the questions themselves are concerned, "who" and "where" can be obtained automatically, but not the "what".

– Who: is defined from the application user profile which can thus give very detailed information about the person's health. But it can be completed due to user ID from the SIM chip too. Each one has a unique International Mobile Equipment Identity (IMEI), linked to a user account at the line operator. Of course, the device could have been lent to somebody else.

– Where: positioning techniques discussed before are used. Of course, there might be a slight difference between the device and his/her owner's positions but we make the assumption that in most cases, they should be close.

– What: user is requested to answer a few questions about his/her health situation in order to allow ESU teams to anticipate the needed equipment. Other questions are asked about the environment: rubbles, specific danger around, etc. No response at all indicates an unknown issue.

– Sending the request: a message is sent to the network (based on the protocols and techniques presented previously) for delivery to stakeholders whatever the answers to the previous step are (even in the "no response" case).

From a technical point of view, difficulties are still the same as those mentioned for the early detection system. The UI aspects are even more important to build an efficient system to gather answers for the characterization process. Actually, software libraries provided by mobile OS makers support all the requested features. The timed question–response feature is easy to implement: in case no input is given after a certain amount of time, the information is sent that the user has not replied which can indicate a problematic situation.

9.4.2.3. *Ethical concerns*

From the ethical point of view, a major issue is pointed here: can the system override user settings to assist him/her? Indeed, there are situations where this is necessary to have the system work correctly. Here are three examples:

– If the sound has been muted, should the software be given the authorization to unmute it in order to use voice communication?

– If privacy settings deny the application to use GPS and/or other sensors, should they be ignored to allow it to accomplish its duty?

– Even worse: to help ESU assist a fainted person, can they take the control of a smartphone, for instance, to give additional information?

There is no simple answer to these questions. Societal studies have to be led in order to find a compromise between user control and system efficiency. This takes part in a global reflection about ethical, legal and social (ELS) aspects, which is discussed by social sciences communities. [RIZ 14] is an example of what can be analyzed for a given situation.

9.4.3. *Victim and population on-site assistance*

We will follow the same path to present this system: first a description, then the various concerns it raises.

9.4.3.1. *Description*

This system aims at enabling a communication channel between risk managers and people located in the area where a disaster strikes to provide on-site assistance to individuals. It relies on any remaining networking infrastructure available and can transfer any type of data: text of course, but audio and even images if needed. Video, for its part, maybe too resource-demanding (from energy and networking points of view) given its relevancy. The system also takes the appearance of a mobile application, like the two other ones, communicating with a server. It may of course be the same application which plays the various roles.

It allows sharing three categories of information:

– Stakeholders-to-population information: this consists of general information concerning the whole population, for instance a forthcoming threat like the aftershock than can be expected after an earthquake. This information is broadcasted, that is to say there is no condition for a person in the area to receive it.

– Stakeholders-to-a person information: it consists of information that is sent to a particular person regarding its situation: he/she may not be a victim but has some specific needs:

- nearest safe place (or how to be extracted from the danger zone);

- nearest place to have water and food;

- nearest place to find accommodation;

- survival advice;

This information may automatically be sent, given the information acquired from the owner's devices when using the previous systems. Thus, it eases the pressure on risk experts work who do not have to answer repetitive requests manually anymore.

– Person-to-stakeholders information: this allows us to manage special requests triggered from users that are too specific. It may concern a limited amount of situations: for instance, one person reporting the situation to another (who does not have a smartphone or is not able to use it). Further test simulation with the system should give the most probable use-cases in the near future.

We could also include a person-to-person information transfer system, but it may be less useful regarding the risk-management operations. Moreover, general-purpose applications like the Serval Project [GAR 11] already exist for device-to-device personal communication when the normal infrastructure is lacking.

9.4.3.2. *Technical concerns*

All systems presented in this chapter were described starting from the user device point of view. But in all situations there is an information system, that is to say a server receiving information from the devices. It was not a particular issue until this system because the communication was rather simple: the server role was to store data sent by devices and report it to stakeholders, eventually processed to calculate statistics and send them to a geographical information system (GIS) to display a map. This is an easy task given the resources available nowadays in servers. This is why we have not focused on this aspect of the systems until now.

Indeed, this particular system has higher needs in term of processing: an individual response must be calculated for every single user, according to his/her very own data. Operations include:

– a constraint solver: to compute a recommendation taking into account all the data available about a user, in particular regarding his/her profile, not only the environment he/she currently is;

– a navigation unit: to guide the user, most certainly a pedestrian, to a particular point (for safety or supply) in relation to the constraint solver.

To achieve this at a relatively low computing cost, techniques from the simulation and videogames industry can be used, for example for real-time path finding [LAM 04]. Of course, algorithms have to be adapted, as the model does not match completely reality, without mentioning the fact that users are not tracked in real time, in our proposition.

Nevertheless, the demand for the server side is higher than before and it must be anticipated. However, there are solutions, as servers do not have to be physically in the area the disaster strikes. Moreover, the cloud computing paradigm allows us to increase processing capacity on-the-go, by aggregating physical machines into a virtual one to face the increasing workload.

9.4.3.3. *Difficulties*

The energy issue is even more important for this on-site assistance system because the demand increase because of the number of peripherals that have to work at the same time: network interfaces, always having the screen on to display a map and directions, audio indications, positioning interfaces, etc. In addition to this, the time during the battery drain is high because devices have to work during the entire time the user is looking for a particular place.

There is currently no solution that could limit energy consumption in this particular context. We must take the risk of having the mobile terminals (and more particularly the smartphone) power off before the user reaches his goal.

Network aspects are a less important matter, although congestion can occur because of the multiple one-to-one communications that can occur in a Wi-Fi-Direct or LTE D2D mesh network. The real trouble here is the lack of feedback given the fact that those standards are new. Large experiments still have to be done to check their robustness in harsh conditions.

9.4.3.4. *Ethical concerns*

In this context, ethical concerns are less critical than the previous system. General issues about privacy remain, because the application has access to the user profile which can contain very personal information.

A solution could consist of creating a file at a country scale level. However, constraints exist about the legal frame that must exist to support such an action. Quite surprisingly, it is easier for private firms like Google or Microsoft to gather and especially use publicly information about people than for states, without mentioning the costs.

9.4.4. *Information gathering, dissemination and exploitation*

As for the previous systems, we will firstly present a description of the goals concerning this topic and then discuss technical and non-technical trends and difficulties. This subsection presents lower level concerns compared to previous ones: the discussion will be centered on some specific technical issues.

9.4.4.1. *Description*

Conventional communication networks are the first systems that fail during disasters [PAL 12]. This results in poor level of initial response. The majority of existing applications uses some form of centralized processing, communication or storage. They suffer from the common communication network failure. However, alternatives for usable communication exist. One of them consists of leveraging the opportunistic and collaborative network formed by surrounding mobile devices. As we have seen earlier, more mobile phones are likely to be present in the terrain disaster than expert specialized devices. They have computing, storage and communication capabilities. The association of multiple devices coupled with wireless communication makes it possible to build or support distributed applications or systems. We will focus in the following on the possibilities such a collaborative network offers to information gathering, dissemination and exploitation.

9.4.4.2. *Technical concerns*

Each participant contributes to the establishment and the maintenance of a distributed application or system. However, challenges are inherent to such an association [ROT 06]. To mention some of them, we can cite the following:

– wireless communications are often faulty: connectivity between devices is weak; exchanged data can be lost;

– topology is dynamic: devices can move; some of them can leave while some join the network;

– communication delay or latency is not null and can fluctuate even with the same participants;

– wireless communication is unsecure: exchanged data can be intercepted, altered in transit or recovered by unintended entities;

– control is distributed: any operation in the network is realized with the coordination of the members;

– network is often partitioned or not fully connected: a device may be or become unreachable by other devices;

– operations are concurrent: multiple devices can modify the same data concurrently.

9.4.4.2.1. Information gathering

The collection of gathered data represents a valuable shared common good [PAL 12]. Studies demonstrate the willingness of users to participate in building the collection [AL 01]. This collaborative gathering of data while resulting in a rich source of information, may introduce conflicts and inconsistency. One illustration of a conflict is when two or more pieces of incoming data report inconsistent values for a piece of information that permits a single value at most. That is, for example, the case when a user receives contradictory danger levels for the current place. One behavior is to consider the most recent information and discard the others. Dynamics (in terms of latency, topology or partitioning) in the network favor the occurrence of conflicts. Dissemination of outdated information illustrates such a situation. Conflict resolution and consistency preservation mechanisms, are, then, of capital importance: they result in either a new value of the information combining all or partial conflicting ones or an existing value that should be considered as correct by all participating devices.

Conflict resolution is commonly encountered in a distributed database. A participant runs a partial or a full replica of the database. The major issue consists of dealing with multiple successive conflicting (and non-conflicting) versions of a piece of data: how to guarantee data consistency while allowing concurrent read and write operations. Several solutions exist:

– Optimistic replication: this approach maximizes the availability of the system in the presence of network partitioning. A replica may run an operation (e.g. data modification, removal, etc.) without requiring immediate synchronization, coordination or consensus with other replicas. Changes are reported later on. Thus, at a given time, the value of a piece of data may not be the same on different replicas. However, in the long run, the data eventually converges to the same value when all changes are reported. There is no *a priori* conflict avoidance (e.g. by setting a lock). A conflict is resolved at the time it is detected. Bayou [TER 95] is one of the first replicated databases proposed for the mobile computing environment. A device designated as a primary replica for a particular data item commits all the write attempts. An application defines how a conflict is characterized and how to resolve it for the system to work autonomously. The main drawback of Bayou is that the primary replica becomes a single point of failure. Constant data availability (CODA) [COD 87] is a distributed file system based on optimistic replication and designed for disconnected operation in mobile computing. A support for the Linux Kernel exists starting from version 2.6.

– Multi-version concurrency control (MVCC): this is a control scheme permitting concurrent access to a distributed database. Successive versions are stored in the database. Old versions are kept but marked as obsolete. A new version is visible only when the related update has been completed. It means that a read results in a snapshot of the database regardless of ongoing updates. Thus, a write operation does not require locking previous versions of the data: reading never blocks writing and *vice versa*. However, maintaining multiple versions of data incurs costs in terms of storage. Periodic purges or archiving of obsolete versions are necessary to save space. CouchDB [COU 05] and its derivative CouchBase

[COU 11] are flagship MVCC-based databases. They are document-oriented databases: stored information is semi-structured in that document contains both data and schema. Multiple instances replicate their data through HTTP-based synchronization protocols. CouchDB and CouchBase were designed with offline use in mind. An embedded version of CouchBase named CouchBase Lite is available for mobile.

– Operational transformation (OT): this addresses synchronous real time collaborative applications. Participants can view and edit the same document at the same time. Each participant locally maintains a replica of the shared document. OT ensures the consistency over the multiple replicas. Local modifications, also called operations, are expected to be highly responsive: they are performed immediately [KUM 10]. Operations are, then, propagated to other participants in a timely manner. They will eventually converge to a version reflecting the intention of editors. A participant can edit any part of the document. A document must be linearly addressable. A received remote operation is transformed against previous operations before being executed locally. The new form preserves the intention of the editor and ensures the convergence over all participants. Suppose two users A and B that collaborate on a document containing the string "abc". User A inserts the character "z" at position 0 resulting to the operation OpA = Insert(0, "z"). User B deletes the character "c" with the operation OpB = Delete(2). We will consider processing on A when receiving OpB, however, the mechanic is the same on B. OpB is transformed to OpB' = Delete(3) facing the previous execution of OpA on A. Finally, OpB' is executed on A and results in the string "zab". Asynchronous and non-real time situations limit the usage of OT. Indeed, a high latency of operations propagation blurs the intention of multiple editors. A suitable use case of OT in disaster management is, for example, a fully connected network composed by nearby devices. Collaborative editing applications such as Google Docs [DOC 06], Apache Wave [WAV 09] and Etherpad [ETH 08] leverage OT.

9.4.4.2.2. Information dissemination

We will focus in this section on ways to capitalize on mobile phone networks to convey information from a source to a destination. In a classical situation, a mobile phone is a terminal node: it is either the source of communication or a final destination. The network infrastructure transports the data from a terminal node to another terminal node. In the considered infrastructure-less network of mobile phones, some devices may be out of communication range of others. Given the absence of infrastructural nodes, mobile devices have to be in charge themselves of relaying and routing functions.

Routing is widely studied in mobile *ad hoc* networks (MANETs). A MANET is an autonomous self-configured network composed of mobile nodes over wireless links. All nodes act as routers, discover and maintain a route to each node of the network. Coordination between nodes is necessary for conveying information to its destination. Routing protocols have been designed for MANETs. To mention two of them, we can cite the optimized link state routing protocol (OLSR) and better approach to mobile *ad hoc* networking (BATMAN). The key concept of OLSR [CLA 03] lies in the use of multi-point relays (MPR) that relay information. MPRs are chosen among nodes in such a way that any node of a network is covered by at least one MPR. MPRs exchange regularly routing information. BATMAN [JOH 08] is intended to replace OLSR. It does not maintain the full route to the destination: each node along the route only maintains the information about the next link through which you can find the best routes. The Serval Project [GAR 11] relies on BATMAN. Routing protocols for MANETs are usually designed for general-purpose application. Maintaining routes in permanence is costly and may not be necessary in all cases.

Opportunistic networks are an alternative solution. A device may communicate with another device even if a complete path connecting them is missing. An intermediate device can be involved in an opportunistic manner to relay data if it allows us to come closer to the recipient. Thus, data transport is done step by step according to this scheme to the final destination. Opportunistic networks exploit node mobility. In this perspective, the Twimight system [LEG 11] offers a microblogging service similar to Twitter, dedicated to emergency and crisis situations.

9.4.4.2.3. Information exploitation

In a collaborative network, information is received from different sources. It is essential to identify and authenticate the sender for security concerns. It is also vital to certify and track back received information to the sender or the data collector. Incorrect information, or doubts about the veracity of information can lead to very poor resource allocation, which in turn can cause needless suffering and loss of life [PAL 12]. From a practical point of view, asymmetric cryptography represents a response to identification and authentication issues. Pretty Good Privacy (PGP), is one the most widely used programs for privacy and authentication. A user generates a private key and the associated public key. He sends the public key to his contacts and keeps the private key confidential. User's contacts use the public key to encrypt information intended for him and to authenticate information coming from him. Conversely, the private key is used by the user to sign issued information and decrypt received information. PGP comprises a "Web of trust" feature that allows establishing the authenticity of the binding between a public key and its owner. For this aim, other users can certify the public key associated with the owner. The Networking and Cryptography library (NaCl) is another candidate to authentication for mobile phone environments [NAC 08]. It avoids various types of cryptographic disaster suffered by previous cryptographic libraries and uses shorter public keys compared to PGP.

Considering the possibility for participants to share information securely, we will focus now on the form in which information is exchanged and how applications will use it. Information can be structured, unstructured or semi-structured. Structured information means that it follows a fixed pre-defined formal data model. The data model explicitly defines the structure of data. It makes data processing easier. For example, any profile information must include the fields "name" and "email". Unstructured information does not conform to a pre-defined data model nor is it organized in a pre-defined manner. Semi-structured information does not follow a pre-defined formal data model but includes markers that describe its structure. Information is self-described. It benefits from the structure-oriented processing similar to structured information and exhibits the flexibility of structure defined at runtime. Semi-structured information is described by markup language. EXtensible Markup Language (XML) and JavaScript Object Notation (JSON) represent two popular markup languages.

An application can exploit the flexibility of semi-structured information. Its design can even be data-centric. That choice simplifies the development and maintenance of the application over a collaborative network. Indeed, it benefits from underlying information gathering and replication as presented earlier. RAVEN is an implementation of this approach [PAL 12a]. It is a framework which makes it possible to build applications for collaborative editions. RAVEN offers developers compile time tools, which only use the schema to generate all database handling components, edit and list user interfaces. It also provides a schema editing application as a portion of the framework in order to allow users to define a new database on the phone at runtime. CouchApps constitute another mature and production grade alternative. They are web applications, represented as JSON document, served directly from CouchDB. A CouchApp uses web technologies (HTML/CSS/Javascript) and can be replicated by CouchDB.

9.4.4.3. *Ethical concerns*

When participating in a collaborative network, a mobile phone can process, store or relay information of other devices in addition to its own information. Those functions drain energy and decrease the lifetime of the device. It is essential to ensure that this cost matches the willingness of the device owner to allocate resources for the common good. The system must also guarantee the security of the data and the communication while ensuring the privacy of the users.

On the other hand, user information can be conveyed by intermediary devices. Users should be aware of the burden of transporting that information. As resources (battery, bandwidth, etc.) are constrained in the whole network, communication must be limited to vital information. Even if this policy is ultimately achieved via technical means, communications are triggered on the user's initiative.

9.5. Conclusion/perspectives

In this chapter, we attempted to provide complementary means of managing a crisis by relying on mobile consumer devices as a personal assistant for the population. The idea is to gather information and communicate with people in an efficient way to ease stakeholders and risk professional duty.

To achieve this, lots of technical concerns must be taken into account, but scientific and technological progress is regularly made to address the issues: for example, as far as the energy question is concerned, one of the major ones, it is a necessity for device manufacturers to create more efficient batteries and there is no doubt they will.

Non-technical aspects could actually raise problems that may dramatically slow down the development of such a solution if they are not taken into account at a very early stage. This is why we put emphasis on the fact that all the actors should collaborate as soon as terrain experiments are ready to be launched.

Ethical concerns also emerge from this proposition: researchers must integrate what people are ready to do and what they are willing to let specialists do with their private data, in order to receive assistance in case of emergency. That is why experiments must also be led with ELS specialists to have quantitative as well as qualitative feedback.

Such experiments need support from local and national organization (such as civil security, coastguards, municipal officials, etc.) to be conducted in the most realistic condition. It also allows people to work together by making every actor aware of what the others are doing. It is a non-technical aspect but it remains essential regardless of the system used.

9.6. Bibliography

[AL 01] AL-AKKAD A., ZIMMERMANN A., "User study: involving civilians by smart phones during emergency situations", *8th International ISCRAM Conference*, pp. 1–10, 2011.

[BHA 15] BHAT A.P., MESHRAM N.S., DHOBLE S.J. *et al.*, "Microcontroller based low cost earthquake monitoring using lab-view", *International Symposium on Ultrasonics*, vol. 22, no. 24, pp. 376–380, 2015.

[CAM 12] CAMERON M.A., POWER R., ROBINSON B. *et al.*, "Emergency situation awareness from twitter for crisis management", *Proceedings of the 21st International Conference Companion on World Wide Web*, Lyon, ACM, pp. 695–698, April 2012.

[CAM 13] CAMPS-MUR D., GARCIA-SAAVEDRA A., SERRANO P., "Device to device communications with Wi-Fi direct: overview and experimentation", *IEEE Wireless Communications*, vol. 20, no. 3, pp. 96–104, 2013.

[CHA 11] CHANDRAPPA R., GUPTA S., KULSHRESTHA U.C., "Predicting disaster: Asian scenario", *Coping with Climate Change*, Springer Berlin Heidelberg, pp. 149–154, 2011.

[CLA 03] CLAUSEN T., JACQUET P., Optimized link state routing protocol (OLSR), RFC 3626, 2003.

[COD 87] http://www.coda.cs.cmu.edu/, 1987.

[COU 05] http://couchdb.apache.org/, 2005.

[COU 11] http://www.couchbase.com/, 2011.

[DOC 06] https://docs.google.com/, 2006.

[ETH 08] http://etherpad.org/, 2008.

[FEN 13] FENG J., Device-to-device communications in LTE-advanced network, PhD Thesis, Télécom Bretagne, Université de Bretagne-Sud, 2013.

[GAR 11] GARDNER-STEPHEN P., The serval project: practical wireless ad-hoc mobile telecommunications, Flinders University, Adelaide, South Australia, Tech. Rep, 2011.

[GOA 15] GOASDUFF L., RIVERA J., *Press Release March 2015*, Gartner Group, Egham, 2015.

[GPP 13] 3GPP WG SA2, Technical specification 22.278, Service requirements for the evolved packet system, 2013.

[GPP 14] 3GPP WG SA1, Technical specification 22.468, Group communication system enablers for LTE (GCSE_LTE), 2014.

[HUA 12] HUANG R., SONG W.-Z., XU M. *et al.*, "Real-world sensor network for long-term volcano monitoring: design and findings", *IEEE Transactions on Parallel and Distributed Systems*, vol. 23, no. 2, pp. 321–329, 2012.

[JOH 08] JOHNSON D., NTLATLAPA N., AICHELE C., "A simple pragmatic approach to mesh routing using BATMAN", *2nd IFIP International Symposium on Wireless Communications and Information Technology in Developing Countries*, pp. 1–10, 2008.

[KHA 13] KHAN W.Z., XIANG Y., AALSALEM M.Y. *et al.*, "Mobile phone sensing systems: a survey", *Communications Surveys & Tutorials*, IEEE, vol. 15, no. 1, pp. 402–427, 2013.

[KUM 10] KUMAWAT S., KHUNTETA A., "A survey on operational transformation algorithms: challenges, issues and achievements", *International Journal of Computer Applications*, vol. 3, no. 12, pp. 30–38, 2010.

[LAM 04] LAMARCHE F., DONIKIAN S., "Crowd of virtual humans: a new approach for real time navigation in complex and structured environments", *Computer Graphics Forum*, Blackwell Publishing, Inc, pp. 509–518, 2004.

[LAN 10] LANE N.D., MILUZZO E., LU H., *et al.*, "A survey of mobile phone sensing", *Communications Magazine*, IEEE, vol. 48, no. 9, pp. 140–150, 2010.

[LAW 14] LAWRENCE J.F., COCHRAN E.S., CHUNG A. *et al.*, "Rapid earthquake characterization using MEMS accelerometers and volunteer hosts following the M 7.2 Darfield, New Zealand, earthquake", *Bulletin of the Seismological Society of America*, vol. 104, no. 1, pp. 184–192, 2014.

[LEF 14] LEFEUVRE F., TANZI T.J., "Radio science's contribution to disaster emergencies", *The Radio Science Bulletin*, vol. 348, pp. 37–46, 2014.

[LEG 11] LEGENDRE F., HOSSMANN T., SUTTON F. *et al.*, "30 years of wireless *ad hoc* networking research: what about humanitarian and disaster relief solutions? What are we still missing?", *Proceedings of the 1st International Conference on Wireless Technologies for Humanitarian Relief (ACWR)*, Kerala, India, 2011.

[MIN 15] MINSON S.E., BROOKS B.A., GLENNIE C.L. *et al.*, "Crowd sourced earthquake early warning", *Science Advances*, vol. 1, no. 3, pp. 1–7, 2015.

[NAC 08] http://nacl.cr.yp.to/, 2008.

[PAL 12] PALMER N.O., Smartphones: a platform for disaster management, PhD Thesis, Vrije Universiteit, 2012.

[PAN 15] PANAITOPOL D., MOUTON C., LECROART B. *et al.*, "Recent advances in 3GPP Rel-12 standardization related to D2D and public safety communications", *Computing Research Repository*, 2015.

[RAD 14] RADIANTI J., DUGDALE J. *et al.*, "Smartphone sensing platform for emergency management", *arXiv preprint arXiv:1406.3848*, 2014.

[ROT 06] ROTEM-GAL-OZ A., "Fallacies of distributed computing explained", RGO Architects, White paper arXiv: 1505.07140, available at http://www.rgoarchitects.com/Files/fallacies.pdf, 2006.

[RIZ 14] RIZZA C., PEREIRA A.G., "Building a resilient community through social network: ethical considerations about the 2011 Genoa floods", *11th International ISCRAM Conference*, pp. 289–293, 2014.

[SEB 14] SEBASTIEN O., HARIVELO F., SEBASTIEN D., "Using general public connected devices for disasters victims location", *General Assembly and Scientific Symposium (URSI GASS), 2014 XXXIth*, URSI, pp. 1–4, 6–23 August 2014.

[TAN 09] TANZI T.J., LEFEUVRE F., "L'apport des radios sciences à la gestion des catastrophes", *Journées Scientifiques Propagation et Télédétection URSI–France*, pp. 401–428, Paris, France , 24–25 March 2009.

[TAN 14] TANZI T.J., SEBASTIEN O., HARIVELO F., "Toward a collaborative approach for disaster management using radio science technologies", *Radio Science Bulletin*, no. 348, p. 25, 2014.

[TER 12] TERPSTRA T., DE VRIES A., STRONKMAN R. *et al.*, "Toward a real time Twitter analysis during crises for operational crisis management", *Proceedings of the 9th International ISCRAM Conference*, Vancouver, Canada, April pp. 1–9, 2012.

[WAV 09] https://incubator.apache.org/wave/, 2009.

[WIF 09] WI-FI ALIANCE, Wi-Fi Peer-to-Peer (P2P) Technical Specification v1.0, December 2009.

[ZOR 10] ZORN S., ROSE R., GOETZ A. *et al.*, "A novel technique for mobile phone localization for search and rescue applications", *International Conference on Indoor Positioning and Indoor Navigation (IPIN),* IEEE, pp. 1–4, 2010.

10

How to Ensure Quality Standards in Emergency Management Systems

10.1. Introduction

An emergency can be defined as "an unexpected situation deriving from an incident which puts life in hazard as well as material goods and/or the natural environment, and which requires immediate response" [WAY 08, ALT 12, FED 15b].

In essence, an emergency has a dynamic and unpredictable character, and to confront, it is usually necessary to involve multiple agencies, which should combine efforts and interact before, during and after the occurrence of the disaster. Integration, coordination and cooperation are essential processes for operational

---

Chapter written by Marcelo ZAMBRANO V., Manuel ESTEVE and Carlos PALAU.

success and are closely related to communication and information systems (CISs) and public safety networks (PSNs), through which communication is possible with all parts involved, and quality, sharing and distribution of the required information is guaranteed. Emergency management (EM) deals with all the actions and measures necessary to effectively prevent and respond to disasters and emergency situations. EM concerns the actions required for response as well as with prevention, recovery and mitigation [ALT 12, EST 15, BRI 14]. One of the key features of EM is its dependency regarding the resources, scope and promptness of the response.

An emergency management system (EMS) is the tool with which to implement the management of an emergency. It is the set of resources and processes within a common organization framework which allows the development, maintenance and improvement of the capabilities implied in quality EM [FED 14a, BRI 08]. Systemic treatment of GE has allowed us to improve responses in the presence of a threat, risk or disaster; through a continuous, orderly and methodological process, optimization of available resources and the involvement of multiple agencies providing to system, maximizing the different capabilities for a comprehensive response. The PSNs are a critical element within EMSs, to allow effective interaction and communication between the agencies involved and active participation of the community, at the moment to prevent and addressing a damaging incident.

The quality of EM is strongly dependent on the quality of its associated EMS and it is based upon the system's inner capabilities and of each of its components to meet the requirements and needs of all the parts implied, regarding the procedure to respond or prevent disasters. Quality EM allows minimizing the risk or the impact of destructive events and makes a dramatic difference in human casualties and damage to material and environmental goods in the area where the emergency occurs.

10.1.1. *Problem*

The series of natural disasters which have occurred over the last few years have revealed a great deal of inconsistencies in the area of management of emergency situations. Hurricane Katrina (USA, 2005), the earthquake in Port Au Prince (Haiti, 2010), forest fires in Cortes de Pallas and Andilla (Spain, 2012) and the floods in Tiflis (Georgia 2015) are but some instances of these shortcomings. According to statistics from United Nations Office for Disaster Risk Production (UNISDR), over one million human lives have been lost due to natural disasters and two and a half million people affected, with economic losses of over one trillion dollars over the last 12 years [UNI 13].

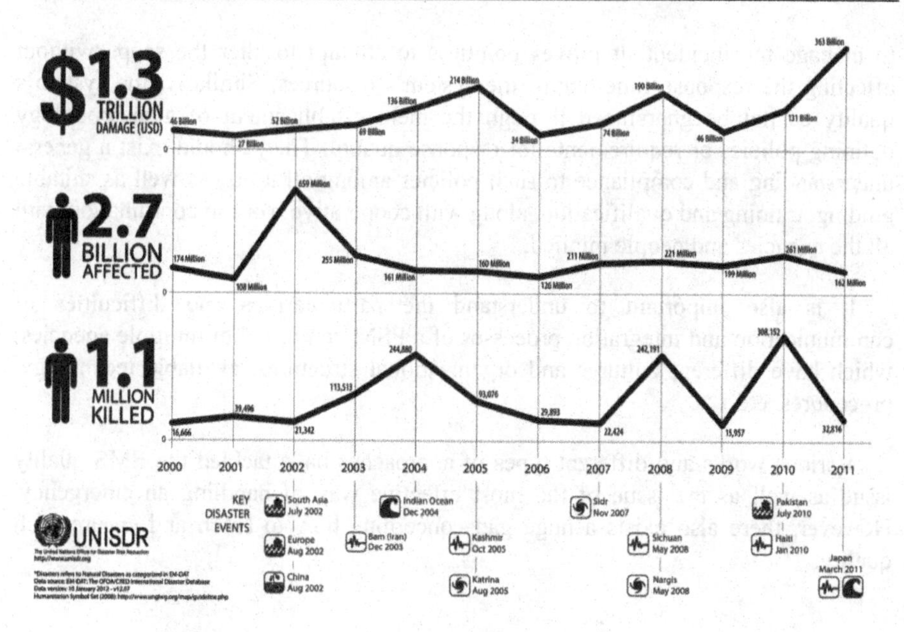

Figure 10.1. *The economic and human impact of disasters in the last 12 years [UNI 13]*

Despite the huge efforts and resources involved in dealing with them, such incidents have brought damage and unacceptable losses for the social environments and geographical areas involved. Nowadays, technological advances and a sharper awareness on the issue have fostered a certain degree of improvement in EM, but serious issues are still present and need to be solved.

The Federal Emergency Management Agency in the USA (FEMA) states that "issues in the quality of response to disaster or damaging incidents does not concern the lack of resources or tactical problems in most cases, but the quality of management" [FED 14a, UNI 12]. The British Standard Institution (BSI) in the standard for Disaster and Emergency Management Systems, affirms that "there are two types of emergencies: the ones which are managed by the organization and the ones which get to manage the organization" [BRI 08].

From our point of view, the problem lies in the lack of knowledge concerning the interrelations and dependencies between the parameters determining the quality of an EMS and allow its assurance. We cannot apply an isolated or individualized treatment of the parameters to determine the quality of an EMS; understanding that the variation or modification of any of them will exert changes on the other factors. For instance the scope of a disaster's management relies on the resources available to manage the incident. It proves pointless to attempt to alter the scope without affecting the response time and/or the system's resources. Similarly, the system's quality cannot be guaranteed through the mere establishment of a methodology defining policies or requirements for response quality. There should exist a general understanding and compliance to such policies among all users, as well as suitable guiding, training and qualification, along with cooperative work in coordination with all the agencies and people implied.

It is also important to understand the particularities and difficulties of communication and integration processes of a PSN, composed of multiple agencies, which have different cultures and organizational structures, available technology, procedures, etc.

Various works and different types of approaches have tackled the EMS quality issue as well as the issue of the most effective way of handling an emergency. However, there also exists a huge gap concerning how to reach and ensure such quality.

10.1.2. *Scheme*

This chapter provides a vision of an EMS under the perspective of quality. We define and specify a framework within which the parameters determining quality are present, and consider the requirements for the development, maintenance and improvement of the capabilities ensuring the quality in the management of an emergency. It delivers a suitable analysis tool serving as a starting point for the definition either of vulnerable spaces or likely to be improved within the processes and structures of an EMS.

It starts by understanding the EMS as a real-time system, where efficiency depends on communication and synergy between its elements that enable to meet the objectives and response time planned, and wherein the quality will be defined as the meeting point to house the best possible response of system, for all parties involved.

No accomplishment is possible without the fulfillment of scope and time of response. In order to achieve that, it is vital for the EMS to possess the capabilities to adapt itself and be able to respond at any time to whatever kind of disaster, of whatever magnitude under any circumstance.

The framework is based upon four principles:

– Establishment of parameters and interrelations determining the quality of an EMS. The "quality triangle for project management" has been taken a reference, one of whose variants derives into what is known as "agile triangle".

– Ensuring and improving the quality of the EMS. It is based on the Total Quality Management, framed within Joiner's triangle for continuous quality ensuring and improvement.

– An integrated and continued EMS. Should consider all the necessary processes before, during and after the occurrence of a disaster, and the existence of a PSN to allow the participation of multiple agencies required for a quality response. The regulations and recommendations of FEMA and the International Organization of Standardization (ISO) have been taken as a reference.

– Agility as a critical capability of an EMS. As emergency environments are similar to military scenarios for the dynamic and hostiles features of the operational environments. The system's abilities to adapt itself and respond effectively to disasters of any kind and magnitude is fundamental. We have based ourselves upon the papers of the Command and Control Research Program for the US's Defense of Department (CCRP), which lays great emphasis on agility as one of the main features of an EMS for effective response.

These four guidelines converge into a model which summarizes the factors, dependencies among themselves and the relationship with the general structure of an EMS.

10.2. Background

10.2.1. *Emergency management*

Broadly speaking, the word "management" is defined a "set of interrelated processes aimed at coordinating diverse strategies as well as efforts to effectively move on towards the achievement of goals" [SAN 03].

According to the International Strategy for Reduction of Disasters Organization (EIRD), an EM consists of the organization of and management of resources and responsibilities for the handling of all the aspects of emergencies, specially preparedness, response and recovery [EST 15].

ISO's regulation 22320 for EM promotes the implication of the whole community in the application and development of measures for the prevention and effective response to all kinds of disasters, both involuntary or intentional incidents [ASO 13].

The office of United Nations High Commissioner for Refugees (UNHCR) defines EM as the organization of capabilities and resources toward coping with threats against life and welfare [ALT 12].

From a different viewpoint, the BSI mentions Crisis Management as an organization's capacity to handle destructive incidents [BRI 14].

The International Association of Emergency Managers (IAEM) defines EM as the executive function in charge of creating the legal framework through which communities have their vulnerability to different risks lessened and stand up to disasters [ISO 15a]. FEMA also sticks to this definition.

In any case, the aim of EM is to diminish human losses as well as damage to property and environment, and to protect the whole community from any hazard or threat [FED 15b].

In accordance with the aforementioned, it could be said that EM:

> *Creates the framework to help the whole community prevent and stand up to emergency situations of any kind, with the aim of reducing fatalities, material and environmental damage.*

EM has evolved significantly over the last few years. We have come from non-organized public action lacking specialized institutions, usually preceded by the government's reaction capabilities on an improvisation basis, to a continued and planned EM, taking into account coordinated collaborative action of multiple specialized institutions, based on agility and quality EMS [IZU 09].

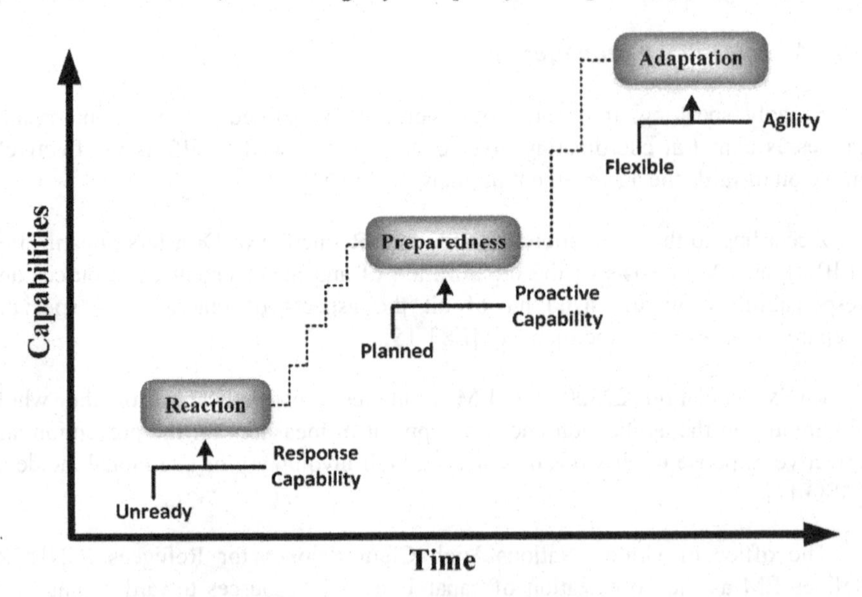

Figure 10.2. *Emergency management evolution*

EM comprises a whole range of topics related to how to better prevent and face disasters. It involves other concepts such as management of incidents, disasters, risks, civil protection and others. Not only does it address response phases to disasters, but also to the whole range of actions designed for preparedness, mitigation and recovery of damage after destructive incidents, in previous as well as in subsequent phases.

10.2.1.1. *Phases of emergency management*

Due to its complexity, EM is often described through phases which make analysis and understanding easier. Quality EM requires actions before, during and after destructive incidents of any kind, at any time and under any circumstance. The final phase of a previous phase stands as the initial phase of the following phase, thus defining an endless and continuous cycle of interrelated processes. One of the main processes of each phase is the planning of the next, which allows the control and assessment of the results obtained in the phase itself.

Many international organizations in the field of EM and security represent EM through a cycle which groups processes into four phases: preparedness, response, recovery and mitigation. This scheme has become widely accepted and shows some variations in some cases, but the essential model remains the same [BAI 10]:

– *Preparedness*: actions related to the construction and improvement of the necessary means to prevent or stop the occurrence of a disaster. This phase is associated to phases such as prevention and protection. All response plans for the main existing threats should be created at this phase.

– *Response*: actions required immediately after the occurrence of an incident, with the aim of saving lives, protecting property and the natural environment, and meeting basic human needs. One of the main goals of this phase is to obtain stabilization and control of the situational.

– *Recovery*: actions required to help a community affected by a destructive incident recover and return to a normal or enhanced status.

– *Mitigation*: actions developing a solid ground which should diminish or eliminate the risk for persons and goods and/or lessen the effects of damaging incidents [FED 14b, BAI 10].

10.2.2. *Emergency management systems*

A system is a "set of interrelated organized elements interacting towards the achievement of a goal". Such a goal could not be reached if the elements making up the system were acting in isolation.

An EMS must provide its users with the required capacities to confront emergencies in the best possible way. The quality of the system is related to such capabilities for reaching its goals.

FEMA defines an integrated emergency management system (IEMS) as the framework for an effective management of emergencies and comprises actions on every level of government, private sectors and mass media, through coordination, management and control of disasters, irrespectively of type, origin, size and complexity [BOB 13]. One of the key components of IEMS is the incident command system (ICS). It is known as the methodology which allows the integration of facilities, assets, processes and communications within the same organization structure with the aim of delivering a coordinated response [FED 14a].

The BSI establishes a disaster and emergency management system (DEMS) as the tool which allows the prevention and control of potential disasters and emergencies, thus enabling the people in charge to find suitable acceptable solutions [BRI 08].

The ISO defines a command and control system (C2S) as the system sustaining the effective management of resources at the time of an emergency of all resources available, on the preparedness phase, at response time and during follow-up and/or recovery [ASO 13]. It is relevant to mention that the PSN and the integration, coordination and cooperation processes between the agencies involved, are established as parallel and complementary processes to C2S.

A general approach could say that an EMS is:

> *A set of processes and resources integrated within a common organization structure with the aim of creating, developing and maintaining the capacities for preparedness, protection, mitigation, response and recovery of the whole community before an emergency.*

An EMS is a tool which allows the effective realization of EM. It allows the guiding, coordination and command over the efforts of all parts involved by integrating resources and processes within the same organizational structure with the aim of fulfilling the goals outlined for the EM.

The EMSs have improved the management of an emergency or disaster, by increasing the capabilities available to prevent and deal a damaging incident. They consider an integral response, in which different agencies interact in a coordinated and cooperative manner, providing the full range of capabilities and resources required for an agile and effective response.

Due to the multidisciplinary character of EM, an EMS must allow, through PSN, integration and interaction of the multiple agencies involved with the aim of obtaining the best possible response, in accordance with the environment and the scale and type of risk or disaster. Such circumstances, however necessary, introduce additional problems such as coordination, cooperation and the onset of the required command and control processes so that the people involved can work together and add up their efforts. EMSs should prop up those additional components, specific for EM, by creating synergies and the necessary organizational structure for providing a quality response.

Coordination is the way in which the various institutions involved work together to fulfill a common goal. Without coordination, organizations have difficulties in identifying a common goal in incident response and in accepting strategic implementation [DOB 12].

Cooperation is an agreement for acting in a coordinated way on the basis of agreed goals and/or values. Cooperation must be assessed, prepared, defined and tested beforehand on the basis of risk analysis. This will create opportunities for the planning of effective and economical response options. Cooperation can share or reduce costs as well as improve the continuity and recovery of economic activity [DOB 12].

Command and control processes are implemented and executed through C2S with the aim of building up an organization structure and enough control and information for decisions and actions to be executed in an agile and effective way. C2Ss base their operation capacity on standardization, well-defined command chains and the establishment of a common situation awareness. For this purpose, the exchange, distribution and availability of quality information is a critical requirement [ALB 06].

10.2.3. *Command, control and information systems (C2IS)*

The concept of command and control (C2) was created inside the army and refers to the exercise of authority, command and monitoring by the assigned operational command, over the chosen forces for the accomplishment of a specific mission [DEP 10].

Military environments present several stress situations where the system's agility and the quality of the response are of critical relevance. Such features have been transferred to EM taking into consideration the similarity with the requirements of the environment. C2Ss both in emergency and military application domains must be designed to fulfill the mission's goals while operating in critical environments.

One of the main drawbacks of developing of a C2S with a civil orientation compared with C2S on a military environment is the additional ingredient of heterogeneity and the lack of specific training of organizations. The military C2S are hierarchical and are tightly coupled, while civil C2S are often organized in multiple agencies, making them weakly coupled; its structures take a longer time to establish and communications are more complex [REP 15]. It should also be mentioned that civil environments usually require a lower level of security in communications and information, which simplifies standardization and scalability of your PSN, with the corresponding impact on the system's agility.

We can define C2 for civil environments as:

> *The exercise of authority, command and follow-up up by the person in charge over the resources and staff available for the fulfillment of goals.*

C2S is a relevant part of the core of an EMS and defines the organization structure and the required processes to carry out the processes in C2 by supporting the officer in charge of decision making, order transmission and supervising that all orders are enforced.

Communication and information systems (CIS) are one of the main components of C2S, so it is common to refer to the Command and Control Information Systems (C2IS). They are responsible for ensuring information flow and its quality through all levels and agencies of PSNs, thus creating a common situation awareness of the operations theatre, an effective transfer of decisions and an agile and effective execution of all the processes.

During emergencies, production and circulation of timely and transparent information contributes to generating confidence and credibility on EMS [ORG 09]. Communications are a critical factor in achieving and ensuring a quality response for EMS, so the PSN should have high reliability and availability.

In the case of EMS, is also it required that the CIS allow communications with its environment, enabling the delivery of relevant information to the outside of the system, as well as the exchange of information with other CISs. The PSN is the tool that provides the infrastructure and organization needed to allow the integration of the CISs of the involved agencies, communication and exchange of information between them and the whole community.

10.2.4. *Quality in emergency management systems*

"Quality" is a concept with no exact definition as it relies on the orientation and environment of the product or service. From the product's point of view, quality is a measurable parameter which allows comparison. From the user point of view, it is linked to the capabilities of the product or service to meet the needs and expectations of users. From a production point of view, it refers to the fulfilling of the product's design specification [SAN 03].

The ISO gives us a general definition intended to offer wide coverage in the whole range of fields of its regulation and oriented to all kinds of organizations. Quality is defined as "the extent to which a set of inherent features will meet the requirements" [ISO 05].

Ishikawa makes a definition of quality from the point of view of quality ensuring. He defines quality as the equivalent to client's satisfaction. He proposes a systemic treatment where the product's quality implies quality in all and every one of the elements which make up the production system: processes, equipment, human resources, objectives and so on [HOY 05].

Respectively, the US Department of Defense defines quality as "doing the right thing from the first time, always aspiring for improvement and user satisfaction".

It can be affirmed that quality for EM is defined as:

> *The system's capacities to satisfy the needs or requirements of all parts involved and their perception of such satisfaction when the time comes to prevent or confront a risk or disaster.*

An EMS should develop, maintain and enhance the capabilities for prevention, protection, mitigation, response and recovery of its internal and external users and of the whole community in order to confront incidents effectively.

The scope response time and resources available for an EMS are outlined by the government. However, the system's quality is determined in a collective way by all parties involved. From the Government's point of view, quality is defined by the fulfillment of the scope and response time set during the design phase. From the point of view of internal users, quality defines itself by their perception as concerns the system's ability to satisfy their needs and requirements. From the perspective of external users and people affected, quality will have to do with the system's ability to mitigate the damage and losses produced by a destructive incident.

One of the most common challenges when designing systems is to find a common meeting point between all parts involved to defining quality, where we can design and develop an EMS that allows the whole community to take part, deliver adequate responses to those affected, provides suitable tools to internal users and fulfill the proposed scope and goals, in accordance with the resources available and the environment's conditions.

There are several methodologies that address the issue of quality management, one of the most successful and widely used today is Total Quality Management (TQM).

10.2.4.1. *Total quality management (TQM)*

TQM is the management philosophy which has its foundations in a balanced satisfaction of needs and expectations of all parts involved as the best way to reach the desired goals [EUS 11]. Understood as a philosophy it should be accepted by all parts and applied over the processes in the organization. This is why the word "total" appears in its definition.

It is based upon eight principles [GAI 07, ISO 05]:

– *User focused:* understanding the present and future needs and expectations of users, meeting them and enhance their perception. Internal and external users can be distinguished.

– *Leadership:* it is the board's responsibility to create and maintain the atmosphere and synergy enabling the personnel to become involved in the organization's goals.

– *People's implication:* people are the most important resource of an organization. It is the implication of the human force which will allow them to pursue the corporative goals.

– *Process-based:* results are more easily obtained when resources and actions are managed as processes.

– *Focus on management as a system:* interrelated processes should be understood as a system.

– *Continuous improvement:* must be a permanent goal for the organization.

– *Focus on fact-based decision-making:* effective decisions are based on information analysis.

– *Mutually beneficial relationships with all parties implied:* relationships based on mutual benefit will improve the capacities for value creation.

TQM is based on the control and constant search of continuous improvement of all the processes and the system itself. To satisfy this premise, it resorts to the PDCA cycle (Plan, Do, Check, Act) created by Deming, who proposes a dynamic cycle which repeats itself continuously and controls the results of each one of the processes and subsequently acting in the appropriate way to achieve continuous enhancement [CEN 15, GRI 95, ISO 08].

The process is described in four phases [DIA 05, MOJ 15]:

– *Plan:* set the goals and required process to obtain the desired results, in accordance with the requirements of all parts involved and the organization's policy.

– *Do:* implement previously planned processes toward reaching of goals.

– *Check:* follow-up and assessment of processes and results in relation to policies, goals and requirements, briefing those in charge on the results achieved.

– *Act:* carry out actions oriented toward the improvement of performance in each of the processes in a permanent way.

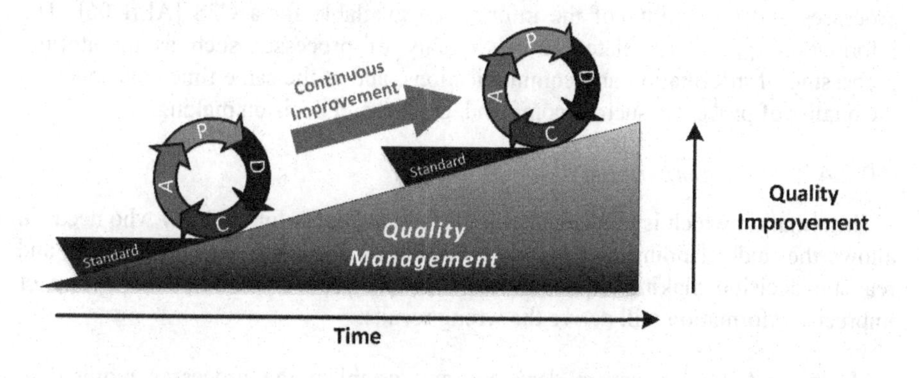

Figure 10.3. *PDCA cycle*

TQM rises as a response to the requirements of organizations in order to operate in dynamic volatile markets due to globalization and a staggering technological development. These dynamic features can be compared to those we find in emergency environments, that is why part of the proposals in this chapter refer to the application of TQM in EMS.

A systemic and process-based approach to EM facilitates its study by dividing it into smaller parts, which are easier to understand and analyze. The quality of an EMS depends on the quality of its resources and processes. The available information and processes such as C2 and those required for integral response have a special influence on the system quality.

10.2.4.2. *Process quality*

In general, the quality of a process is based on the quality of material and human resources involved in its execution, process knowledge of the staff involved and quality of the organizational structure that allows those resources to interact.

The staff must have the infrastructure and equipment to fulfill its functions, as well as the skills and training to permit them to develop and use the capabilities required for a quality EM. The community should have information and training about the risk in their environment and the procedures to prevent and/or deal with them.

The processes of coordination and cooperation necessary for integral response have special relationship with the PSN, which must allow the interconnection of CISs of all agencies involved, provide reliable communications and ensure the distribution of information throughout the entire EMS.

As regards C2 processes, there is mutual dependence between the quality of processes and the quality of the information available for a C2S [ALB 06]. The information quality is related to the quality of processes such as monitoring, processing of information and communications, and at the same time contributes to the quality of processes such as command, control and decision making.

10.2.4.3. *Information quality*

Information which is precise, intelligible and available for anybody who needs it, allows the understanding and proper execution of processes; and fast, timely and realistic decision making. It is obvious that any process starting with wrong or imprecise information will derive the wrong results.

In order to make correct decisions and establish the processes required to manage an emergency, we must get a clear understanding of the context in which operations are conducted. This situational awareness is based on the quality of information available.

Information quality is the one which is obtained through the appropriate management of such information. The tool which enables us to reach that goal is the information and communications system (CIS).

The CIS includes the set of processes and resources which enable the quality and flow of information in any direction at any time. One of the main functions of a CIS is to offer support to command actions in decision making, as well as the effective transfer of such decisions to every level in the organization [LAP 11].

Quality of information is assessed according to four parameters [SPA 12]:

– *Availability:* it is the key factor since the information is useless unless it is available. It is closely linked to the PSN services and maintains an inverse relationship with information security.

– *Presentation:* information must be understandable, which implies conciseness and a clear format.

– *Context:* information must be updated regularly and be a reflection of reality.

– *Intrinsic:* refers to the accurateness and reliability of sources.

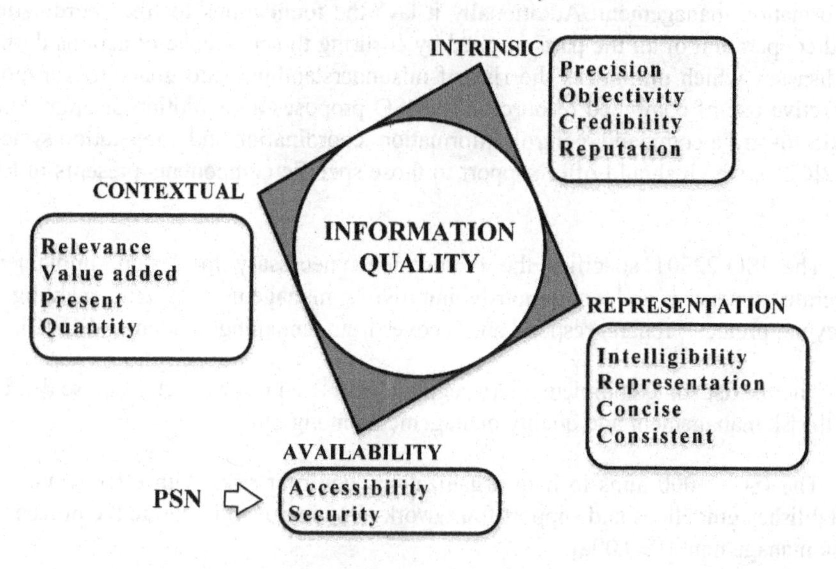

Figure 10.4. *Information quality*

10.2.5. *International Standardization Organization (ISO) and emergency management*

The ISO is formed by the International Standardization Organizations' (ISOs) government and non-government organizations of one hundred and seventy-three countries, organized into a series of technical sub-committees in charge of developing the regulations which will contribute to the enhancement of commerce, information exchange and technology transfer.

The ISO-223 is the technical committee in charge of developing the regulations concerning social security, improving the management capabilities before a crisis and reducing the risk of disaster [EST 15]. It shows a vision of wide scope, addressed to covering all possible damaging incidents which should crop up for any kind of organization, as well to promoting a best practices and strategies policy, proactive as well as reactive [ISO 15b].

The ISO-22399 shows the general guidelines for EMS design allowing for the preparedness before the occurrence of incidents as well as operational continuity during and before the incident [ISO 07].

The ISO-22320 describes the best practices for the establishment of organizational structures, C2 processes, decision making support processes and information management. Additionally it lays the foundations for the coordination and cooperation of all the parts implied by ensuring the coherence of actions during a disaster, which minimizes the risk of misunderstandings and guarantees a more effective use of combined resources. The ISO proposes an evolution or upgrade of C2IS toward a command, control, information, coordination and cooperation system (C2IC2S), which should offer support to those specific components presents in EM [ASO 13].

The ISO-22301 specifies the requirements necessary to design, implement, operate, supervise and continuously improve a management system enabling to prevent, protect, prepare, respond and recover from damaging incidents [ISO 12].

There exist some complementary regulations related to EM such as those dealing with risk management and quality management among others.

The ISO-31000 aims to help organizations confront risks with effectiveness. It establishes guidelines and support framework allowing us to integrate the process of risk management [ISO 09a].

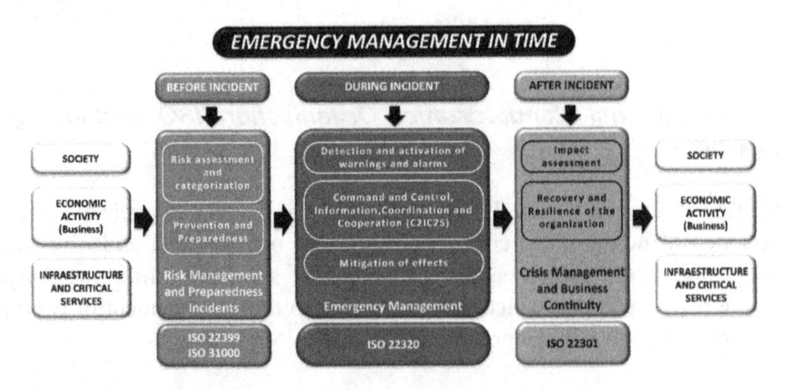

Figure 10.5. *ISO emergency management*

Concerning quality, the ISO makes its proposal out of the family of regulations ISO-9000, developed by the ISO-176 technical committee in charge of management and ensuring quality.

The ISO-9000 family refers the requirements or guidelines related to quality management and presently it is one of the most important references on worldwide [BEL 09].

They suggest a model for quality management clearly focused on goal achievement, under the basic principles of a balanced satisfaction of the requirements and needs of all parts involved.

The guidelines established in the ISO-9000 regulations rest upon the eight general principles of TQM described in point 3.2.4.1.

The ISO-9000 describes the principles of quality management and specifies terminology for quality management systems [ISO 05].

The ISO-9001 specifies the requirements for quality management systems which apply to all organizations intending to show their capacity for providing products or services meeting the requirements of their users as well as the set of regulations they should comply with. ISO 9001 introduces a process-based model for quality management where users play a fundamental role [ISO 08].

The ISO-9004 provides the guidelines which take into account both effectiveness and efficiency of the quality management system. The aim of this regulation is to improve the performance in the organization, client satisfaction and other parts implied [ISO 15b].

The ISO establishes the PDCA cycle in all its regulations, as one of the foundations for the control and continued improvement of processes.

10.2.6. *Federal Emergency Management Agency (FEMA)*

It is the Agency of the American Government whose aim is to "support all citizens and response organizations to develop, maintain and improve their preparedness, protection, response, recovery and mitigation capacities before any kind of menace" [FED 15a].

They draw proposals from the concept of an Integrated Emergency Management System (IEMS), on the principles of an integrated EM which should consider all parts involved as a single entity and enabling the assurance and improvement of the response capabilities before disasters [FED 83].

The IEMS concept was developed and implemented on the basis of two main systems [FED 14a]:

– National Preparedness System (NPS) lays out the guidelines, programmes and processes which permit the whole community to fulfill with their National Security Plan.

– National Incident Management System (NIMS) outlines recommendations, concepts, terminology and organizational processes which should enable an effective management of incidents in a collaborative way.

Its core points are the command and management processes of incidents, supported by the resources management, information and communication, as well as the preparedness processes. Command and management are realized through an Incident Command System (ICS), a Multiagency Coordination System and a Public Information System [FED 14a].

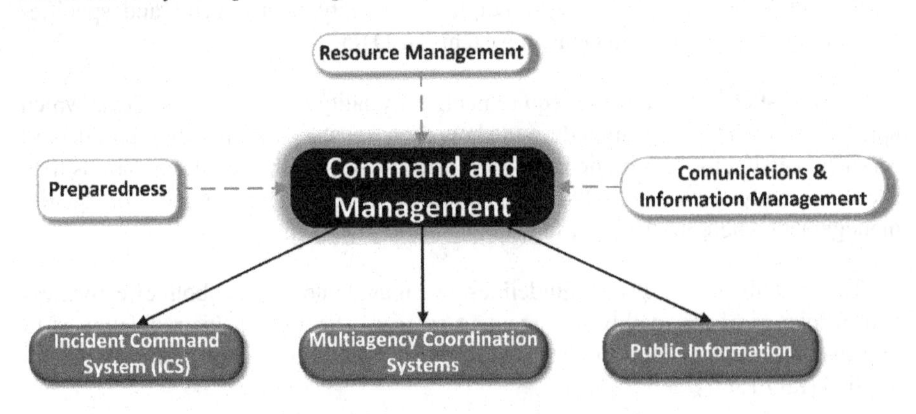

Figure 10.6. *NIMS components*

10.2.6.1. *Incident command system-oriented*

The ICS was developed during the 1970s after a number of catastrophic fires in California, US. The investigations found that the problems in response were due more to a wrong management than any other reason [FED 14a].

An ICS is "a combination of facilities, equipment, personnel, protocols, procedures and communications, operating under a common organizational structure with the aim of administering the assigned resources and effectively fulfill the desired goals of an event, incident or operation" [UNI 12].

The ICS aims ensuring the safety of first responders as well as in achieving the goals and an effective use of resources. It does not replace nor modify the organizational structures of the implied agencies, rather it is a tool enabling integration, coordination and cooperation of its resources and processes, through its management under a single organizational structure.

One of its main features is that it is modular, scalable and flexible and its size and deployment will be dependent on the magnitude and complexity of the disaster. The command and the operations are functional elements critical to the organizational structure of the ICS. Planning, design and finance are functions initially assumed by the command, and only will deploy in the case where the system is extended enough to require it. According to the case, the command could delegate other functions such as security, public information and links to the external environment.

The ICS model has been taken as a reference for the development of great number of EMSs all over the world and is widely used in the USA, United Kingdom, Canada, Brazil and is recommended by the UN. The ICS is based on fourteen principles grouped in six big blocks according to the features of the processes [UNI 12].

Features	*Principles*
Standardization	Common Terminologies
Command	Establishment and transfer command Chain of Command and Unity of Command Unified of command
Planning and Organization Structure	Management by Objectives Incident Action Planning Modular Organization Manageable Span of Control Comprehensive Resources management Incident Facilities and Locations
Communications and Information Management	Integrated Communications Information and Intelligence Management
Professionalism	Accountability Dispatch/Deployment

Figure 10.7. *Principles of ICS – FEMA*

Among the most relevant features of the model developed by the FEMA for an EMS, we can mention:

– model dimensioned according to the environment and capabilities of the USA;

– emphasis on the standardization of language in order to prevent confusion and communication problems;

– integrated system enabling different agencies involved on the EM and the community, operate in coordinated and cooperative form. The term "whole community" is introduced as key to an effective management of the event;

– scalable and flexible;

– emphasis on decentralization as a basic concept for the system's agility;

– introduction of the word "incident" with the purpose of widening the scope of the system and offer coverage to the different kinds of damaging event which may occur;

– well-defined chains of command where each unit reports to a single person in charge.

10.3. Methods

In order to outline a model which allows us to ensure quality, we have to start by understanding what quality means for an EMS.

The quality of any system is a function of the quality of each of its parts, therefore it will depend on the quality of its processes and its resources.

For the specific case of an EMS, quality could be understood as the point where the best alternative is located to satisfy the needs and expectations of all parts implied (government, internal and external users, etc.). The system must comply with all the parameters and service standards defined during the system's design and its operative planning. Additionally it should also possess the necessary resources to create, develop and improve the required capabilities to assist its internal and external users in accordance with the proposed scope and time response.

The proposed model is based on compliance with the parameters that limit the quality of EMS, and assurance of this quality at all times of the operating cycle. It is important to take into account that EMS must be dimensioned and personalized according to the specific conditions of its environment, and in terms of its operation, the involvement of multiple agencies is required.

10.3.1. *Quality parameters*

A project's quality is defined by three interdependent parameters: resources, scope and time. The scope should be defined based on available resources and the planned response time; resources should be allocated in relation to the scope and time response required; and the response time will be consistent with scope and available resources. All three parameters and their interrelationships can be represented by a triangle whose interior shows the project's quality. This triangle is known as the "triangle of project management", "quality" or "iron triangle" [DIS 12, PEA 06].

The idea of a triangle suggests that none of these parameters can be modified without affecting the other two, and shortcomings in one of them cannot be replaced by sufficiency in another.

Generally, two of the parameters stay fixed and the third is set in function of the former two, according to environment conditions. For example, the available resources may be fixed as well as the project's scope, and the required time can be determined from on that basis.

Quality is defined by the inscribed circle area and not by the triangle's area. The idea would be to find an appropriate balance in the triangle in order to obtain the best possible quality by means of optimizing resources, maximizing of scope and minimizing of the required time. In Figure 10.8, we can visualize that the quality is the same in the first two triangles (equal areas of inscribed circles), though some oversizing in the resources and scope sides of the project can be perceived in the second case. It can also be observed that triangles number one and number two are similar and, in spite that proportions between parameters, are the same, and the quality in the third one is lower.

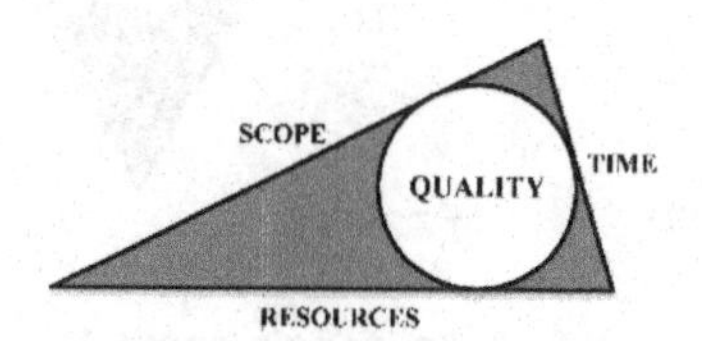

Figure 10.8. *Quality triangle*

Projects are basically characterized by using resources, being unique and having a specific duration [PRO 15]. Such features can be associated to the features displayed in emergency or disasters. Every emergency has its own temporal, geographical and typological features. In order to manage it, it is necessary to have the resources and a planning as regards to the scope and response time.

In the case of EM, the scope is related to the services offered, the geographical coverage of these services and with threats and risks that can be managed. The response time is defined by the maximum time which it is required to respond to a threat or disaster. The resources relate to the elements or means (material, technological, financial and human) an EMS must possess to meet the planned response parameters.

According to the stated above, the quality triangle can be applied to the EMSs by taking into account that in order to obtain quality responses it is imperative to count with the necessary resources, meet the required scope and minimize response time.

There is a scheme known as the "*agile triangle*" to give its agility to the "*quality triangle*".

A single parameter is fixed, leaving the other two dynamic according to different circumstances. When applied to an EMS, the fixed parameters are the available resources, while system's scope and response time will vary according to type and magnitude of event, and the operations environment. Thus, the importance of planning relies on its correspondence with scope, response time, and available or assigned resources to EMS.

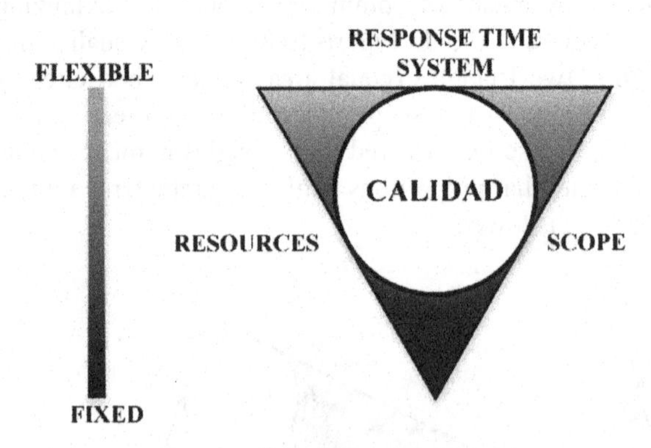

Figure 10.9. *Agile triangle applied to an EMS*

10.3.2. *Ensuring quality and continuous improvement*

Joiner summarizes the TQM framework into three interrelated elements making up a triangle: culture of quality, scientific approach and all one team [LOP 04, TIM 99]:

– *Quality Culture*: excellence should be sought in every aspect of an organization. The executive board should be obsessed with ensuring quality and comprehend its relevance for the organization and be adopted as part of the organizational culture. The stakeholders' voice should be listened to at all times, and the system's efforts must focus on meeting their needs and expectations.

– *Scientific Approach*: actions and decisions should be based on data analysis. Processes must be geared toward making realistic decisions known, understood, controlled and continuously improved. The information must be reliable and available at all times and any node on the PSN.

– *All One Team*: it is generally the most difficult element to attain. All agencies and users (internal and external) of EMS should be part of a PSN that allows them to be treated as a single entity of interconnected elements, that work together in a coordinated and cooperative way, to obtain the desired results. The unified command should create the environment and synergy that allows all parts involved to direct their efforts toward the prevention, protection, response, recovery and mitigation of the effects of a damaging incident. The staff should have the sufficient training and coaching to make smart and suitable decisions, which gives confidence in its performance, and permits it to overcome the limitations imposed by a management based on the inspection.

This theory is propped up on the interdependence and harmony between all three components of the triangle. It affirms that one of the main causes of failure in management initiatives stems from not recognizing such interdependencies. As an example we can mention that when the concept "the client is always right" became fashionable, no significant improvements were registered, because those initiatives were not propped up by data, and the processes were not understood.

We are proposing the application of the concepts of TQM to EMS with the aim of ensuring the system's quality. In order to achieve and ensure quality responses, an EMS should follow the principles established by Joiner's three fundamental concepts. In the same way as in the previous case, shortages of one component cannot be replaced with abundance in the other.

There exists a fourth element which reinforces all three sides in the triangle and makes it into a pyramid. The fourth element is Information and Communication Technologies (ICT) which offer a solid background for the ensuring and improvement of the whole system [ARE 06].

A GE of quality, should identify, control and deal with threats, risks and disasters that arise or may arise in its social core; this requires for continuous and permanent processes to manage it at all times.

10.3.3. *Integral and continued emergency management*

FEMA, in its National Response Framework describing document, alludes to the relationship between functional and temporal phases of NIMS: "preparedness is necessary for protecting, protecting for response and response is necessary for recovery starting from mitigation" [FED 14b].

Quality EM should contemplate the whole set of interrelated processes with social and national security before, during and after the occurrence of a disaster. That is why an EMS is required with an integral and continued response capability, with a predictive, reactive and adaptive orientation, which should be able to confront the dynamism and unpredictability of an emergency [IZU 09].

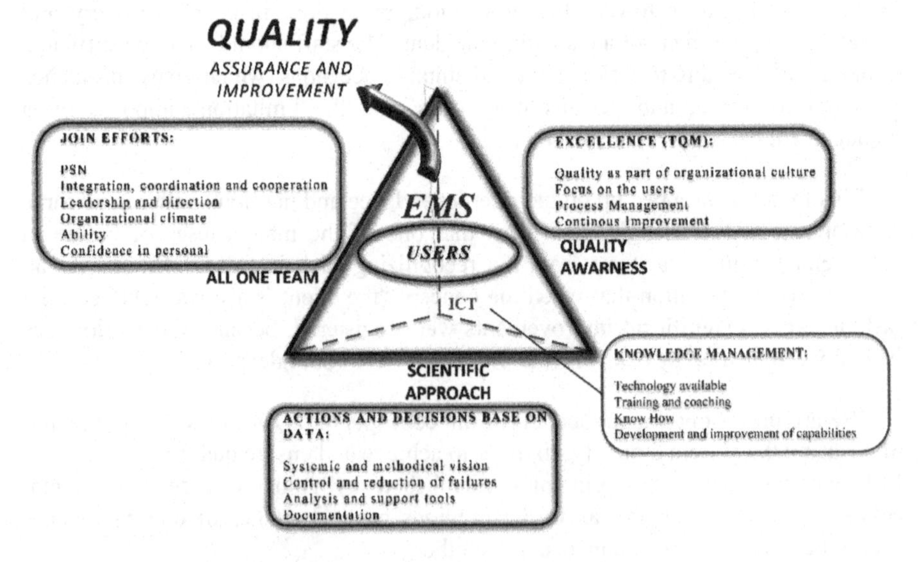

Figure 10.10. *Joiner's triangle*

The EMS should be able to develop, maintain and improve the capabilities of all agencies, personnel and social core involved in order to prevent, protect, respond, recover and mitigate the effects of a disaster.

The participation of multiple agencies creates enough diversity and specificity of knowledge and capabilities to deliver an integral response to issues, thus enabling suitable responses to diverse incidents and environments [UPO 94]. Qualification and training staff and the whole community acquire special relevance as the means for capabilities develop of people.

Agencies involved in the GE depend on the environmental conditions of each country or region. The PSN is the main artery through which agencies are integrate to EMS, enabling the communication and exchange of information between them and the community.

Ensuring quality and the continuous improvement of processes for dynamic environments, as is the case of EM, is linked not only to improving satisfaction and expectations users', but also related to the system's agility to maintain quality levels before the occurrence of different types of incidents and dynamic environments.

10.3.4. *Agility*

Due to an emergency's dynamism and unpredictability, an EMS's capacity to effectively prevent and respond to any type of incident and adapt itself to whatever environment is absolutely imperative.

Agility is defined as the set of capacities enabling an organization or system to effectively confront the possible changes which may occur [ALB 10].

As concerns an EMS, it can be defined as the promptness and effectiveness of the system to adapt and respond to sudden unexpected changes which may arise in the operations environment before a damaging event of any type or magnitude.

Agility is a key feature of an EMS, and its quality depends on it. We cannot talk about quality without mentioning the fulfillment of scope and time response planned. The quality of an EMS is related to its effectiveness, which in turn is related to the quality of human resources, information, available technology and organizational structure of the system:

– agility is associated to the capacities of flexibility and innovation [LOP 08]. Flexibility is defined as an organization's capacity to adapt to new situations, without these changes meaning significant expenses in terms of time, cost, effort or performance [UPO 94]. Innovation is linked to an organization's creative capacity to generate new ideas enabling changes in the processes or structures with the aim of improving or adapting responses to circumstances [UPO 94]. Innovation and flexibility hold a direct connection to the quality of human resources. Knowledge, abilities and skills of the staff, in a higher or lesser degree, will allow for the adaptation of the operational processes to the dynamics of a changing environment [LOP 08], as well as the possibility to make fast decisions at the edge of the network [ATK 07];

– information quality contributes to the agility of all processes in the EMS. The information should be understandable, reliable, accurately describe the operating environment and be available to all agencies and users at any time. This allows making quick, timely and appropriate decisions, improving situation awareness throughout the PSN and reducing delays and errors in the execution of processes;

– current technologies have made it possible to overcome spatial boundaries and bring civil and military operations over to global environments, where response times are critical. The integration and cooperation between agencies and systems from different countries or regions, related to EM, is essential. Cloud computing, high speed data transmission, deployable networks, virtual and augmented reality, the internet of things, drones, big data, and social networks, among many others, are clear examples of the advantages offered by actual technologies, and its contribution with agility and minimization of response times a EMS;

– a suitable organizational structure allows an agile deployment of resources to the place and at the time they are required, as well as scaling the PSN in accordance with the type and magnitude of the incident. The EMS should provide a integral response, which has the resources and cooperation of agencies required to a response commensurate to the needs and requirements of all parts involved.

Standardization and decentralization are fundamental processes within the organizational structure of an EMS. Standardization facilitates communication and understanding between the various agencies that are part of the PSN. Decentralization facilitates the distribution of resources, eliminate bottlenecks and encourages decision-making at the edges of the network (power to the edge). These features translate into the greater agility and effectiveness of a system.

10.4. Conceptual model

Here we present a conceptual model which describes and synthesizes the aforementioned concepts, providing a framework for ensuring quality.

Among the most significant considerations, we can note:

– the model provides a vision of EMSs from the outside as a unique entity, and an integral systemic vision from its inside;

– an EMS should consider the environment's conditions prior to its dimensioning, design and operation. Such premises will be determining the threats and risks to be prevented and mitigated, as well as the factors holding an influence on the response and recovery: geographical conditions, government's policies, social development, technology. Technology is an input and output variable, it must take into account both available technology within the system and in the external environment;

– as for quality ensuring, all four guidelines previously described should be taken into account: quality parameters, quality ensuring, integral and continued EM, and agility;

– conceptually, an EMS should be oriented toward meeting the needs and expectations of all parties involved. From a practical point of view, we should find the possible better quality, in accordance with the available resources, scope and time response planned, type of disaster and present environment;

– additionally to C2 processes, integration-coordination-cooperation processes should be taken into account. The ISO's model has been taken as reference, in which a C2 C2S covers the whole range of processes required. Standardization and decentralization are features required for the organizational structure of an EMS;

– an EMS, through its PSN, should permit for the interaction with its environment, by delivering relevant information to external users and enabling communication with the rest of involved agencies, as well as the exchange of information with other EMSs. Feedback from external sources should be possible, and, once analyzed and verified, should improve and update situation awareness.

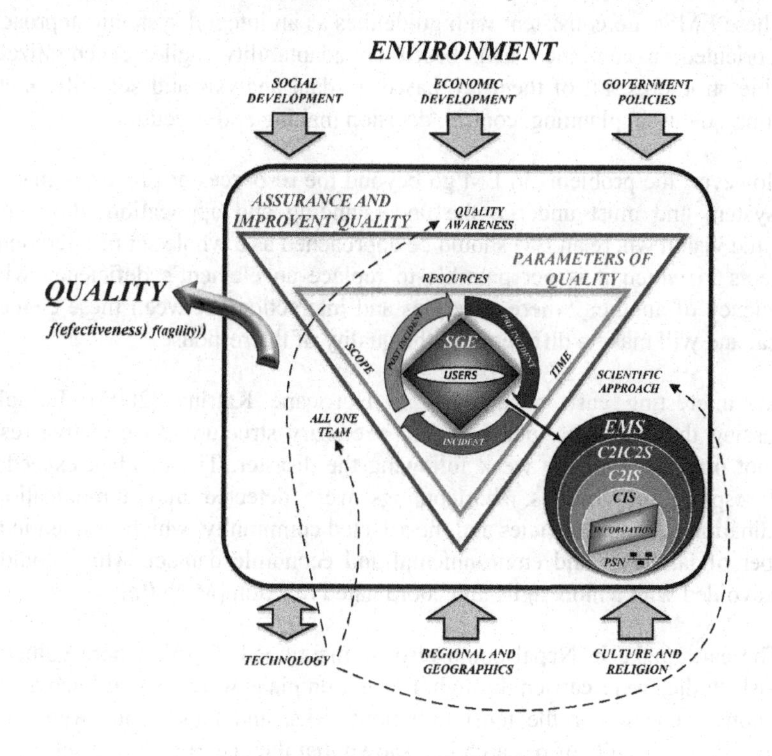

Figure 10.11. *Conceptual model for EMS quality assurance*

10.5. Discussion

An EMS is the tool that ensures the suitable performance and agility of operations in preventing and coping with a disaster; the command and control of resources and processes, as well as effective integration and coordination of agencies should be involved in the response.

The quality assurance as to the response of an EMS is a determining factor, since this depends on minimizing human, environmental and material losses, as well as the stability and continuity of the operations of the social nucleus in question.

There are several success cases that support this quality model for EMSs. Undoubtedly, the one developed by the FEMA is one of the most relevant and stands as a reference for this chapter. Similarly, we can mention the Disaster and Emergency Management System (DEMS) of United Kingdom, or the Swiss Civil Protection System (CPS), which have shown significant improvements at the time to prevent or deal with different types of damaging incidents.

These EMSs are consistent with guidelines as an integral systemic approach and user oriented; a continued EM, based on adaptability, agile, decentralized and scalable structures; all of them are based on data analysis and scientific methods, enabling quality in planning, control, decision making and execution.

However, the problems in EM go beyond the resources or processes making up the system and must undergo the understanding and application of an integral systemic vision where an EM should be approached as a whole set of interconnected elements in which it is not possible to replace an element's deficiency with the sufficiency of another, where synergies and interactions between these elements is critical and will make a difference in the quality of the response.

An interesting case of analysis is Hurricane Katrina (2005). In spite of possessing the resources, planning and necessary structure, an effective response was not possible until the week following the disaster. The incident exceeded the local response capabilities, shortcomings were detected in communication and coordination between agencies and the affected community, which resulted in a high number of fatalities, and environmental and economic damage which could have been avoided with a more agile and coordinated reaction [MOF 08].

The earthquake in Nepal could also be mentioned (2015), where vulnerability and risk studies were carried through. Prevention plans were also outlined as well as a response scheme for the most important risks, and these plans were properly socialized. The results of research had shown that the occurrence of a seism of huge magnitude was imminent and this information was known. However, lack of resources, sustainable development policies and a social awareness did not allow such schemes to be brought into practice. The type and location of the damages also confirms the results concerning vulnerabilities and hazards present. Deficiencies in the prevention, protection, response, recovery and mitigation processes, allowed for the loss of more four thousand people and over seven thousand individual affected [PER 15].

10.6. Bibliography

[ALB 06] ALBERTS D.S., RICHARD E.H., *Understanding Command and Control*, CCRP, Washington, 2006.

[ALB 10] ALBERTS D.S., *The Agility Imperative: Précis*, CCRP, 30 March 2010.

[ALT 12] ALTO COMISIONADO DE LAS NACIONES UNIDAS PARA LOS REFUGIADOS, Manual para situaciones de emergencia, ACNUR Internal report, 2012.

[ARE 06] ARENAS P., EDGAR A., La gestión del conocimiento: el triángulo de Joiner o la pirámide de Joiner-Arenas, available at http://www.entorno-empresarial.com/archivo/articulo/372/la-gestion-del-conocimiento-el-triangulo-joiner-o-la-piramide-joiner-arenas, 2006.

[ASO 13] ASOCIACIÓN ESPAÑOLA DE NORMALIZACIÓN Y CERTIFICACIÓN, UNE-ISO22320-Protección y seguridad de los ciudadanos/Gestión de emergencias/Requisitos para la respuesta a incidentes, AENOR Internal report, Madrid, 2013.

[ASO 15] ASOCIACIÓN ESPAÑOLA DE NORMALIZACIÓN Y CERTIFICACIÓN, Gestión de Emergencias, available at http://www.aenor.es/documentos/certificacion/folletos/articulo_gestion_emergencias.pdf, 2015.

[ATK 07] ATKINSON S.R., JAMES MOFFAT Y., *The Agile Organization. From Informal Networks to Complex Effects and Agility*, CCRP, 2007.

[BAI 10] BAIRD M.E., The "Phases" of Emergency Management, Intermodal Freight Trnasportation Intitute (ITFI), University of Memphis, January 2010.

[BOB 13] BOBROWSKY P.T., *Encyclopedia of Natural Hazards,* Springer, London, 2013.

[BRI 08] BRITISH STANDARDS INSTITUTION, Disasters and Emergency Management Systems, British Standards Limited, London, 2008.

[BRI 14] BRITISH STANDARDS INSTITUTION, Crisis Management – Guidance and Good Practice, British Standards Limited, London, 2014.

[CEN 15] CENTRO DE TECNOLOGÍA Y MANUFACTURA AVANZADA SENA, "Ciclo de Deming: PHVA", *Auditoria de la Calidad en Salud,* available at http://auditoriadelacalidadensalud.jimdo.com/curso-de-gerencia/plataforma-estrategica/, 2015.

[DEP 10] DEPARTMENT OF DEFENSE OF UNITED STATES, Dictionary of Military and Associated Terms (JP 1-02), NIPRNET Internal report, 2010.

[DIA 05] DÍAZ M., MARÍA E., HEYLIN Y. *et al.*, Diseño y documentación de un Sistema de Gestión de la Calidad basado en la Norma ISO 9001:2000, Internal report, 2005.

[DIS 12] DISHNO D., Project management skills for all careers, Internal report, 2012.

[DOB 12] DÖBBELING E.P., "Emergency management – global best practice for an incident response system", *ISO Focus+*, vol 3, no. 5, pp. 13–15, 2012.

[EST 15] ESTRATEGIA INTERNACIONAL PARA LA REDUCCIÓN DE DESASTRES, Estrategia Internacional para la reducción de desastres – Las Américas, EIRD, available at http://www.eird.org/esp/terminologia-esp.htm, 2015.

[EUS 11] EUSKALIT, *Calidad Total: Principios y Modelos de Gestión*, EPE, 2011.

[FED 83] FEDERAL EMERGENCY MANAGEMENT AGENCY, "Integrated emergency management system", *Hazard Analysis for Emergency Management*, September 1983.

[FED 04] FEDERAL/PROVINCIAL/TERRITORIAL NETWORK ON EMERGENCY PREPAREDNESS AND RESPONSE, National framework for health emergency management, Guideline for Program Development, 2004.

[FED 14a] FEDERAL EMERGENCY MANAGEMENT AGENCY, Introduction to incident command system IS-100.b., available at: www.training.fema.gov., 2014.

[FED 14b] FEDERAL EMERGENCY MANAGEMENT AGENCY, National protection framework, available at: www.training .fema.gov., 2014.

[FED 15a] FEDERAL EMERGENCY MANAGEMENT AGENCY, About FEMA, available at http://www.fema.gov/about-agency, 2015.

[FED 15b] FEDERAL EMERGENCY MANAGEMENT AGENCY, Fundamentals of emergency management IS-0230.d., availabale at: www.training .fema.gov., 2015.

[GAI 07] GAITÁN R., LINDA K., Diseño de un modelo de gestíon de calidad basado en los modelos de excelencia y el enfoque de gestión por procesos, PhD Thesis, Universidad del Norte, 2007.

[GIB 10] GIBSON C.A., MICHAEL TARRANT Y., "A 'Conceptual Models' approach to organisational resilience", *The Australian Journal of Emergency Management*, vol. 25, no. 2, pp. 6–12, 2010.

[GRI 95] GRIMA C., PEDRO, TROT-MARTORELL L.J., *Técnicas para la Gestión de Calidad*, Díaz de Santos S.A., Madrid, 1995.

[HER 06] HERNÁNDEZ M., PMBOK en la Gestión de Proyectos, Escuela Politécnica Nacional, Quito, 2006.

[HOY 10] HOYER R.W., HOYER Y.B.Y., "¿Qué es calidad?", *Quality Progress*, vol. 34, no. 7, pp. 53–62, 2001.

[INT 15a] INTERNATIONAL ASSOCIATION OF EMERGENCY MANAGERS, International Association of Emergency Managers – Spain, available at http://www.iaem.es/?page_id=35, 2015.

[ISO 05] INTERNATIONAL ORGANIZATION FOR STANDARDIZATION, ISO9000: Quality management system –Terminology, avialable at: www.iso.org, 2005.

[ISO 07] INTERNATIONAL ORGANIZATION FOR STANDARDIZATION, ISO22399: Societal Security, available at: https://www.iso.org/obp/ui/#iso:std:iso:pas:22399:ed-1:v1:en, 2007.

[ISO 08] INTERNATIONAL ORGANIZATION FOR STANDARDIZATION, ISO9001: Quality Management System – Requirements, available at www.iso.org, 2008.

[ISO 09a] INTERNATIONAL ORGANIZATION FOR STANDARDIZATION, ISO31000: Risk Management, available at http://solomantenimiento.blogspot.com.es/2012/05/norma-iso-310002009-gestion-del-riesgo.html, 2009.

[ISO 09b] INTERNATIONAL ORGANIZATION FOR STANDARDIZATION, Quality management system terminology – principles to improve performance, available at: www.iso.org, 2009.

[ISO 12] INTERNATIONAL ORGANIZATION FOR STANDARDIZATION, ISO22301: Business Continuity, available at https://www.iso.org/obp/ui/#iso:std:iso:22301:ed-1:v2:en, 2012.

[ISO 15b] INTERNATIONAL ORGANIZATION FOR STANDARDIZATION, ISO/TC 223, available at http://www.isotc223.org/Published-Standards/, 2015.

[IZU 09] IZU B., MIGUEL J., "De la protección civil a la gestión de emergencias", *Revista Aragonesa de Administración Pública N.35,* Zaragoza, 2009.

[LÓP 04] LÓPEZ L., LORENA, SÁNCHEZ S. *et al.*, "Pensamiento Estadístico para los Empresarios del Siglo XXI", *Industrial,* vol. 25, no. 1, pp. 3–9, 2004.

[LÓP 08] LÓPEZ C., ÁLVARO, VALLE C., RAMÓN, "Capital humano, prácticas de gestión y agilidad empresarial: ¿están relacionadas?" *Revista Europea de Dirección y Economía de la Empresa,* vol. 17, no. 2, pp. 155–178, 2008.

[MOF 08] MOFFAT J., To response to hurricane katrina: a case study of changing C2 maturity, Defence Science and Technology Laboratory, Farnborough, 2008.

[MOJ 15] MOJONNIER T., Business theory – has Toyota lost its continuous improvement Mojo?, available at http://businesstheory.com/toyota-lost-continuous-improvement-mojo/, Junio de 2015.

[ORG 09] ORGANIZACIÓN MUNDIAL DE LA SALUD, Gestión de la información y comunicación en emergencias y desastres, Panamá, 2009.

[PEA 06] PEARLSON, KERI, SAUNDERS, *et al.*, *Managing and Using Information Systems: A Strategic Approach,* John Wiley & Son, 2006.

[PER 15] PÉREZ-ORTIZ C., LUIS E., ¿Qué faltó en Nepal?, available at http://site.edinssed.com/que-falto-en-nepal/, 2015.

[PRO 15] PROJECT MANAGEMENT INSTITUTE, Project Management Institute, available at http://www.pmi.org/, 2015.

[REP 10] REPETO A.J.M., ESPÍNDOLA NAHUEL, "Interoperabilidad y Comunicaciones Utilizando P2P en Sistemas de Comando y Control para Emergencias y Catastrofes", *Congreso argentino de Ciencias de la Computación*, Caba, pp. 102–111, 2010.

[SAN 03] SANGÜESA S., MARTHA, Manual de Gestión de Calidad, Universdad de Navarra, Pamplona, 2003.

[SMI 05] SMITH E.A., *Effects Based Operations. Aplying Network Centrics Warfare in Peace, Crisis, and War*, CCRP, Washington, 2005.

[SPA 12] SPANEVELLO, FABIÁN A.., "IQ: Calidad de la Información", *Revista de Publicaciones Navales*, pp. 48–55, no. 710, 2012.

[TIM 99] TIMBERLAKE A., Quality, Quantity, Cost: Which is Your Driver?, Intitute of Paper Science and Technology, Atlanta, Georgia, 1999.

[UNI 12] UNITED STATES AGENCY INTERNATIONAL DEVELOPMENT, *Curso Básico Sistema de Comando de Incidentes,* USAID, 2012.

[UNI 13] UNITED NATIONS INTERNATIONAL STRATEGY FOR DISASTER REDUCTION, The Economic and Human Impact of Disaster in the Last 10 Years, UNISDR, 2013.

[UPO 94] UPON D., "The management of manufacturing flexibility", *California Management Review,* vol. 36, no. 2, pp. 72–89, 1994.

[WAY 08] WAYNE B., "Guide to emergency management and related terms, definitions, concepts, acronyms,organizations, programs, guidance, executive orders & legislation", FEMA, 2008.

List of Authors

Olayinka ADIGUN
Kingston University
UK

Gianmarco BALDINI
European Commission
Joint Research Centre
Ispra
Italy

Konstantia BARBATSALOU
CISUC-DEI
University of Coimbra
Portugal

Bert BOUWERS
Rohill Technologies B.V.
Hoogeveen
Netherlands

Jérôme BROUET
Alcatel-Lucent International
Boulogne Billancourt
France

Daniel CÂMARA
Institut Mines-Telecom–Telecom
ParisTech
LTCI UMR 5141 CNRS
Paris
France

Theofilos CHRYSIKOS
University of Patras
Greece

Luís CORDEIRO
OneSource
Coimbra
Portugal

Manuel ESTEVE
Distributed Real-Time Systems Lab
at Communications Department
Universitat Politècnica de València
Spain

Riccardo FEDRIZZI
CREATE-NET Research Center
Trento
Italy

Hugo FONSECA
OneSource
Coimbra
Portugal

Federico FROSALI
Selex ES
Rome
Italy

Panagiotis GALIOTOS
University of Patras
Greece

Francesco GEI
Università degli Studi di Firenze
Italy

Eric GEORGEAUX
AIRBUS Defence & Space
Toulouse
France

Karina Mabell GOMEZ CHAVEZ
RMIT University
Melbourne
VIC
Australia

Leonardo GORATTI
CREATE-NET Research Center
Trento
Italy

Christophe GRUET
AIRBUS Defence & Space
Toulouse
France

Fanilo HARIVELO
Laboratoire d'Informatique et de Mathématiques
University of Réunion
France

Oliver C. IBE
Department of Electrical and Computer Engineering
University of Massachusetts
USA

Jean ISNARD
International Union of Radio Science
Paris
France

David JELENC
University of Ljubljana
Slovenia

Branko KOLUNDZIJA
University of Belgrade
Serbia

Jernej KOS
University of Ljubljana
Slovenia

Alexandros LADAS
Kingston University
UK

Ram Gopal LAKSHMI NARAYANAN
Verizon Labs
San Jose, CA
USA

Philippe LASSERRE
Alcatel-Lucent International
Boulogne Billancourt
France

Etienne LEZAACK
Belgium Federal Police
Belgium

Georgios MANTAS
Instituto de Telecomunicações
Aveiro
Portugal

Dania MARABISSI
Università degli Studi di Firenze
Italy

Hugo MARQUES
Instituto de Telecomunicações
Lisbon
Portugal

Luigia MICCIULLO
Università degli Studi di Firenze
Italy

François MONTAIGNE
AIRBUS Defence & Space
Toulouse
France

Edmundo MONTEIRO
CISUC-DEI
University of Coimbra
Portugal

Wilmuth MÜLLER
Fraunhofer Institut für Optronik
Systemtechnik und Bildauswertung
Karlsruhe
Germany

Lirida NAVINER
Télécom Paris Tech
Paris
France

Navid NIKAEIN
Mobile Communications Department
EURECOM
Sophia Antipolis
France

Andy NYANYO
Airwave Solutions Limited
West Berkshire
UK

Dragan OLCAN
University of Belgrade
Serbia

Dejene Boru OLJIRA
Karlstad University
Sweden

Carlos PALAU
Distributed Real-Time Systems Lab
at Communications Department
Universitat Politècnica de València
Spain

David PALMA
OneSource
Coimbra
Portugal

Cristina PÁRRAGA NIEBLA
DLR German Aerospace Center
Institute of Communications and Navigation
Bavaria
Germany

Luís PEREIRA
Instituto de Telecomunicações
Aveiro
Portugal

Guy PHILIPPE
AIRBUS Defence & Space
Toulouse
France

Christos POLITIS
Kingston University
UK

Xavier PONS-MASBERNAT
AIRBUS Defence & Space
Toulouse
France

Tinku RASHEED
CREATE-NET Research Center
Trento
Italy

Roberto RIGGIO
CREATE-NET Research Center
Trento
Italy

Jonathan RODRIGUEZ
Instituto de Telecomunicações
Aveiro
Portugal

Jean-Christophe SCHIEL
AIRBUS Defence & Space
Toulouse
France

Olivier SEBASTIEN
Laboratoire d'Informatique et de Mathématiques
University of Réunion
France

Paulo SIMÕES
CISUC-DEI
University of Coimbra
Portugal

Bruno SOUSA
OneSource
Coimbra
Portugal

Gary STERI
European Commission
Joint Research Centre
Ispra
Italy

Tullio Joseph TANZI
Institut Mines-Telecom – Telecom ParisTech
LTCI UMR 5141 CNRS
Paris
France

Denis TRČEK
University of Ljubljana
Slovenia

Nuwan WEERASINGHE
Kingston University
UK

Peter WICKSON
Airwave Solutions Limited
West Berkshire
UK

Marcelo ZAMBRANO V.
Distributed Real-Time Systems Lab at Communications Department
Universitat Politècnica de València
Spain

Daniel ZERBIB
AIRBUS Defence & Space
Toulouse
France

Index

L, M, N, O

P, Q, S

T, U, V, W

Printed in the United States
By Bookmasters